Communications and Control Engineering

El-Kébir Boukas

Control of Singular Systems with Random Abrupt Changes

 Springer

Prof. E.K. Boukas
Mechanical Engineering Department
École Polytechnique de Montréal
Succirsale Centre-Ville H3C
3A7 Montreal, Quebec
Canada
email: *El-Kebir.Boukas@polymtl.ca*

ISBN: 978-3-540-74344-6   e-ISBN: 978-3-540-74345-3

DOI 10.1007/978-3-540-74345-3

Communications and Control Engineering Series ISSN: 0178-5354

Library of Congress Control Number: 2007934281

Cover design: LE-TEX Jelonek, Schmidt & Vöckler GbR, Leipzig
Typesetting and Production: LE-TEX Jelonek, Schmidt & Vöckler GbR, Leipzig

Printed on acid-free paper

9 8 7 6 5 4 3 2 1

springer.com

Dedicated to my wife Saida and my kids Imane, Ibtissama and Anas

Preface

Singular systems also referred to as descriptor systems, implicit systems, generalized state-space systems, differential-algebraic systems or semi-state systems (see [43, 76]) represents an interesting class of dynamical systems since it combines differential equations and algebraic equations and generalizes the linear time invariant model which is extensively used in linear control theory. This class of systems has been used to model varieties of systems like economics, chemical processes, mechanics, electrical systems, etc. Singular systems have attracted a lot of researchers from the mathematical and control communities and a great number of fundamental notions and results in control and systems theory based on linear time-invariant systems with state space representation have been successfully extended to the class of singular systems. For more details on what it has been done on this subject we refer the reader to [3, 4, 51, 75, 122, 123, 127, 50, 109, 112, 122, 124, 126], and the references therein. More specifically, we tackled the stability and the stabilizability of the class of linear continuous-time singular systems and many results have been reported in the literature in the LMI setting. Among these results we quote those of [14]. The robust stability and the robust stabilization have also been studied. Many types of uncertainties have been considered among them we quote the norm bounded form, the linear fractional transformation (LFT, which generalizes the norm bounded type) and polytopic form.

The $\mathcal{H}_\infty$ control and the filtering problems have also been treated and interesting results were reported in the literature. For the $\mathcal{H}_\infty$ control, under the assumptions that the external disturbances have finite power or finite energy, a control law is designed to guarantee that the closed-loop system is regular, impulse-free and stable and at the same time assures the disturbance rejection with a given level $\gamma > 0$. For the filtering problem, the objective is to design a dynamical estimator that estimates the state vector that can be used in the control to make the closed-loop of the singular system regular, impulse-free and stable.

Some of the practical systems are stochastic or more specifically we can model them by a class of stochastic systems driven by continuous-time Markov chains that we will refer to as the class of stochastic systems with abrupt changes. This class of systems is more appropriate to model many practical systems, where random failures and repairs and sudden environment changes may occur. For more detail on what it has been done on the subject, we refer the reader to [27], [96] and the references therein. This class of systems has also attracted a lot of researchers from both mathematical and control community. Many results on stochastic stability and stochastic stabilization have been reported in the literature. For more details on these results we refer the reader to [29, 26, 49, 60, 94] and the references therein, where different approaches have been used. The $\mathcal{H}_\infty$ control problem was investigated in [45, 105], where sufficient conditions for the solvability of this problem was proposed. When time delays appear in a Markovian jump system, the results on stability analysis and $\mathcal{H}_\infty$ control were also reported in [31], [24] and [25] for different types of time delays. For more detail on Markovian jumping systems with time delay, we refer the reader to [14, 27] and the references therein.

Our goal in this volume is to combine the class of singular systems with the one of systems with abrupt changes which will give the class of systems that we will refer to as the class of stochastic singular systems with abrupt changes. Our main objective in this volume is to treat the stability and the stabilization problems using different techniques. We will also handle the filtering problem.

The rest of this book is organized as follows. In Chap. 1, the different problems are stated and the necessary assumptions are given. Chapter 2 deals with the stability problem of the class of singular systems with random abrupt changes and LMI conditions are developed to check if a given system of this class of systems is piecewise regular, impulse-free and stochastically stable. The robust stability problem is also considered. Chapter 3 treats the stabilization problem and its robustness. State space controllers are considered and LMI design approaches are developed. Chapter 4 deals with the $\mathcal{H}_\infty$ control for the class of singular piecewise deterministic systems. In Chap. 5, the static output stabilization is tackled and LMI results are developed for the class of singular systems with random abrupt changes. The robust case is also considered. Chapter 6 deals with observer-based output stabilization for the class of Markovian jump singular systems. In Chapt. 7 the filtering problem is considered and design procedures are proposed in the LMI formalism to solve the the $\mathcal{H}_\infty$ filtering problem. In Chap. 8, the guaranteed cost control problem is tackled and LMI results are developed to synthesize the state feedback controller that makes the closed-loop piecewise regular, impulse-free and stochastically stable and at the same time guaranteed an upper bound for the cost for all admissible uncertainties. In Chap. 9, the mixed $\mathcal{H}_2/\mathcal{H}_\infty$ control is tackled and design procedure is developed to synthesize a state feedback controller. Finally, Chap. 10 provides some tools that can be used to solve all the LMI conditions. It gives an idea to the reader on how to write his program under Mathlab to solve the considered problem of this class of systems.

Contents

List of Figures

List of Tables

Part I

Modeling and problem statements

1

Introduction

Practical systems with random abrupt changes in their dynamics represent a class of systems that has stochastic behavior and which can not be appropriately described by the linear time-invariant model usually used extensively in control theory. These abrupt changes resulted from many causes like failures, repairs, connection and disconnection of some components, etc. Among the systems that have random abrupt changes in their dynamics, we quote those of manufacturing systems, powers systems, telecommunications systems, etc.

To model the behavior of this class of systems, Krasovskii and Lidskii [70, 69] proposed a model that is known in the literature as a Markovian jump systems, piecewise deterministic systems, stochastic hybrid systems and dynamical systems with random abrupt changes. This model is more general which makes it popular in both theoretical and applied research. For this class of systems, the stability and the stabilization problems have been studied and many results have been reported in the literature. Many stabilization techniques have been considered and most of the developed results are in the LMI framework which makes the results powerful and tractable. For more details on what it has been done on the subjects we refer the reader to [27, 65, 95] and the references therein.

Practically not all the systems are normal and it may happen that we can encounter physical systems that can't be modeled by the previous class of systems. In the literature, these systems are referred to as singular systems, descriptor systems, implicit systems, generalized state-space systems, semi-state systems or differential-algebraic systems. Singular systems arise in many practical systems like electrical circuits, power systems, networks, etc. (see for more examples [43] and the references therein). The goal of this chapter is to present the modeling of the class of singular systems with random abrupt changes in the structure and consider the formulation of some of the problems treated earlier in the literature and see how we can extended the previous results to this case.

The rest of this chapter is organized as follows. In Sect. 1.1 some examples of singular systems are presented and their mathematical models are given. Section 1.2 gives the statements of the different problems we will treat in this volume. Section 1.3

covers the solution of the dynamics of singular systems. Section 1.4 and 1.5 gives respectively some useful mathematical concepts and lemmas that will be used in the rest of the volume.

1.1 Examples of Singular Systems

To justify the importance of the class of systems we are considering in this volume, let us consider some practical examples of systems. The first example is the DC motor which represents the actuator that is usually used in the position control ser-vomechanism. It is the mean by which the electrical energy is converted to mechanical energy. In this example, we consider a DC motor driving a load (see Fig. 1.1). If we neglect the DC motor inductance L_m and let $u(t)$, $i(t)$ and $\omega(t)$ denote respectively the voltage of the armature, the current in the armature and the speed of the shaft at time t, based on the basic electrical and mechanic laws we have the following:

$$\begin{cases} u(t) = Ri(t) + K_w w(t) \\ J\dot{w}(t) = K_t i(t) - bw(t) \end{cases} \tag{1.1}$$

where R, K_w, K_t represent respectively the electric resistor of the armature, the electromotive force constant, the torque constant (in the IS unit, both constants are equal), J and b are defined by:

$$J = J_m + \frac{J_c}{n^2} \tag{1.2}$$

$$b = b_m + \frac{b_c}{n^2} \tag{1.3}$$

with J_m and J_c are the moments of inertia of the rotor and the load, and b_m and b_c are the damping ratios of the motor and the load, and n is the gear ratio.

Now if we let $x_1(t) = i(t)$, $x_2(t) = \omega(t)$ and $y(t) = x_2(t)$ we get:

$$\begin{bmatrix} 0 & 0 \\ 0 & J \end{bmatrix} \begin{bmatrix} \dot{x}_1(t) \\ \dot{x}_2(t) \end{bmatrix} = \begin{bmatrix} R & K_w \\ K_t & -b \end{bmatrix} \begin{bmatrix} x_1(t) \\ x_2(t) \end{bmatrix} + \begin{bmatrix} 1 \\ 0 \end{bmatrix} u(t) \tag{1.4}$$

$$y(t) = \begin{bmatrix} 0 & 1 \end{bmatrix} \begin{bmatrix} x_1(t) \\ x_2(t) \end{bmatrix}, \tag{1.5}$$

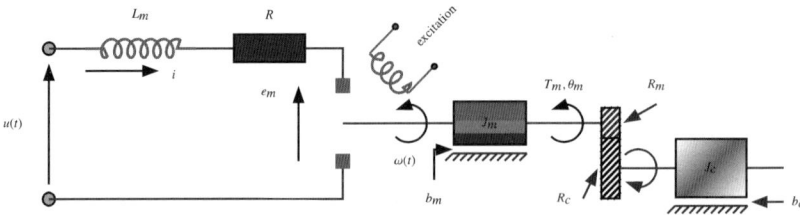

Fig. 1.1. Block diagram of a DC motor

that gives the following standard form:

$$\begin{cases} E\dot{x}(t) = Ax(t) + Bu(t) \\ y(t) = Cx(t), \end{cases} \tag{1.6}$$

where

$$E = \begin{bmatrix} 0 & 0 \\ 0 & J \end{bmatrix}, \ A = \begin{bmatrix} R & K_w \\ K_t & -b \end{bmatrix},$$

$$B = \begin{bmatrix} 1 \\ 0 \end{bmatrix}, \ C = \begin{bmatrix} 0 & 1 \end{bmatrix}.$$

Now if we assume that the load changes randomly and abruptly which we can model by the changes of the inertia, J and if these changes are represented by a continuous-time Markov process $\{r_t, t \geq 0\}$ taking values in a finite set $\mathscr{S} = \{1, 2, \ldots, N\}$, then we will have the form we are considering in this volume:

$$\begin{cases} E(r_t)\dot{x}(t) = A(r_t)x(t) + B(r_t)u(t), \ x(0) = x_0 \\ y(t) = C(r_t)x(t) \end{cases} \tag{1.7}$$

In this case, where r_t occupies the state i, i.e. $r_t = i$, the matrices $A(i)$ $B(i)$ and $C(i)$ are known. This means that the system will switch between different modes randomly.

As a second example, let us consider the electrical circuit of the Fig. 1.2. It consists of an electrical resistance and inductances in parallel. We assume that the switch occupies two positions and it switches from one position to another in a random way that we assume to be modeled by a continuous-time Markov process with finite state space. If we denote the state that models the position of the switch by r_t and by $\mathscr{S} = \{1, 2\}$ the state space and based on Markov process theory, we have:

$$\mathbb{P}\left[r_{t+h} = j | r_t = i\right] = \begin{cases} \lambda_{ij}h + o(h) & \text{when } r_t \text{ jumps from } i \text{ to } j, \\ 1 + \lambda_{ii}h + o(h) & \text{otherwise,} \end{cases} \tag{1.8}$$

where λ_{ij} is the transition rate from mode i to mode j with $\lambda_{ij} \geq 0$ when $i \neq j$ and, $\lambda_{ii} = -\sum_{j=1, j\neq i}^{2} \lambda_{ij}$ and $o(h)$ is such that $\lim_{h\to 0} \frac{o(h)}{h} = 0$.

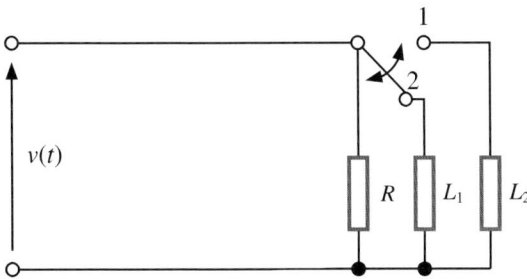

Fig. 1.2. RLC circuit: singular circuit

The corresponding transition matrix is given by:

$$\Lambda = \begin{bmatrix} \lambda_{11} & \lambda_{12} \\ \lambda_{21} & \lambda_{22} \end{bmatrix}.$$

Let us assume that the switch occupies the position r_t at time t and denote respectively by $i_R(t)$, $i_L(t)$ and $i(t)$, the currents passing through the electrical resistance, the inductances L_1 (or L_2 if the switch occupies the position two). Applying now the Kirchhoff laws, we have:

$$L(r_t)\frac{di_L}{dt}(t) = v(t)$$

$$i(t) = i_R(t) + i_L(t)$$

$$Ri_R(t) = v(t),$$

which can be rewritten as follows:

$$L(r_t)\frac{di_L}{dt}(t) = v(t)$$

$$0\frac{di(t)}{dt} = -i(t) + i_R(t) + i_L(t)$$

$$0\frac{di_R(t)}{dt} = -Ri_R(t)v(t)$$

where $L(r_t)$ is defined by:

$$L(r_t) = \begin{cases} L_1 & \text{if } r_t = 1, \\ L_2 & \text{otherwise,} \end{cases}$$

Now if we let:

$$x_1(t) = y_1(t) = i_L(t)$$

$$x_2(t) = y_2(t) = i(t)$$

$$x_3(t) = i_R(t)$$

$$u(t) = v(t),$$

we get the following state space representation:

$$\begin{bmatrix} L(r_t) & 0 & 0 \\ 0 & 0 & 0 \\ 0 & 0 & 0 \end{bmatrix}\begin{bmatrix} \dot{x}_1(t) \\ \dot{x}_2(t) \\ \dot{x}_3(t) \end{bmatrix} = \begin{bmatrix} 0 & 0 & 0 \\ 1 & -1 & 1 \\ 0 & 0 & -R \end{bmatrix}\begin{bmatrix} x_1(t) \\ x_2(t) \\ x_3(t) \end{bmatrix} + \begin{bmatrix} 1 \\ 0 \\ 1 \end{bmatrix}u(t)$$

$$y(t) = \begin{bmatrix} y_1(t) \\ y_2(t) \end{bmatrix} = \begin{bmatrix} 1 & 0 & 0 \\ 0 & 1 & 0 \end{bmatrix}\begin{bmatrix} x_1(t) \\ x_2(t) \\ x_3(t) \end{bmatrix}.$$

This generalized state space representation can be transformed to the following equivalent one by using the fact that $x_3(t) = \frac{1}{R}u(t)$ and by performing some algebraic transformations:

$$\begin{bmatrix} 1 & 0 & 0 \\ 0 & 0 & 0 \\ 0 & 0 & 0 \end{bmatrix} \begin{bmatrix} \dot{x}_1(t) \\ \dot{x}_2(t) \\ \dot{x}_3(t) \end{bmatrix} = \begin{bmatrix} 0 & 0 & 0 \\ 1 & -1 & 0 \\ 0 & 0 & 1 \end{bmatrix} \begin{bmatrix} x_1(t) \\ x_2(t) \\ x_3(t) \end{bmatrix} + \begin{bmatrix} \frac{1}{L(r_t)} \\ \frac{1}{R} \\ \frac{1}{R} \end{bmatrix} u(t)$$

$$y(t) = \begin{bmatrix} y_1(t) \\ y_2(t) \end{bmatrix} = \begin{bmatrix} 1 & 0 & 0 \\ 0 & 1 & 0 \end{bmatrix} \begin{bmatrix} x_1(t) \\ x_2(t) \\ x_3(t) \end{bmatrix}.$$

This model can be put in the general framework of singular systems with random abrupt changes:

$$\begin{cases} E(r_t)\dot{x}(t) = A(r_t)x(t) + B(r_t)u(t), \; x(0) = x_0 , \\ y(t) = C(r_t)x(t). \end{cases} \tag{1.9}$$

Now if we assume that the resistor is given by the following expression:

$$R = R_n + \Delta R ,$$

where R_n is known and $|\Delta R| \leq \rho$, notice in this case that the matrices $E(i)$, $B(i)$ and $C(i)$ are not affected by ΔR.

In this case we can have:

- Polyhedral uncertainty: here since $-\rho \leq \Delta R \leq \rho$, the state matrix $A(i)$ belongs to the following convex hull:

$$\left\{ \begin{bmatrix} 0 & 0 & 0 \\ 1 & -1 & 1 \\ 0 & 0 & -R+\rho \end{bmatrix} , \begin{bmatrix} 0 & 0 & 0 \\ 1 & -1 & 1 \\ 0 & 0 & -R-\rho \end{bmatrix} \right\}.$$

- Norm bounded uncertainty: here since $|\Delta R| \leq \rho$, the state matrix $A(i) + \Delta A(i)$ can be rewritten as:

$$A(i) + \Delta A(i) = \begin{bmatrix} 0 & 0 & 0 \\ 1 & -1 & 1 \\ 0 & 0 & -(R+\Delta R) \end{bmatrix}$$

$$= \begin{bmatrix} 0 & 0 & 0 \\ 1 & -1 & 1 \\ 0 & 0 & -R \end{bmatrix} + \begin{bmatrix} 0 \\ 0 \\ -1 \end{bmatrix} F_A(i) \begin{bmatrix} 0 & 0 & 1 \end{bmatrix}$$

$$= A(i) + D_A(i)F_A(i)E_A(i),$$

with $\|F_A(i)\| \leq \rho$.

As a third example, let us consider a production system producing one part type. We will assume that the produced parts deteriorate with time at a given rate that may depend on the state of the production system. If we denote respectively by $x(t)$, $u(t)$ and $d(t)$ the stock level, the production rate and the demand rate of the system at time t, the stock level is described by the following dynamics (see [15]):

$$\begin{cases} \dot{x}(t) = -\rho(r_t)x(t) + \beta(r_t)u(t) - d(t), \\ x(0) = x_0, \end{cases}$$

where $\rho(r_t)$ is the deteriorating rate of the stock level and $\beta(r_t)$ is defined as follows:

$$\beta(r_t) = \begin{cases} 1 & \text{if the machine is up,} \\ 0 & \text{otherwise.} \end{cases}$$

Let us now assume that the demand rate $d(t)$ is described by the following:

$$d(t) = K_v v(t) + K_w w(t),$$
$$d(0) = d_0,$$

where K_v and K_w are known constants, $v(t)$ is a control variable that represents the advertisement we made to improve the demand rate and $w(t)$ is a disturbance with bounded energy.

Combining the two dynamics we get:

$$\begin{bmatrix} 1 & 0 \\ 0 & 0 \end{bmatrix} \begin{bmatrix} \dot{x}(t) \\ \dot{d}(t) \end{bmatrix} = \begin{bmatrix} -\rho(r_t) & -1 \\ 0 & -1 \end{bmatrix} \begin{bmatrix} x(t) \\ d(t) \end{bmatrix} + \begin{bmatrix} \beta(t) & 0 \\ 0 & K_v \end{bmatrix} \begin{bmatrix} u(t) \\ v(t) \end{bmatrix}$$
$$+ \begin{bmatrix} 0 \\ K_w \end{bmatrix} w(t).$$

The control variables must satisfy the following constraints:

$$\underline{u} \le u(t) \le \bar{u},$$
$$\underline{v} \le v(t) \le \bar{v}.$$

The machine is assumed to occupy three modes with the following transition rates:

$$\Lambda = \begin{bmatrix} \lambda_{11} & \lambda_{12} & 0 \\ \lambda_{21} & \lambda_{22} & \lambda_{23} \\ \lambda_{31} & 0 & \lambda_{33} \end{bmatrix},$$

with $\lambda_{ij} \ge 0$ and $\lambda_{ii} = -\sum_{j=1}^{3} \lambda_{ij}$ for $i = 1, 2, 3$.

Before closing this section let us focus on the comparison between the class of deterministic singular systems and the one considered in this volume. In fact for the deterministic singular systems, it is always possible to put the dynamics of the autonomous system in the following form (see [43]):

$$\begin{cases} E\dot{x}(t) = Ax(t), \\ x(0) = x_0, \end{cases}$$

where E and A have the following forms:

$$x(t) = \begin{bmatrix} x_1(t) \\ x_2(t) \end{bmatrix}, \quad E = \begin{bmatrix} \mathbb{I} & 0 \\ 0 & 0 \end{bmatrix}, \quad A = \begin{bmatrix} A_1 & A_2 \\ A_3 & A_4 \end{bmatrix},$$

that gives

$$\begin{cases} \dot{x}_1(t) = A_1 x_1(t) + A_2 x_2(t), \\ 0 = A_3 x_1(t) + A_4 x_2(t), \end{cases}$$

that we can rewrite in turn as follows when the matrix A_4 is nonsingular:

$$\begin{cases} \dot{x}_1(t) = \left[A_1 - A_2 A_4^{-1} A_3 \right] x_1(t), \\ x_2(t) = -A_4^{-1} A_3 x_1(t), \end{cases}$$

Therefore, if the solution exists, the variables $x_1(t)$ (slow) and $x_2(t)$ (fast) are both continuous in time. Meanwhile for the class of systems we are dealing with the behavior is different and in fact we have discontinuities in the fast variable $x_2(t)$. To show this, let us write similarly the dynamics as follows:

$$\begin{cases} E(r_t)\dot{x}(t) = A(r_t)x(t), \\ x(0) = x_0. \end{cases}$$

where $E(r_t)$ and $A(r_t)$ have the following forms:

$$x(t) = \begin{bmatrix} x_1(t) \\ x_2(t) \end{bmatrix}, \quad E(r_t) = \begin{bmatrix} \mathbb{I} & 0 \\ 0 & 0 \end{bmatrix}, \quad A(r_t) = \begin{bmatrix} A_1(r_t) & A_2(r_t) \\ A_3(r_t) & A_4(r_t) \end{bmatrix},$$

that gives

$$\begin{cases} \dot{x}_1(t) = A_1(r_t)x_1(t) + A_2(r_t)x_2(t), \\ 0 = A_3(r_t)x_1(t) + A_4(r_t)x_2(t). \end{cases}$$

That we can rewrite in turn as follows when the matrix $A_4(i)$ is nonsingular for every mode i:

$$\begin{cases} \dot{x}_1(t) &= \left[A_1(r_t) - A_2(r_t)A_4^{-1}(r_t)A_3(r_t) \right] x_1(t), \\ x_2(t) &= -A_4^{-1}(r_t)A_3(r_t)x_1(t). \end{cases}$$

Now if we denote by τ_k the instant of kth jump of the process $\{r_t, t \geq 0\}$, $x_1(t)$ will be continuous in time while $x_2(t)$ has discontinuities. In fact, if the Markov process jumps from mode i to mode j at time τ_k, we have:

$$x_2(\tau_k^-) = -A_4^{-1}(i)A_3(i)x_1(\tau_k),$$
$$x_2(\tau_k) = -A_4^{-1}(j)A_3(j)x_1(\tau_k),$$

Since $A_4^{-1}(i)A_3(i)$ and $A_4^{-1}(j)A_3(j)$ are in general different which implies the discontinuities in the slow variable every time the Markov process jumps. The values of the state vector at the instant of the kth jump can be defined:

$$x_1(\tau_k) = x_1(\tau_k^-),$$
$$x_2(\tau_k) = x_2(\tau_k^-) + \left[A_4^{-1}(i)A_3(i) - A_4^{-1}(j)A_3(j)\right]x_1(\tau_k),$$

that can be rewritten as follows:

$$\begin{bmatrix} x_1(\tau_k) \\ x_2(\tau_k) \end{bmatrix} = \left[\begin{matrix} \mathbb{I} & 0 \\ \left[A_4^{-1}(i)A_3(i) - A_4^{-1}(j)A_3(j)\right] & \mathbb{I} \end{matrix}\right]\begin{bmatrix} x_1(\tau_k) \\ x_2(\tau_k^-) \end{bmatrix},$$
$$= \Psi_{ij}\begin{bmatrix} x_1(\tau_k) \\ x_2(\tau_k^-) \end{bmatrix}.$$

Based on the different systems we presented and the recent comments, the dynamics that we will consider in this volume is described by the following differential algebraic equations:

$$\begin{cases} E(r_t)\dot{x}(t) = [A(r_t) + \Delta A(r_t)]\,x(t) + [B(r_t) + \Delta B(r_t)]\,u(t) \\ \qquad\qquad + B_w(r_t)w(t)\,, x(0) = x_0\,, \\ y(t) = [C(r_t) + \Delta C(r_t)]\,x(t)\,, \end{cases} \tag{1.10}$$

with discontinuities at each jump of the Markov process $\{r_t, t \geq 0\}$.
    In the rest of this book we will restrict ourselves to norm bounded uncertainty.

1.2 Problem Statements

Let us consider a dynamical singular system with random abrupt changes defined in a fundamental probability space $(\Omega, \mathcal{F}, \mathbb{P})$ and assume that its state equation is described by the following differential algebraic equations:

$$\begin{cases} E(r_t)\dot{x}(t) = A(r_t, t)x(t) + B(r_t, t)u(t)\,, \\ x(0) = x_0\,, \end{cases} \tag{1.11}$$

where $x(t) \in \mathbb{R}^n$ is the state vector, $x_0 \in \mathbb{R}^n$ is the initial state, $u(t) \in \mathbb{R}^m$ is the control input, $\{r_t, t \geq 0\}$ is the continuous-time Markov process taking values in a finite space $\mathcal{S} = \{1, 2, \cdots, N\}$ and describes the evolution of the mode at time t, $E(i) \in \mathbb{R}^{n \times n}$ is a known singular matrix with rank $(E(i)) = n_r \leq n$, $A(r_t, t) \in \mathbb{R}^{n \times n}$ and $B(r_t, t) \in \mathbb{R}^{n \times m}$ are matrices with the following forms for every $i \in \mathcal{S}$:

$$A(i, t) = A(i) + D_A(i)F_A(i)E_A(i)\,,$$
$$B(i, t) = B(i) + D_B(i)F_B(i)E_B(i)\,,$$

with $A(i) \in \mathbb{R}^{n \times n}$, $D_A(i) \in \mathbb{R}^{n \times n_D}$, $E_A(i) \in \mathbb{R}^{n_E \times n}$, $B(i) \in \mathbb{R}^{n \times m}$, $D_B(i) \in \mathbb{R}^{n \times m_D}$ and $E_B(i) \in \mathbb{R}^{m_E \times m}$ are real known matrices with appropriate dimensions, and $F_A(i) \in \mathbb{R}^{n_D \times n_E}$ and $F_B(i) \in \mathbb{R}^{m_D \times m_E}$ are unknown real matrices that satisfy the following:

$$\begin{cases} F_A^\top(i) F_A(i) \leq \mathbb{I}, \\ F_B^\top(i) F_B(i) \leq \mathbb{I}. \end{cases} \tag{1.12}$$

The Markov process $\{r_t, t \geq 0\}$ beside taking values in the finite set $\mathscr{S}$, represents the switching between the different modes and its state equation is described by the following probability transitions:

$$\mathbb{P}\left[r_{t+h} = j | r_t = i\right] = \begin{cases} \lambda_{ij} h + o(h), & \text{when } r_t \text{ jumps from } i \text{ to } j, \\ 1 + \lambda_{ii} h + o(h), & \text{otherwise}, \end{cases} \tag{1.13}$$

where λ_{ij} is the transition rate from mode i to mode j with $\lambda_{ij} \geq 0$ when $i \neq j$ and $\lambda_{ii} = -\sum_{j=1, j \neq i}^{N} \lambda_{ij}$ and $o(h)$ is such that $\lim_{h \to 0} \frac{o(h)}{h} = 0$.

It may happen that the transition matrix has uncertainties beside the uncertainties in the different matrices of the dynamics. In this volume, we will consider also that the matrix Λ belongs to a polytope, i.e.,

$$\Lambda = \sum_{k=1}^{\kappa} \alpha_k \Lambda_k, \tag{1.14}$$

with κ is a positive given integer, $0 \leq \alpha_k \leq 1$ with $\sum_{k=1}^{\kappa} \alpha_k = 1$ and $\Lambda_k \in \mathbb{R}^{N \times N}$ is a known transition matrix and its expression is given by:

$$\Lambda_k = \begin{bmatrix} \lambda_{11}^k & \cdots & \lambda_{1N}^k \\ \vdots & \ddots & \vdots \\ \lambda_{N1}^k & \cdots & \lambda_{NN}^k \end{bmatrix}, \tag{1.15}$$

where λ_{ij}^k keeps the same meaning as before.

Remark 1.2.1 *The uncertainties that satisfy the condition (1.12), (1.14) are referred to as admissible. The uncertainty term, in (1.12), is supposed to depend on the system's mode, r_t.*

Remark 1.2.2 *The matrix $E(i)$ for each $i \in \mathscr{S}$ is supposed to be singular which makes the dynamics (1.11) different from the one usually used to describe the behavior of the time-invariant dynamical systems as it was considered in [27, 65, 95] and the references therein.*

Remark 1.2.3 *Notice that when $E(i)$ for each $i \in \mathscr{S}$ is not singular, (1.11) can be transformed easily to the class of Markovian jump linear systems and the results developed in the literature can be used to check the stochastic stability, to design the appropriate controller that stochastically stabilizes this class of systems and even design the appropriate filter.*

It is well known for the class of singular systems even for the deterministic case that for an arbitrary finite initial condition, the time response of a singular system may exhibit impulsive or non-causal behavior along with the derivatives of these impulses. It is also known that singular systems usually contain three kinds of modes:

- finite dynamic modes,
- infinite dynamic modes,
- and non dynamic modes.

The undesired impulsive behavior in a singular system can be generated by infinite dynamic modes. Sometimes, even if a singular system is impulse-free, it can still have initial finite discontinuities due to inconsistent initial conditions.

Due to the presence of infinite dynamic modes and non dynamic modes, the existence and uniqueness of a solution to a given singular system is not always guaranteed and the system can also have undesired impulsive behavior. So, The definitions of regularity and non-impulsiveness have to be introduced. Therefore, for a singular system, it is important to develop conditions which guarantee that the given singular system is not only stable but also regular and impulse-free.

Definition 1.2.1 *[43]*

i. *System (1.11) is said to be regular if the characteristic polynomial, $\det(sE(i) - A(i))$ is not identically zero for each mode $i \in \mathscr{S}$.*
ii. *System (1.11) is said to be impulse free, if $\deg(\det(sE(i) - A(i))) = rank(E)$ for each mode $i \in \mathscr{S}$.*

Notice that when the system (1.11) is piecewise regular and impulse-free, this implies that the solution exists and moreover it is piecewise impulse-free and unique. For more details we refer the reader to [43, 122] and the references therein.

The first problem that we should tackle is the one of stability and it consists as it is the case for deterministic to determine if the solution of the system (1.11) will remain bounded when the time goes to infinity. In the literature there exists many definitions for stochastic stability for more details we refer the reader to [14, 27, 95] and the references therein. In this volume, we will use the following definition.

Definition 1.2.2 *System (1.11) with $u(t) \equiv 0$ is said to be stochastically stable if there exists a constant $M(x_0, r_0) > 0$ such that the following holds for any initial conditions (x_0, r_0):*

$$\mathbb{E}\left[\int_0^\infty x^\top(t)x(t)|x_0, r_0\right] \le M(x_0, r_0). \tag{1.16}$$

In this volume, in addition to the stochastic stability we would like to check also if the system is piecewise regular and impulse-free.

The concept of robust stability is defined in the same manner. In fact we would like to see if a stochastically stable system will remain stable even if the dynamics are subject to admissible uncertainties. The following definition is used for robust stability.

Definition 1.2.3 *Uncertain system (1.11) is said to be robust stochastically stable (RSS) if there exists a constant $M(x_0, r_0)$ such that (1.16) holds for all admissible uncertainties.*

As it is the case for nominal system (system with uncertainties equal to zero), we would be interested to know if the uncertain system under study is piecewise regular, impulse-free and stochastically stable for all admissible uncertainties. Results either on stability or robust stability in the LMI setting are searched since they are easily tractable using existing tools in the marketplace.

The second control problem of importance is the one of stabilizability. It consists of designing the appropriate controller that makes the closed-loop dynamics regular, impulse-free and stochastically stable and sometimes attains some prescribed performances. Among the structures of controllers that we can use for this purpose depending on the measurability of the state vector and the mode we quote:

1. state feedback controller;
2. $\mathcal{H}_\infty$ state feedback controller;
3. static output feedback controller;
4. dynamic output feedback controller;
5. $\mathcal{H}_\infty$ output feedback controller;
6. mixed $\mathcal{H}_2/\mathcal{H}_\infty$ state or output feedback controller.

The following definition will be used in the rest of this volume for stochastic stabilization.

Definition 1.2.4 *System (1.11) is said to be stochastically stabilizable if there exists a control such that the closed-loop system is stochastically stable.*

When the dynamics have uncertainties, the following definition will be used for robust stabilization.

Definition 1.2.5 *Uncertain system (1.11) is said to be stochastically stabilizable if there exists a control such that the closed-loop system is piecewise regular, impulse-free and stochastically stable.*

Definition 1.2.6 *Nominal system (1.11) is said to be stochastically stabilizable if there exists a control*

$$u(t) = K(r_t)x(t),\tag{1.17}$$

with $K(i)$, $i \in \mathcal{S}$, a constant matrix such that the closed-loop system is piecewise regular, impulse-free and stochastically stable.

More details on these techniques of stochastic stabilization will be given in the subsequent chapters. The third control problem we will consider in this volume is the filtering problem. It consists of estimating the state vector of the systems at each time t. This problem will be covered in Chap. 7.

1.3 Solution for the Dynamics

To facilitate the understanding of the different concepts of this volume, let us transform the dynamics (1.11) to the Jordan form and develop the closed-form solution of this dynamics. For simplicity of presentation, let us assume that the uncertainties are all equal to zero either in the matrix $A(i)$ or the matrix $B(i)$ for all $i \in \mathscr{S}$. In fact, for singular systems, we know that we can find, using Jordan canonical form decomposition, nonsingular matrices $\widehat{M}(i)$ and $\widehat{N}$ for more details of this subject we refer the reader to Dai [43]. The searched transformation will divide the state vector into two components named respectively slow and fast and denoted respectively by $\eta_1(t) \in \mathbb{R}^{n_1}$ and $\eta_2(t) \in \mathbb{R}^{n_2}$ with $n = n_1 + n_2$. This can be obtained by choosing:

$$\widehat{N}\eta(t) = x(t),$$

Using this transformation, the nominal state equation becomes:

$$E(r_t)\widehat{N}\dot{\eta}(t) = A(r_t)\widehat{N}\eta(t) + B(r_t)u(t).$$

If we pre-multiply this equation by $\widehat{M}(i)$ and using the fact that:

$$\widehat{M}(i)E(i)\widehat{N} = \begin{bmatrix} \mathbb{I} & \mathbf{0} \\ \mathbf{0} & N \end{bmatrix},$$

$$\widehat{M}(i)A(i)\widehat{N} = \begin{bmatrix} A_1(i) & \mathbf{0} \\ \mathbf{0} & \mathbb{I} \end{bmatrix},$$

$$\widehat{M}(i)B(i) = \begin{bmatrix} B_1(i) \\ B_2(i) \end{bmatrix}.$$

with $N \in \mathbb{R}^{n_2 \times n_2}$ is a nilpotent matrix (i.e., a square matrix whose eigenvalues are all 0, it is also a square matrix, N, such that $N^k \neq 0$ for $k = 1, \cdots, p-1$ and $N^p = 0$ for some positive integer power p) and $\mathbb{I} \in \mathbb{R}^{n_1 \times n_1}$ is an identity matrix with appropriate dimension, with $n_1 + n_2 = n$, we get:

$$\begin{bmatrix} \mathbb{I} & 0 \\ 0 & N \end{bmatrix}\begin{bmatrix} \dot{\eta}_1(t) \\ \dot{\eta}_2(t) \end{bmatrix} = \begin{bmatrix} A_1(r_t) & 0 \\ 0 & \mathbb{I} \end{bmatrix}\begin{bmatrix} \eta_1(t) \\ \eta_2(t) \end{bmatrix} + \begin{bmatrix} B_1(r_t) \\ B_2(r_t) \end{bmatrix}u(t),$$

which can be rewritten as follows:

$$\dot{\eta}_1(t) = A_1(r_t)\eta_1(t) + B_1(r_t)u(t),$$
$$N\dot{\eta}_2(t) = \eta_2(t) + B_2(r_t)u(t).$$

For the first differential equation, if we let $\tau_1, \tau_2, \ldots, \tau_k, \tau_{k+1}, \ldots$ and $i_1, i_2, \ldots, i_k, i_{k+1}, \ldots$ denote respectively the instant at which the process r_t jumps and the visited modes, the solution in the interval $[\tau_k, \tau_{k+1}]$ is given by (see Boukas [12]):

$$\eta_1(t) = e^{A_1(i_k)(t-\tau_k)}\eta_1(\tau_k) + \int_{\tau_k}^t e^{A_1(i_k)(t-\tau)}B_1(i_k)u(\tau)d\tau.$$

From the other side, we have:

$$\eta_1(\tau_k) = e^{A_1(i_{k-1})(\tau_k - \tau_{k-1})}\eta_1(\tau_{k-1}) + \int_{\tau_{k-1}}^{\tau_k} e^{A_1(i_{k-1})(\tau_k - \tau)}B_1(i_{k-1})u(\tau)d\tau\,.$$

Using this, we can rewrite the solution in the internal $[\tau_k, \tau_{k+1}]$ as follows:

$$\eta_1(t) = e^{A_1(i_k)(t-\tau_k)}\left[e^{A_1(i_{k-1})(\tau_k - \tau_{k-1})}\eta_1(\tau_{k-1}) + \int_{\tau_{k-1}}^{\tau_k} e^{A_1(i_{k-1})(\tau_k - \tau)}B_1(i_{k-1})u(\tau)d\tau \right]$$

$$+ \int_{\tau_k}^{t} e^{A_1(i_k)(t-\tau)}B_1(i_k)u(\tau)d\tau\,.$$

Let $\Psi_1(i_0)$ and $\Psi_2(i_l)$ be defined as follows:

$$\Psi_1(i_0) = e^{A_1(i_k)(t-\tau_k)}e^{A_1(i_{k-1})(\tau_k - \tau_k - 1)}\cdots e^{A_1(i_1)(\tau_2 - \tau_1)}\,,$$

$$\Psi_2(i_l) = e^{A_1(i_l)(t-\tau_l)}\,,$$

and proceeding iteratively we get:

$$\eta_1(t) = \Psi_1(i_0)\left[e^{A_1(i_0)(\tau_1 - \tau_0)}\eta_{10} + \int_{\tau_0}^{\tau_1} e^{A_1(i_0)(\tau_1 - \tau)}B_1(i_0)u(\tau)d\tau \right]$$

$$+ \sum_{l=2}^{k} \Psi_2(i_l)\left[\int_{\tau_{l-1}}^{\tau_l} e^{A_1(i_l)(\tau_l - \tau)}B_1(i_l)u(\tau)d\tau \right]$$

$$+ \int_{\tau_k}^{t} e^{A_1(i_k)(t-\tau)}B_1(i_k)u(\tau)d\tau\,,$$

which is completely determined by the initial conditions and the values of the control in the interval $[\tau_0, t]$, with $\tau_0 = 0$.

For the second algebraic equation, let us assume that the control $u(t)$ is p-times piecewise continuously differentiable. Therefore, by differentiating p-times and at each differentiation we pre-multiply by N, we have:

$$N\eta_2^{(1)}(t) = \eta_2(t) + B_2(r_t)u(t)\,,$$

$$N^2\eta_2^{(2)}(t) = N^1\eta_2^{(1)}(t) + N^1 B_2(r_t)u^{(1)}(t)\,,$$

$$\vdots$$

$$N^p\eta_2^{(p)}(t) = N^{p-1}\eta_2^{(p-1)}(t) + N^{p-1}B_2(r_t)u^{(p-1)}(t)\,,$$

where $\eta_2^{(k)}(t)$ and $u^{(k)}(t)$ stand respectively for the k-times derivative of $\eta_2(t)$ and $u(t)$ respectively.

Using now the fact that $N^p = 0$ and by adding all these equations we get:

$$0 = \eta_2(t) + \sum_{k=1}^{p-1} N^k B_2(i)u^{(k)}(t)\,,$$

which gives in turn:

$$\eta_2(t) = -\sum_{k=1}^{p-1} N^k B_2(i) u^{(k)}(t).$$

These equations are useful for stochastic stability. As it can be seen when $u(t) = 0$ for all $t \geq 0$, the stochastic stability of the system (1.11) is brought to the one of the first differential equation that is related to the stochastic stability of the slow state variable which is linked to the matrix $A_1(i)$, $i \in \mathscr{S}$, since the second gives $\eta_2(t) = 0$ for all $t \geq 0$.

1.4 Stochastic $\mathscr{H}_2$ and $\mathscr{H}_\infty$ Norms

In the subsequent chapters we will use the $\mathscr{H}_2$ and $\mathscr{H}_\infty$ norms. Before defining these norms let us recall some concepts on norms. First of all, let us consider the case of deterministic scalar function of time, i.e., $x(t) : [0, \infty[\rightarrow \mathbb{R}$. This function may represent one component of the state vector. In this case for any $x(t) \in \mathbb{R}$, we have the following norms:

- L_1 norm (total resources):

$$\|x\|_1 = \int_0^\infty |x(t)| dt.$$

- L_2 norm (energy):

$$\|x\|_2 = \left[\int_0^\infty x^2(t) dt \right]^{\frac{1}{2}}.$$

- L_∞ norm (peak value):

$$\|x\|_\infty = \sup_{t \geq 0} |x(t)|.$$

- RMS (root mean square):

$$\|x\|_{rms} = \left[\lim_{T \to \infty} \frac{1}{T} \int_0^T x^2(t) dt \right]^{\frac{1}{2}}.$$

For the case of deterministic vector function of time, i.e., $x(t) : [0, \infty[\rightarrow \mathbb{R}^n$, we have the following norms:

- L_1 norm (total resources):

$$\|x\|_1 = \int_0^\infty \sum_{k=1}^n |x_k(t)| dt,$$

$$= \int_0^\infty \|x(t)\|_1 dt,$$

$$= \sum_{k=1}^n \|x_k\|_1,$$

- L_2 norm (energy):

$$\|x\|_2 = \left[\int_0^\infty x^\top(t)x(t)dt\right]^{\frac{1}{2}} .$$

- L_∞ norm (peak value):

$$\|x\|_\infty = \sup_{t\geq 0} \max_k |x_k(t)|$$
$$= \sup_{t\geq 0} \|x(t)\|_\infty .$$

- RMS (root mean square):

$$\|x\|_{rms} = \left[\lim_{T\to\infty} \frac{1}{T} \int_0^T x^\top(t)x(t)dt\right]^{\frac{1}{2}} .$$

When the vector function of time t is stochastic and modeled by a stationary stochastic process we have the following norm:

- Covariance

$$V(x) = \mathbb{E}\left[x(t)x^\top(t)\right] .$$

- Autocorrelation

$$R(x, \tau) = \mathbb{E}\left[x(t + \tau)x^\top(t)\right] .$$

- Variance

$$\|x\|_{rms} = \left[\mathbb{E}x^\top(t)x(t)\right]^{\frac{1}{2}}$$
$$= \mathbb{E}\left[\operatorname{tr}\left[x(t)x^\top(t)\right]\right]$$
$$= \operatorname{tr}\left[\mathbb{E}x(t)x^\top(t)\right]$$
$$= \operatorname{tr}\left[V(x)\right]$$
$$= \operatorname{tr}\left[R(x, 0)\right] .$$

1.5 Lemmas

In the rest of this volume the following lemmas will be extensively used, it is why we recall them here to facilitate the readability of the book.

Lemma 1.5.1 *(See [27]) Let Y be a symmetric matrix and H, E be given matrices with appropriate dimensions and F satisfying $F^\top F \leq \mathbb{I}$. Then, we have*

(i) For any $\varepsilon > 0$, $HFE + E^\top F^\top H^\top \leq \varepsilon HH^\top + \varepsilon^{-1}E^\top E$.
(ii) $Y + HFE + E^\top F^\top H^\top < 0$ holds if and only if there exists a scalar $\varepsilon > 0$ such that the following holds $Y + \varepsilon HH^\top + \varepsilon^{-1}E^\top E < 0$.

Lemma 1.5.2 *(see [27])*

1. For any $x, y \in \mathbb{R}^n$,

$$\pm 2x^\top y \leq x^\top X x + y^\top X^{-1} y, \tag{1.18}$$

holds for any $X > 0$.

2. For any matrices U and $V \in \mathbb{R}^{n \times n}$ with $V > 0$, we have

$$U V^{-1} U^\top \geq U + U^\top - V. \tag{1.19}$$

Proof: The proof of (1.) is trivial and can be found in the Appendix of [27]. For the proof of (2.), note that since $V > 0$, we have the following:

$$(U - V)V^{-1}(U - V)^\top \geq 0,$$

which yields

$$U V^{-1} U^\top - U V^{-1} V^\top - V V^{-1} U^\top + V \geq 0.$$

This gives the desired results and ends the proof of the lemma.

1.6 Notes

This chapter presented the class of Markovian jump singular systems also known as systems with random abrupt changes. Different concepts where presented and some models were developed to motivate the studies of this book. We have mainly concentrated on square singular systems, but we could mention that we may encounter practical systems that may be modeled by rectangular systems with random abrupt changes. Notice also that some singular systems may exhibit discontinuities at the jumps in the state vector. For more details of this class of systems we refer reader to the works done by Raouf and Boukas ([1] and the references therein).

Part II

Stochastic stability

The stability problem is one of the most important problems in control theory and differential equations. This is due to the fact that the stability is always the first requirement that we should satisfy in each design problem. Roughly speaking, for a given dynamical system, the stability of this system is related to the values taken by the outputs that should take finite values for bounded inputs. The stability problem, either for the deterministic framework or the stochastic one, has attracted many researchers from mathematical and control communities and many results have been reported in the literature. For more details on this subject for normal dynamical systems, we refer the reader to [102, 66] for the deterministic framework and to [95, 27, 65, 14] for the stochastic framework and the references therein.

For the class of singular systems with random abrupt changes, we are considering in this book, only few results have been reported in the literature. Among them, we quote the works reported in [2, 123, 28, 21]. But for the deterministic framework, the subject has attracted a lot of researchers and many interesting results have been reported in the literature. Among them we quote [43, 59, 52, 6, 35, 32, 34, 33, 51, 54, 63, 64, 62, 74, 78, 79, 80, 88, 116, 117, 125, 122, 123] and the references therein. In these references, the stability problems for both the continuous-time and the discrete-time linear singular systems have been tackled and conditions were established. Some references gave LMI conditions for the continuous-time but for the discrete-time case to the best of our knowledge no result in the LMI setting is available up to date. The robust stability problem has also been tackled in the previous references. In most cases, the norm bounded uncertainties are considered.

This part deals with the stochastic stability of the class of systems with random abrupt changes. The concept of stochastic stability that we consider in this volume is the one used in the literature for linear systems with Markovian jumps. Notice that there exist different ways of defining the stochastic stability. For more details, we refer the reader to [60, 95, 27, 65] and the references therein. Our goal is to develop LMI conditions that can help us to check if a given system of the class under study is piecewise regular, impulse-free and stochastically stable. The robust stability problem is also considered and LMI conditions will be developed.

2

Stability

The stability problem is one of the most important problems in control theory and differential equations. This problem, either for the deterministic framework or the stochastic one, has attracted many researchers from mathematical and control communities and many results on the subject have been reported in the literature. For more details on this topic, we refer the reader to [66, 102] for the deterministic framework and to [27, 14, 95, 65] for the stochastic framework and the references therein.

For the class of systems we are considering in this book only few results have been reported in the literature. Among them, we quote the works reported in [2, 21]. But for the deterministic framework, the stability problem has attracted a lot of researchers and many interesting results on the subject have been reported in the literature. Among them we quote [43, 122] and the references therein.

As it was pointed out previously, the class of systems, in addition to inherent impulses, has discontinuities during the instants of the jumps of the process $\{r_t, t \geq 0\}$. The goal of this chapter is to develop results in the LMI setting to check if a given nominal system of the class we are treating in this book is piecewise regular, impulse-free and stochastically stable or specifically regular, impulses-free and stochastically stable between consecutive jumps. The results also are extended for the uncertain linear singular systems with random abrupt changes. Different sub-systems are also discussed and appropriate results are also developed.

2.1 Problem Statement

Let us consider a dynamical singular system with random abrupt changes defined in a fundamental probability space $(\Omega, \mathcal{F}, \mathbb{P})$ and assume that its state equation is described by the following differential algebraic systems:

$$\begin{cases} E(r_t) \dot{x}(t) = A(r_t, t)x(t), \\ x(0) = x_0, \end{cases} \tag{2.1}$$

where $x(t) \in \mathbb{R}^n$ is the state vector, $x_0 \in \mathbb{R}^n$ is the initial state, $\{r_t, t \geq 0\}$ is the continuous-time Markov process taking values in a finite space $\mathscr{S} = \{1, 2, \cdots, N\}$

and describes the evolution of the mode at time t, $E(i) \in \mathbb{R}^{n \times n}$ is a known singular matrix with rank $(E(i)) = n_r \leq n$ for all $i \in \mathscr{S}$, $A(i, t) \in \mathbb{R}^{n \times n}$ is a matrix with the following form for every $i \in \mathscr{S}$:

$$A(i, t) = A(i) + D_A(i) F_A(i) E_A(i),$$

with $A(i) \in \mathbb{R}^{n \times n}$, $D_A(i) \in \mathbb{R}^{n \times n_D}$, and $E_A(i) \in \mathbb{R}^{n_E \times n}$, are real known matrices with appropriate dimensions, and $F_A(i) \in \mathbb{R}^{n_D \times n_E}$ is an unknown real matrix that satisfies the following:

$$F_A^\top(i) F_A(i) \leq \mathbb{I}. \tag{2.2}$$

The Markov process $\{r_t, t \geq 0\}$ beside taking values in the finite set $\mathscr{S}$, represents the switching between the different modes and its behavior in time is described by the following probability transitions:

$$\mathbb{P}\left[r_{t+h} = j \,|\, r_t = i\right] = \begin{cases} \lambda_{ij} h + o(h), & \text{when } r_t \text{ jumps from } i \text{ to } j, \\ 1 + \lambda_{ii} h + o(h), & \text{otherwise,} \end{cases} \tag{2.3}$$

where λ_{ij} is the transition rate from mode i to mode j with $\lambda_{ij} \geq 0$ when $i \neq j$ and $\lambda_{ii} = -\sum_{j=1, j \neq i}^{N} \lambda_{ij}$ and $o(h)$ is such that $\lim_{h \to 0} \frac{o(h)}{h} = 0$.

Notice that, as we mentioned in the previous chapter, we will assume here when it is necessary that the transition matrix, Λ, belongs to a polytope, i. e.,

$$\Lambda = \sum_{k=1}^{K} \alpha_k \Lambda_k, \tag{2.4}$$

where κ is a positive given integer, $0 \leq \alpha_k \leq 1$ with $\sum_{k=1}^{K} \alpha_k = 1$ and Λ_k is a known transition matrix and its expression is given by:

$$\Lambda_k = \begin{bmatrix} \lambda_{11}^k & \cdots & \lambda_{1N}^k \\ \vdots & \ddots & \vdots \\ \lambda_{N1}^k & \cdots & \lambda_{NN}^k \end{bmatrix}, \tag{2.5}$$

where λ_{ij}^k keeps the same meaning as previous.

Remark 2.1.1 *The uncertainties satisfying the condition (2.2), (2.4) are referred to as admissible. The uncertainty term, in (2.2), is supposed to depend on the system's mode, r_t.*

Remark 2.1.2 *The matrix $E(i)$, for each $i \in \mathscr{S}$ is supposed to be singular which makes the dynamics (2.1) different from the one usually used to describe the behavior of the time-invariant dynamical systems as it is the normal practice. Notice that when $E(i)$, for each $i \in \mathscr{S}$ is not singular, (2.1) can be transformed easily to the class of Markov jump linear systems and the results developed in the literature (see Mariton [95], Boukas and Liu [27], Boukas [14] and the references therein), can be used to check the stochastic stability, of this class of systems.*

Definition 2.1.1 *System (2.1) is said to be stochastically stable (SS) if there exists a constant $M(x_0, r_0) > 0$ such that the following holds for any initial conditions (x_0, r_0):*

$$\mathbb{E}\left[\int_0^\infty x^\top(t)\, x(t)\,|\, x_0,\, r_0\right] \leq M(x_0,\, r_0). \tag{2.6}$$

The robust stochastic stability is defined in a similar manner.

Definition 2.1.2 *Uncertain system (2.1) is said to be robust stochastically stable (RSS) if there exists a constant $M(x_0, r_0)$ such that (2.6) holds for all admissible uncertainties.*

The goal of this chapter is to develop LMI-based stability conditions for system (2.1). The nominal and the uncertain cases are considered. Since in the case of uncertain systems, our results are only sufficient, therefore the emphasis is made only on the sufficient conditions even for nominal systems, these conditions are also necessary for nominal systems.

2.2 Stability of Singular Systems

Let us now consider the class of systems described by (2.1) and assume that all the uncertainties on the state matrix are equal to zero and study the stochastic stability of the nominal system. Our concern is to establish LMI conditions that can be used to check if a given dynamical system belonging to the class of systems we are considering in this chapter is piecewise regular, impulse-free and stochastically stable. The following theorem states the first result on stability of such class of systems.

Theorem 2.2.1 *The nominal singular system with random abrupt changes (2.1) is piecewise regular, impulse-free and stochastically stable if there exists a set of non-singular matrices $P = (P(1), \cdots, P(N))$, with $P(i) \in \mathbb{R}^{n \times n}$, such that the following set of coupled LMIs holds for each $i \in \mathscr{S}$:*

$$P^\top(i)\,A(i) + A^\top(i)\,P(i) + \sum_{j=1}^{N} \lambda_{ij}\, E^\top(j)P(j) < 0, \tag{2.7}$$

with the following constraints:

$$E^\top(i)P(i) = P^\top(i)E(i) \geq 0,\, \forall i \in \mathscr{S}. \tag{2.8}$$

Proof: Under the conditions of the theorem, we will first show the regularity and absence of impulses in the system (2.1) between two consecutive jumps. By (2.7), the following holds for each $i \in \mathscr{S}$:

$$P^\top(i)A(i) + A^\top(i)P(i) + \sum_{j=1}^{N} \lambda_{ij}E^\top(j)P(j) < 0. \tag{2.9}$$

Now, choose two nonsingular matrices $\widehat{M}(i)$ and $\widehat{N}$ such that:

$$\widehat{M}(i)E(i)\widehat{N} = \begin{bmatrix} I & 0 \\ 0 & 0 \end{bmatrix},$$

and write

$$\widehat{M}(i)A(i)\widehat{N} = \begin{bmatrix} \widehat{A}_1(i) & \widehat{A}_2(i) \\ \widehat{A}_3(i) & \widehat{A}_4(i) \end{bmatrix}, \quad \widehat{M}^{-\mathsf{T}}(i)P(i)\widehat{N} = \begin{bmatrix} \widehat{P}_1(i) & \widehat{P}_2(i) \\ \widehat{P}_3(i) & \widehat{P}_4(i) \end{bmatrix}.$$

Then, by (2.8), it can be shown that $\widehat{P}_2(i) = 0$. Pre- and post-multiplying (2.9) by $\widehat{N}^{\mathsf{T}}$ and $\widehat{N}$, respectively, we have

$$\begin{bmatrix} \star & \star \\ \star & \widehat{A}_4^{\mathsf{T}}(i)\widehat{P}_4(i) + \widehat{P}_4^{\mathsf{T}}(i)\widehat{A}_4(i) \end{bmatrix} < 0, \tag{2.10}$$

where $\star$ will not be used in the following development. Then, by (2.10), we have

$$\widehat{A}_4^{\mathsf{T}}(i)\widehat{P}_4(i) + \widehat{P}_4^{\mathsf{T}}(i)\widehat{A}_4(i) < 0,$$

which implies that $\widehat{A}_4(i)$ is nonsingular (i. e., $\widehat{A}_4^{-1}(i)$ exists for each $i \in \mathscr{S}$). Therefore, nominal system (2.1) is piecewise regular and impulse-free.

Next, we will show the stochastic stability. Since system (2.1) is piecewise regular and impulse-free, for any $i \in \mathscr{S}$, we can choose nonsingular matrices $\widetilde{M}(i)$ and $\widetilde{N}$ such that

$$\widetilde{M}(i)E(i)\widetilde{N} = \begin{bmatrix} I & 0 \\ 0 & 0 \end{bmatrix}, \quad \widetilde{M}(i)A(i)\widetilde{N} = \begin{bmatrix} \widetilde{A}_1(i) & 0 \\ 0 & I \end{bmatrix}.$$

Write the matrix $\widetilde{M}^{-\mathsf{T}}(i)P(i)\widetilde{N}$ as follows:

$$\widetilde{P}(i) = \widetilde{M}^{-\mathsf{T}}(i)P(i)\widetilde{N} = \begin{bmatrix} \widetilde{P}_1(i) & \widetilde{P}_2(i) \\ \widetilde{P}_3(i) & \widetilde{P}_4(i) \end{bmatrix}.$$

Then, for any $i \in \mathscr{S}$, nominal system (2.1) becomes equivalent to the following one:

$$\dot{\xi}_1(t) = A_1(i)\xi_1(t),$$
$$0 = \xi_2(t),$$

where

$$\xi(t) = \begin{bmatrix} \xi_1(t) \\ \xi_2(t) \end{bmatrix} = \widetilde{N}^{-1}x(t).$$

Now, let us choose the following Lyapunov functional:

$$V(x(t), r_t) = x^{\mathsf{T}}(t)E^{\mathsf{T}}(r_t)P(r_t)x(t),$$

Let $\mathscr{L}$ be the weak infinitesimal operator of the random process $\{(x(t), r_t), t \geq 0\}$. Then, if at time t, $x(t) = x$ and $r_t = i$, $i \in \mathscr{S}$, the infinitesimal operator emanating from the point (x, i) at time t is given by:

$$\mathscr{L}V(x(t), i) = \lim_{h \to 0} \frac{1}{h} \mathbb{E}\left[V(x(t+h), r_{t+h}) - V(x, i)|x(t) = x, r_t = i\right]$$

$$= \dot{x}^\top(t)E^\top(i)P(i)x(t) + x^\top(t)E^\top(i)P(i)\dot{x}(t)$$

$$+ \sum_{j=1}^{N} \lambda_{ij} x^\top(t)E^\top(j)P(j)x(t) \,.$$

Using the fact that $E^\top(i)P(i) = P^\top(i)E(i) \geq 0$, we get:

$$\mathscr{L}V(x(t), i) = x^\top(t)\left[A^\top(i)P(i) + P^\top(i)A(i) + \sum_{j=1}^{N} \lambda_{ij} E^\top(j)P(j)\right]x(t) \,,$$

which can be rewritten as follows:

$$\mathscr{L}V(\mathbf{x}(t), i) = x^\top(t)\Psi(i)x(t) \,,$$

where $\Psi(i) = P^\top(i)A(i) + A^\top(i)P(i) + \sum_{j=1}^{N} \lambda_{ij} E^\top(j)P(j)$.

Thus,

$$\mathscr{L}V(x(t), i) \leq -\min_{i \in \mathscr{S}} \lambda_{min}\left[-\Psi(i)\right] x^\top(t)x(t) \,.$$

By Dynkin's formula, we obtain

$$\mathbb{E}[V(x(t), i)] - V(x_0, r_0) = \mathbb{E}\left[\int_0^t \mathscr{L}V(x(s), r_s)ds\right] \,,$$

$$\leq -\min_{i \in \mathscr{S}}\{\lambda_{min}(-\Psi(i))\}\mathbb{E}\left[\int_0^t x^\top(s)x(s)ds|(x_0, r_0)\right] \,,$$

which implies, in turn,

$$\min_{i \in \mathscr{S}}\{\lambda_{min}(-\Psi(i))\}\mathbb{E}\left[\int_0^t x^\top(s)x(s)ds|(x_0, r_0)\right]$$

$$\leq V(x(0), r_0) - \mathbb{E}[V(x(t), i)]$$

$$\leq V(x(0), r_0) \,.$$

This yields that

$$\mathbb{E}\left[\int_0^t x^\top(s)x(s)ds|(x_0, r_0)\right] \leq \frac{V(x(0), r_0)}{\min_{i \in \mathscr{S}}\{\lambda_{min}(-\Psi(i))\}} \,,$$

holds for any $t > 0$. Letting t goes to infinity implies that

$$\mathbb{E}\left[\int_0^\infty x^\top(s)x(s)ds|(x_0, r_0)\right] \,,$$

is bounded by a constant $T(x_0, r_0)$ given by:

$$M(x_0, r_0) = \frac{V(x(0), r_0)}{\min_{i \in \mathscr{S}}\{\lambda_{min}(-\Psi(i))\}},$$

which implies that (2.1) is stochastically stable.

Remark 2.2.1 *Notice that when $E(i) = \mathbb{I}$ for $i \in \mathscr{S}$, the results of this theorem are the one for stochastic stability of the class of Markovian jump linear systems. Therefore, the results of this theorem is a generalization of those on stochastic stability in Boukas [14].*

Notice that the conditions of Theorem 2.2.1 can be transformed to get equivalent conditions. In the next paragraphs, we will develop new equivalent conditions. First of all, notice that the last term of the condition (2.7) can be rewritten as follows:

$$\sum_{j=1}^{N} \lambda_{ij} E^\top(j) P(j) = \lambda_{ii} E^\top(i) P(i) + \sum_{j=1,j\neq i}^{N} \lambda_{ij} E^\top(j) P(j).$$

Now, if we assume that there exists an $\varepsilon_P > 0$ such that the following holds:

$$E^\top(i) P(i) \leq \varepsilon_P \left[P(i) + P^\top(i) \right], \tag{2.11}$$

we get the following results.

Corollary 2.2.1 *Let ε_P be a given positive scalar. The nominal singular system with random abrupt changes (2.1) is piecewise regular, impulse-free and stochastically stable if there exists a set of nonsingular matrices $P = (P(1), \cdots, P(N))$, with $P(i) \in \mathbb{R}^{n\times n}$, such that the following set of coupled LMIs holds for each $i \in \mathscr{S}$:*

$$P^\top(i) A(i) + A^\top(i) P(i) + \lambda_{ii} E^\top(i) P(i)$$

$$+ \sum_{j=1,j\neq i}^{N} \varepsilon_P \lambda_{ij} \left[P(j) + P^\top(j) \right] < 0, \tag{2.12}$$

with the following constraints:

$$\varepsilon_P \left[P(i) + P^\top(i) \right] \geq E^\top(i) P(i) = P^\top(i) E(i) \geq 0. \tag{2.13}$$

Remark 2.2.2 *Notice that the scalar ε_P can be chosen mode-dependent and therefore get less conservative results.*

Corollary 2.2.2 *Let $\varepsilon_P = (\varepsilon_P(1), \cdots, \varepsilon_P(N))$ be a given set of positive scalars. The nominal singular system with random abrupt changes (2.1) is piecewise regular, impulse-free and stochastically stable if there exists a set of nonsingular matrices $P = (P(1), \cdots, P(N))$, with $P(i) \in \mathbb{R}^{n\times n}$, such that the following set of coupled inequalities holds for each $i \in \mathscr{S}$:*

$$P^\top(i) A(i) + A^\top(i) P(i) + \lambda_{ii} E^\top(i) P(i)$$

$$+ \sum_{j=1,j\neq i}^{N} \varepsilon_P(j) \lambda_{ij} \left[P(j) + P^\top(j) \right] < 0, \tag{2.14}$$

with the following constraints:

$$\varepsilon_P(i)\left[P(i) + P^\top(i)\right] \geq E^\top(i)P(i) = P^\top(i)E(i) \geq 0. \tag{2.15}$$

Some direct conservative results can be obtained if we chose the following Lyapunov function:

$$V(x_t, r_t) = x^\top(t)E^\top(r_t)Px(t),$$

with P mode-independent with the following assumption:

$$E^\top(i)P \leq \varepsilon_P(i)\left[P + P^\top\right],$$

we get the following results:

Corollary 2.2.3 *Let $\varepsilon_P = (\varepsilon_P(1), \cdots, \varepsilon_P(N))$ be a given set of positive scalars. The nominal singular system with random abrupt changes (2.1) is piecewise regular, impulse-free and stochastically stable if there exists a nonsingular matrix $P \in \mathbb{R}^{n \times n}$ such that the following set of coupled LMIs holds for each $i \in \mathscr{S}$:*

$$P^\top A(i) + A^\top(i)P + \lambda_{ii}E^\top(i)P$$

$$+ \sum_{j=1, j\neq i}^{N} \varepsilon_P(j)\lambda_{ij}\left[P + P^\top\right] < 0, \tag{2.16}$$

with the following constraints:

$$\varepsilon_P(i)\left[P + P^\top\right] \geq E^\top(i)P = P^\top E(i) \geq 0, \tag{2.17}$$

When the parameter ε_P is chosen mode-independent, the results become:

Corollary 2.2.4 *Let ε_P be a given positive scalar. The nominal singular system with random abrupt changes (2.1) is piecewise regular, impulse-free and stochastically stable if there exists a nonsingular matrix $P \in \mathbb{R}^{n \times n}$ such that the following set of coupled LMIs holds for each $i \in \mathscr{S}$:*

$$P^\top A(i) + A^\top(i)P + \lambda_{ii}E^\top(i)P + \sum_{j=1, j\neq i}^{N} \varepsilon_P \lambda_{ij}\left[P + P^\top\right] < 0, \tag{2.18}$$

with the following constraints:

$$\varepsilon_P\left[P + P^\top\right] \geq E^\top(i)P = P^\top E(i) \geq 0. \tag{2.19}$$

Remark 2.2.3 *Using the fact that $\lambda_{ii} = -\sum_{j=1, j\neq i}^{N} \lambda_{ij}$ the condition (2.18) becomes*

$$P^\top A(i) + A^\top(i)P + \lambda_{ii}\left[E^\top(i)P - \varepsilon_P\left[P + P^\top\right]\right] < 0.$$

Since $\lambda_{ii} \leq 0$ and $\left[E^\top(i)P - \varepsilon_P\left[P + P^\top\right]\right] < 0$ this condition gives:

$$P^\top A(i) + A^\top(i)P < 0.$$

Now, as a second result on the stochastic stability of system (2.1) if we assume that there exists an $\varepsilon(i) > 0$ such that the following holds:

$$E^\top(i)P(i) \leq \varepsilon(i)\left[P^\top(i)P(i)\right], \tag{2.20}$$

we get the following results.

Corollary 2.2.5 *Let $\varepsilon = (\varepsilon(1), \cdots, \varepsilon(N))$ be a given set of positive scalars. The nominal singular system with random abrupt changes (2.1) is piecewise regular, impulse-free and stochastically stable if there exists a set of nonsingular matrices $P = (P(1), \cdots, P(N))$, with $P(i) \in \mathbb{R}^{n \times n}$, such that the following set of coupled matrix inequalities holds for each $i \in \mathscr{S}$:*

$$P^\top(i)A(i) + A^\top(i)P(i) + \lambda_{ii}E^\top(i)P(i)$$

$$+ \sum_{j=1, j\neq i}^{N} \varepsilon(j)\lambda_{ij}\left[P^\top(j)P(j)\right] < 0, \tag{2.21}$$

with the following constraints:

$$\varepsilon(i)\left[P^\top(i)P(i)\right] \geq E^\top(i)P(i) = P^\top(i)E(i) \geq 0. \tag{2.22}$$

Remark 2.2.4 *We can transform the conditions of the previous corollary into an LMI setting. Letting $X(i) = P^{-1}(i)$ and pre- and post-multiplying (2.21) respectively by $X^\top(i)$ and $X(i)$, we get:*

$$A(i)X(i) + X^\top(i)A^\top(i) + \lambda_{ii}X^\top(i)E^\top(i)$$

$$+ \sum_{j=1, j\neq i}^{N} \varepsilon(j)\lambda_{ij}X^\top(i)\left[X^{-\top}(j)X^{-1}(j)\right]X(i) < 0.$$

By defining $S_i(X)$ and $X_i(X)$ as follows:

$$S_i(X) = \left[\sqrt{(\lambda_{i1})}X^\top(i), \cdots, \sqrt{(\lambda_{ii-1})}X^\top(i), \sqrt{(\lambda_{ii+1})}X^\top(i), \cdots, \sqrt{(\lambda_{iN})}X^\top(i)\right],$$

$$X_i(X) = diag\left[X^\top(1) + X(1) - \varepsilon(1)\mathbb{I}, \cdots, X^\top(i-1) + X(i-1) - \varepsilon(i-1)\mathbb{I},\right.$$

$$\left. X^\top(i+1) + X(i+1) - \varepsilon(i+1)\mathbb{I}, \cdots, X^\top(N) + X(N) - \varepsilon(N)\mathbb{I},\right],$$

we can get the following:

$$\begin{bmatrix} J(i) & S_i(X) \\ S_i^\top(X) & -X_i(X) \end{bmatrix} < 0, \tag{2.23}$$

with $J(i) = A(i)X(i) + X^\top(i)A^\top(i) + \lambda_{ii}X^\top(i)E^\top(i)$.

For (2.22) we get:

$$\varepsilon(i)\mathbb{I} \geq X^\top(i)E^\top(i) = E(i)X(i) \geq 0. \tag{2.24}$$

This development gives the following results.

Corollary 2.2.6 *The nominal singular system with random abrupt changes (2.1) is piecewise regular, impulse-free and stochastically stable if there exists a set of non-singular matrices $X = (X(1), \cdots, X(N))$, with $X(i) \in \mathbb{R}^{n \times n}$ and a set of positive scalars $\varepsilon = (\varepsilon(1), \cdots, \varepsilon(N))$, such that the following set of coupled matrix inequalities holds for each $i \in \mathscr{S}$:*

$$\begin{bmatrix} J(i) & S_i(X) \\ S_i^{\top}(X) & -X_i(X) \end{bmatrix} < 0, \tag{2.25}$$

where $J(i) = A(i)X(i) + X^{\top}(i)A^{\top}(i) + \lambda_{ii}X^{\top}(i)E^{\top}(i)$, with the following constraints:

$$\varepsilon(i)\mathbb{I} \geq X^{\top}(i)E^{\top}(i) = E(i)X(i) \geq 0. \tag{2.26}$$

We can also establish another conditions for the class of systems we are dealing with to be piecewise regular impulse-free and stochastically stable. In fact, notice that the term $E^{\top}(i)P(i)$ can be rewritten as follows:

$$E^{\top}(i)P(i) = \frac{1}{2}E^{\top}(i)P(i) + \frac{1}{2}E^{\top}(i)P(i),$$

and since $E^{\top}(i)P(i) = P^{\top}(i)E(i)$, we get:

$$E^{\top}(i)P(i) = \frac{1}{2}E^{\top}(i)P(i) + \frac{1}{2}P^{\top}(i)E(i),$$

using now Lemma 1.5.1, we get:

$$E^{\top}(i)P(i) \leq \frac{1}{4}\varepsilon(i)\mathbb{I} + \varepsilon^{-1}(i)E^{\top}(i)P(i)P^{\top}(i)E(i),$$

for any $\varepsilon(i) > 0$.

Therefore we have the following results.

Corollary 2.2.7 *The nominal singular system with random abrupt changes (2.1) is piecewise regular, impulse-free and stochastically stable if there exist a set of nonsingular matrices $P = (P(1), \cdots, P(N))$, with $P(i) \in \mathbb{R}^{n \times n}$, and a set of positive scalars $\varepsilon = (\varepsilon(1), \cdots, \varepsilon(N))$ such that the following set of coupled matrix inequalities holds for each $i \in \mathscr{S}$:*

$$P^{\top}(i)A(i) + A^{\top}(i)P(i) + \lambda_{ii}E^{\top}(i)P(i)$$
$$+ \sum_{j=1, j \neq i}^{N} \lambda_{ij} \left[\frac{1}{4}\varepsilon(j)\mathbb{I} + \varepsilon^{-1}(j)E^{\top}(j)P(j)P^{\top}(j)E(j) \right] < 0, \tag{2.27}$$

with the following constraints:

$$E^{\top}(i)P(i) = P^{\top}(i)E(i) \geq 0. \tag{2.28}$$

Remark 2.2.5 *Notice that if we define:*

$$S_i(P) = \left[\sqrt{\lambda_{i1}} E^\top(1)P(1), \cdots, \sqrt{\lambda_{ii-1}} E^\top(i-1)P(i-1), \right.$$
$$\left. \sqrt{\lambda_{ii+1}} E^\top(i+1)P(i+1), \cdots, \sqrt{\lambda_{iN}} E^\top(N)P(N) \right],$$
$$\mathcal{X}_i(\varepsilon) = diag\left[\varepsilon(1)\mathbb{I}, \cdots, \varepsilon(i-1)\mathbb{I}, \varepsilon(i+1)\mathbb{I}, \cdots, \varepsilon(N)\mathbb{I} \right],$$

then we have:

$$\sum_{j=1,j\neq i}^{N} \lambda_{ij}\varepsilon^{-1}(j)E^\top(j)P(j)P^\top(j)E(j) = S_i(P)\mathcal{X}_i^{-1}(\varepsilon)S_i^\top(P).$$

Using this and Schur complement we get the following set of LMIs:

$$\begin{bmatrix} J(i) & S_i(P) \\ S_i^\top(P) & -\mathcal{X}_i(\varepsilon) \end{bmatrix} < 0,$$

where $J(i) = A^\top(i)P(i) + P^\top(i)A(i) + \lambda_{ii}E^\top(i)P(i) + \sum_{j=1,j\neq i}^{N} \frac{1}{4}\lambda_{ij}\varepsilon(j)\mathbb{I}$, *which combined with the following constraints:*

$$E^\top(i)P(i) = P^\top(i)E(i) \geq 0, \forall i \in \mathscr{S},$$

give the required conditions to solve to check if the system is piecewise regular, impulse-free and stochastically stable.

Remark 2.2.6 *For the assumptions that give us the last two corollaries, we can establish similar conservative results to the ones we established for the first assumption. We will omit this and we let this as an exercise for the reader.*

2.3 Robust Stability

Let us now concentrate on the robust stability of our system and develop sufficient conditions which guarantee that the closed-loop state equation of the singular system with random abrupt changes will be piecewise regular, impulse-free and stochastically stable for all admissible uncertainties. For this purpose, using the results of Theorem 2.2.1, the system will be piecewise regular, impulse free and stochastically stable if there exists a set of nonsingular matrices $P = (P(1), \cdots, P(N))$, such that the following set of coupled inequality matrices holds for each $i \in \mathscr{S}$:

$$P^\top(i)A(i) + A^\top(i)P(i) + P^\top(i)D_A(i)F_A(i)E_A(i) + \lambda_{ii}E^\top(i)P(i)$$
$$+ E_A^\top(i)F_A^\top(i)D^\top(i)P(i) + \sum_{j=1,j\neq i}^{N} \lambda_{ij}E^\top(j)P^\top(j) < 0,$$

with the following constraints:

$$E^\top(i)P(i) = P^\top(i)E(i) \geq 0.$$

Using Lemma 1.5.1, for any $\varepsilon_A(i) > 0$, $i \in \mathscr{S}$ we have:

$$P^\top(i)D_A(i)F_A(i)E_A(i) + E_A^\top(i)F_A^\top(i)D^\top(i)P(i)$$
$$\leq \varepsilon_A^{-1}(i)P^\top(i)D_A(i)D_A^\top(i)P(i) + \varepsilon_A(i)E_A^\top(i)E_A(i).$$

Using now this inequality and Schur complement, we get the following results for the robust stochastic stability.

Theorem 2.3.1 *The uncertain singular system with random abrupt changes (2.1) is piecewise regular, impulse-free and stochastically stable if there exist a set of nonsingular matrices $P = (P(1), \cdots, P(N))$, with $P(i) \in \mathbb{R}^{n \times n}$ and a set of positive scalars $\varepsilon_A = (\varepsilon_A(1), \cdots, \varepsilon_A(N))$ such that the following set of coupled LMIs holds for each $i \in \mathscr{S}$ and for all admissible uncertainties:*

$$\begin{bmatrix} J(i) & P^\top(i)D_A(i) \\ D_A^\top(i)P(i) & -\varepsilon_A(i)\mathbb{I} \end{bmatrix} < 0, \tag{2.29}$$

where

$$J(i) = P^\top(i)A(i) + A^\top(i)P(i) + \varepsilon_A(i)E^\top(i)E_A(i)$$
$$+ \lambda_{ii}E^\top(i)P(i) + \sum_{j=1, j \neq i}^{N} \lambda_{ij}E^\top(j)P(j).$$

with the following constraints:

$$E^\top(i)P(i) = P^\top(i)E(i) \geq 0, \tag{2.30}$$

As we did previously for the nominal case with the different assumptions on how to get an upper bound of the term $E^\top(i)P(i)$, we can easily establish the results of the following corollaries.

Corollary 2.3.1 *Let $\varepsilon_P = (\varepsilon_P(1), \cdots, \varepsilon_P(N))$ be a given set of positive scalars. The uncertain singular system with random abrupt changes (2.1) is piecewise regular, impulse-free and stochastically stable if there exist a set of nonsingular matrices $P = (P(1), \cdots, P(N))$, with $P(i) \in \mathbb{R}^{n \times n}$ and a set of positive scalars $\varepsilon_A = (\varepsilon_A(1), \cdots, \varepsilon_A(N))$, such that the following set of coupled matrix inequalities holds for each $i \in \mathscr{S}$ and for all admissible uncertainties:*

$$\begin{bmatrix} J(i) & P^\top(i)D_A(i) \\ D_A^\top(i)P(i) & -\varepsilon_A(i)\mathbb{I} \end{bmatrix} < 0, \tag{2.31}$$

where

$$J(i) = P^\top(i)A(i) + A^\top(i)P(i) + \varepsilon_A(i)E_A^\top(i)E_A(i)$$
$$+ \lambda_{ii}E^\top(i)P(i) + \sum_{j=1, j \neq i}^{N} \varepsilon_P(j)\lambda_{ij}\left[P(j) + P^\top(j)\right],$$

with the following constraints:

$$\varepsilon_P(i)\left[P(i) + P^\top(i)\right] \geq E^\top(i)P(i) = P^\top(i)E(i) \geq 0. \tag{2.32}$$

Now if we use the fact that $E^\top(i)P(i) \leq \varepsilon(i)P^\top(i)P(i)$, for any $\varepsilon(i) > 0$, we get the following results.

Corollary 2.3.2 *The uncertain singular system with random abrupt changes (2.1) is piecewise regular, impulse-free and stochastically stable if there exist a set of non-singular matrices $P = (P(1), \cdots, P(N))$, with $P(i) \in \mathbb{R}^{n \times n}$ and sets of positive scalars $\varepsilon_A = (\varepsilon_A(1), \cdots, \varepsilon_A(N))$ and $\varepsilon = (\varepsilon(1), \cdots, \varepsilon(N))$ such that the following set of coupled matrix inequalities holds for each $i \in \mathscr{S}$ and for all admissible uncertainties:*

$$\begin{bmatrix} J(i) & P^\top(i)D_A(i) \\ D_A^\top(i)P(i) & -\varepsilon_A(i)\mathbb{I} \end{bmatrix} < 0, \tag{2.33}$$

where

$$J(i) = P^\top(i)A(i) + A^\top(i)P(i) + \varepsilon_A(i)E_A^\top(i)E_A(i)$$
$$+ \lambda_{ii}E^\top(i)P(i) + \sum_{j=1, j \neq i}^{N} \varepsilon(j)\lambda_{ij}\left[P^\top(j)P(j)\right],$$

with the following constraints:

$$\varepsilon(i)\left[P^\top(i)P(i)\right] \geq E^\top(i)P(i) = P^\top(i)E(i) \geq 0. \tag{2.34}$$

As we did for the nominal case we can get the following results.

Corollary 2.3.3 *The uncertain singular system with random abrupt changes (2.1) is piecewise regular, impulse-free and stochastically stable if there exist a set of nonsingular matrices $X = (X(1), \cdots, X(N))$, with $X(i) \in \mathbb{R}^{n \times n}$ and sets of positive scalars $\varepsilon_A = (\varepsilon_A(1), \cdots, \varepsilon_A(N))$ and $\varepsilon = (\varepsilon(1), \cdots, \varepsilon(N))$ such that the following set of coupled matrix inequalities holds for each $i \in \mathscr{S}$ and for all admissible uncertainties:*

$$\begin{bmatrix} J(i) & P^\top(i)D_A(i) & S_i(X) \\ D_A^\top(i)P(i) & -\varepsilon_A(i) & 0 \\ S_i^\top(X) & 0 & -X_i(X) \end{bmatrix} < 0, \tag{2.35}$$

where

$$J(i) = P^\top(i)A(i) + A^\top(i)P(i) + \varepsilon_A(i)E_A^\top(i)E_A(i)$$
$$+ \lambda_{ii}E^\top(i)P(i),$$

with the following constraints:

$$\varepsilon(i)\mathbb{I} \geq X^\top(i)E^\top(i) = E(i)X(i) \geq 0. \tag{2.36}$$

By using now the fact that:

$$E^\top(i)P(i) \leq \frac{1}{4}\varepsilon(i)\mathbb{I} + \varepsilon^{-1}(i)E^\top(i)P(i)P^\top(i)E(i),$$

for any $\varepsilon(i) > 0$, we get the following results.

Corollary 2.3.4 *The uncertain singular system with random abrupt changes (2.1) is piecewise regular, impulse-free and stochastically stable if there exist a set of non-singular matrices $P = (P(1), \cdots, P(N))$, with $P(i) \in \mathbb{R}^{n \times n}$ and sets of positive scalars $\varepsilon_A = (\varepsilon_A(1), \cdots, \varepsilon_A(N))$ and $\varepsilon = (\varepsilon(1), \cdots, \varepsilon(N))$ such that the following set of coupled matrix inequalities holds for each $i \in \mathscr{S}$ and for all admissible uncertainties:*

$$\begin{bmatrix} J(i) & P^\top(i)D_A(i) \\ D_A^\top(i)P(i) & -\varepsilon_A(i)\mathbb{I} \end{bmatrix} < 0, \tag{2.37}$$

where

$$J(i) = P^\top(i)A(i) + A^\top(i)P(i) + \varepsilon_A(i)E_A^\top(i)E_A(i) + \lambda_{ii}E^\top(i)P(i)$$
$$+ \sum_{j=1, j\neq i}^{N} \lambda_{ij}\left[\frac{1}{4}\varepsilon(j)\mathbb{I} + \varepsilon^{-1}(j)E^\top(j)P(j)P^\top(j)E(j)\right],$$

with the following constraints:

$$E^\top(i)P(i) = P^\top(i)E(i) \geq 0. \tag{2.38}$$

Remark 2.3.1 *Notice that if we define:*

$$S_i(P) = \left[\sqrt{\lambda_{i1}}E^\top(1)P(1), \cdots, \sqrt{\lambda_{ii-1}}E^\top(i-1)P(i-1),\right.$$
$$\left.\sqrt{\lambda_{ii+1}}E^\top(i+1)P(i+1), \cdots, \sqrt{\lambda_{iN}}E^\top(N)P(N)\right],$$
$$X_i(\varepsilon) = diag\left[\varepsilon(1)\mathbb{I}, \cdots, \varepsilon(i-1)\mathbb{I}, \varepsilon(i+1)\mathbb{I}, \cdots, \varepsilon(N)\mathbb{I}\right],$$

then we have:

$$\sum_{j=1, j\neq i}^{N} \lambda_{ij}\varepsilon^{-1}(j)E^\top(j)P(j)P^\top(j)E(j) = S_i(P)X_i^{-1}(\varepsilon)S_i^\top(P).$$

Using this and Schur complement we get the following set of LMIs:

$$\begin{bmatrix} J(i) & P^\top(i)D_A(i) & S_i(P) \\ D_A^\top(i)P(i) & -\varepsilon_A(i)\mathbb{I} & 0 \\ S_i^\top(P) & 0 & -X_i(\varepsilon) \end{bmatrix} < 0,$$

where $J(i) = A^\top(i)P(i) + P^\top(i)A(i) + \lambda_{ii}E^\top(i)P(i) + \varepsilon_A(i)E_A^\top(i)E_A(i) + \sum_{j=1, j\neq i}^{N} \frac{1}{4}\lambda_{ij}\varepsilon(j)\mathbb{I}$, *which combined with the following constraints:*

$$E^\top(i)P(i) = P^\top(i)E(i) \geq 0, \forall i \in \mathscr{S},$$

give the required conditions to check if the system is piecewise regular, impulse-free and stochastically stable. This result is given by the following corollary.

Corollary 2.3.5 *The uncertain singular system with random abrupt changes (2.1) is piecewise regular, impulse-free and stochastically stable if there exist a set of non-singular matrices $P = (P(1), \cdots, P(N))$, with $P(i) \in \mathbb{R}^{n \times n}$ and sets of positive scalars $\varepsilon_A = (\varepsilon_A(1), \cdots, \varepsilon_A(N))$ and $\varepsilon = (\varepsilon(1), \cdots, \varepsilon(N))$ such that the following set of coupled matrix inequalities holds for each $i \in \mathscr{S}$ and for all admissible uncertainties:*

$$\begin{bmatrix} J(i) & P^\top(i)D_A(i) & S_i(P) \\ D_A^\top(i)P(i) & -\varepsilon_A(i)\mathbb{I} & 0 \\ S_i^\top(P) & 0 & -\mathcal{X}_i(\varepsilon) \end{bmatrix} < 0,$$

(2.39)

where

$$J(i) = P^\top(i)A(i) + A^\top(i)P(i) + \varepsilon_A(i)E_A^\top(i)E_A(i) + \lambda_{ii}E^\top(i)P(i)$$

$$+ \sum_{j=1, j \neq i}^{N} \lambda_{ij}\frac{1}{4}\varepsilon(j)\mathbb{I},$$

with the following constraints:

$$E^\top(i)P(i) = P^\top(i)E(i) \geq 0.$$

(2.40)

Remark 2.3.2 *As we did for the nominal case for the three assumptions that gave us bounds on the term $E^\top(i)P(i)$, we can also establish similar conservative results for the uncertain case.*

Before closing this subsection, let us consider the effect of the uncertainties of the transition matrix besides the uncertainties on the matrices of the dynamics. In this case, the results are given by the following theorems for the different approaches.

Theorem 2.3.2 *The uncertain singular system with random abrupt changes (2.1) is piecewise regular, impulse-free and stochastically stable if there exist a set of nonsingular matrices $P = (P(1), \cdots, P(N))$, with $P(i) \in \mathbb{R}^{n \times n}$ and a set of positive scalars $\varepsilon_A = (\varepsilon_A(1), \cdots, \varepsilon_A(N))$ such that the following set of coupled LMIs holds for each $i \in \mathscr{S}$ and for all admissible uncertainties:*

$$\begin{bmatrix} J(i) & P^\top(i)D_A(i) \\ D_A^\top(i)P(i) & -\varepsilon_A(i)\mathbb{I} \end{bmatrix} < 0,$$

(2.41)

where

$$J(i) = P^\top(i)A(i) + A^\top(i)P(i) + \varepsilon_A(i)E^\top(i)E_A(i)$$

$$+ \sum_{k=1}^{\kappa} \sum_{j=1}^{N} \lambda_{ij}^k E^\top(j)P(j),$$

with the following constraints:

$$E^\top(i)P(i) = P^\top(i)E(i) \geq 0.$$

(2.42)

Using the upper bounds we presented earlier for the term

$$\sum_{k=1}^{\kappa} \sum_{j=1}^{N} \lambda_{ij}^k E^{\top}(j)P(j),$$

we get the following corollaries.

Corollary 2.3.6 *Let* $\varepsilon_P = (\varepsilon_P(1), \cdots, \varepsilon_P(N))$ *be a given set of positive scalars. The uncertain singular system with random abrupt changes (2.1) is piecewise regular, impulse-free and stochastically stable if there exist a set of nonsingular matrices* $P = (P(1), \cdots, P(N))$, *with* $P(i) \in \mathbb{R}^{n \times n}$ *and a set of positive scalars* $\varepsilon_A = (\varepsilon_A(1), \cdots, \varepsilon_A(N))$, *such that the following set of coupled matrix inequalities holds for each* $i \in \mathscr{S}$ *and for all admissible uncertainties:*

$$\begin{bmatrix} J(i) & P^{\top}(i)D_A(i) \\ D_A^{\top}(i)P(i) & -\varepsilon_A(i)\mathbb{I} \end{bmatrix} < 0, \tag{2.43}$$

where

$$J(i) = P^{\top}(i)A(i) + A^{\top}(i)P(i) + \varepsilon_A(i)E_A^{\top}(i)E_A^{\top}(i)$$
$$+ \sum_{k=1}^{\kappa} \lambda_{ii}^k E^{\top}(i)P(i) + \sum_{k=1}^{\kappa} \sum_{j=1,j\neq i}^{N} \lambda_{ij}^k \varepsilon_P(j) \left[P(j) + P^{\top}(j) \right],$$

with the following constraints:

$$\varepsilon_P(i) \left[P(i) + P^{\top}(i) \right] \geq E^{\top}(i)P(i) = P^{\top}(i)E(i) \geq 0. \tag{2.44}$$

When the term $E^{\top}(i)P(i)$ is bounded by $\varepsilon(i)\left[P^{\top}(i)P(i) \right]$, we get:

Corollary 2.3.7 *The uncertain singular system with random abrupt changes (2.1) is piecewise regular, impulse-free and stochastically stable if there exist a set of nonsingular matrices* $P = (P(1), \cdots, P(N))$, *with* $P(i) \in \mathbb{R}^{n \times n}$ *and sets of positive scalars* $\varepsilon_A = (\varepsilon_A(1), \cdots, \varepsilon_A(N))$ *and* $\varepsilon = (\varepsilon(1), \cdots, \varepsilon(N))$ *such that the following set of coupled matrix inequalities holds for each* $i \in \mathscr{S}$ *and for all admissible uncertainties:*

$$\begin{bmatrix} J(i) & P^{\top}(i)D_A(i) \\ D_A^{\top}(i)P(i) & -\varepsilon_A(i)\mathbb{I} \end{bmatrix} < 0, \tag{2.45}$$

where

$$J(i) = P^{\top}(i)A(i) + A^{\top}(i)P(i) + \varepsilon_A(i)E_A^{\top}(i)E_A^{\top}(i)$$
$$+ \sum_{k=1}^{\kappa} \lambda_{ii}^k E^{\top}(i)P(i) + \sum_{k=1}^{\kappa} \sum_{j=1,j\neq i}^{N} \lambda_{ij}^k \varepsilon(j) \left[P^{\top}(j)P(j) \right],$$

with the following constraints:

$$\varepsilon(i) \left[P^{\top}(i)P(i) \right] \geq E^{\top}(i)P(i) = P^{\top}(i)E(i) \geq 0. \tag{2.46}$$

When the term $E^\top(i)P(i)$ is bounded by $\left[\frac{1}{4}\varepsilon(i)\mathbb{I} + \varepsilon^{-1}(j)E^\top(i)P(i)P^\top(i)E(i)\right]$, we get:

Corollary 2.3.8 *The uncertain singular system with random abrupt changes (2.1) is piecewise regular, impulse-free and stochastically stable if there exist a set of non-singular matrices $P = (P(1), \cdots, P(N))$, with $P(i) \in \mathbb{R}^{n \times n}$ and sets of positive scalars $\varepsilon_A = (\varepsilon_A(1), \cdots, \varepsilon_A(N))$ and $\varepsilon = (\varepsilon(1), \cdots, \varepsilon(N))$ such that the following set of coupled matrix inequalities holds for each $i \in \mathscr{S}$ and for all admissible uncertainties:*

$$\begin{bmatrix} J(i) & P^\top(i)D_A(i) \\ D_A^\top(i)P(i) & -\varepsilon_A(i)\mathbb{I} \end{bmatrix} < 0, \qquad (2.47)$$

where

$$J(i) = P^\top(i)A(i) + A^\top(i)P(i) + \varepsilon_A(i)D_A(i)D_A^\top(i) + \sum_{k=1}^{\kappa} \lambda_{ii}^k E^\top(i)P(i)$$

$$+ \sum_{k=1}^{\kappa} \sum_{j=1, j \neq i}^{N} \lambda_{ij}^k \left[\frac{1}{4}\varepsilon(j)\mathbb{I} + \varepsilon^{-1}(j)E^\top(j)P(j)P^\top(j)E(j)\right],$$

with the following constraints:

$$E^\top(i)P(i) = P^\top(i)E(i) \geq 0. \qquad (2.48)$$

Remark 2.3.3 *Notice that the conditions of this corollary can be put in the LMI setting as it was done previously.*

Remark 2.3.4 *The constraints $E^\top(i)P(i) = P^\top(i)E(i)$ may be difficult to solve with some commercial LMI toolboxes like LMI-Toolbox of MATLAB. To overcome this we can use the following LMI condition that may approximate this constraint:*

$$\left[E^\top(i)P(i) - P^\top(i)E(i)\right]^\top \left[E^\top(i)P(i) - P^\top(i)E(i)\right] \leq \beta \mathbb{I},$$

that gives the following LMI:

$$\begin{bmatrix} -\beta \mathbb{I} & \left[E^\top(i)P(i) - P^\top(i)E(i)\right]^\top \\ \left[E^\top(i)P(i) - P^\top(i)E(i)\right] & -\mathbb{I} \end{bmatrix} \leq 0. \qquad (2.49)$$

Therefore, the solution of our problem is brought to the minimization of β subject to the appropriate LMIs of the ones we developed plus (2.49) and $\beta \geq 0$ that we should minimize.

Our goal in this chapter was to study the stochastic stability and its robustness of the class of linear singular systems with random abrupt changes. Using the Lyapunov theory we developed LMI conditions that can be used to check if a given system is piecewise regular, impulse-free and stochastically stable. A certain number of results have been developed to show their effectiveness, let us now in the next section present some numerical examples to see how we can solve the different LMIs we developed for stability and the robust stability for a given system.

2.4 Numerical Examples

In this section, we will provide some numerical examples to show the effectiveness of the results of this chapter.

Example 2.4.1 *As a first example, let us consider a dynamical singular systems with two modes with the following data:*

- *mode # 1:*

$$E(1) = \begin{bmatrix} 1.0 & 0.0 & 0.0 \\ 0.0 & 1.0 & 0.0 \\ 0.0 & 0.0 & 0.0 \end{bmatrix}, \ A(1) = \begin{bmatrix} -1 & 0 & 1 \\ 0 & 0 & 1 \\ 0 & -1 & -1 \end{bmatrix}.$$

- *mode # 2:*

$$E(2) = \begin{bmatrix} 1.0 & 0.0 & 0.0 \\ 0.0 & 1.0 & 0.0 \\ 0.0 & 0.0 & 0.0 \end{bmatrix}, \ A(2) = \begin{bmatrix} -1 & 0 & 1 \\ 0 & 0 & 1 \\ 0 & -1.2 & -1 \end{bmatrix}.$$

The switching between the three modes is supposed to be described by the following transition rates:

$$\Lambda = \begin{bmatrix} -1 & 1 \\ 1.1 & -1.1 \end{bmatrix}.$$

Solving LMIs (2.7)-(2.8) gives the following solution:

$$P(1) = \begin{bmatrix} 0.7139 & -0.1185 & 0.0 \\ -0.1185 & 1.0898 & 0.0 \\ 0.3513 & 0.4970 & 0.4914 \end{bmatrix},$$

$$P(2) = \begin{bmatrix} 0.7093 & -0.1298 & 0.0 \\ -0.1298 & 1.1011 & 0.0 \\ 0.3074 & 0.4047 & 0.4883 \end{bmatrix}.$$

The two matrices are nonsingular matrices, which implies that conditions of the first theorem are satisfied. Therefore, according to Theorem 2.2.6, the system under study is piecewise regular, impulse-free and stochastically stable.

Solving LMIs (2.25)-(2.26) gives the following solution:

$$\varepsilon(1) = 1.0033$$
$$\varepsilon(2) = 1.0039$$

$$P(1) = \begin{bmatrix} 0.8451 & 0.0322 & 0.0 \\ 0.0322 & 0.9178 & 0.0 \\ 0.1008 & -0.2214 & 0.9441 \end{bmatrix},$$

$$P(2) = \begin{bmatrix} 0.8602 & 0.0257 & 0.0 \\ 0.0257 & 0.9090 & 0.0 \\ 0.1304 & -0.2278 & 1.0162 \end{bmatrix}.$$

The two matrices are nonsingular matrices, which implies that conditions of the first theorem are satisfied. Therefore, according to Corollary 2.2.1, the system under study is piecewise regular, impulse-free and stochastically stable.

Solving LMIs (2.27)-(2.28) gives the following solution:

$$\varepsilon(1) = 1.2106$$
$$\varepsilon(2) = 1.2217$$
$$P(1) = \begin{bmatrix} 0.5425 & -0.0391 & 0.0 \\ -0.0391 & 0.7882 & 0.0 \\ 0.2627 & 0.4256 & 0.5400 \end{bmatrix},$$
$$P(2) = \begin{bmatrix} 0.5445 & -0.0486 & 0.0 \\ -0.0486 & 0.8111 & 0.0 \\ 0.2297 & 0.3441 & 0.5303 \end{bmatrix}.$$

The two matrices are nonsingular matrices, which implies that conditions of the first theorem are satisfied. Therefore, according to Corollary 2.2.1, the system under study is piecewise regular, impulse-free and stochastgically stable.

A simulation of this system using MATLAB gives the behavior of the different states in function of time t as illustrated in Fig. 2.1.

Example 2.4.2 *Consider a singular linear system with three modes, i. e., $\mathscr{S} = \{1, 2, 3\}$, and assume that its state equation is described by (2.1) and its data is given by:*

- *mode # 1:*

$$E(1) = \begin{bmatrix} 1.0 & 0.0 & 0.0 \\ 0.0 & 1.0 & 0.0 \\ 0.0 & 0.0 & 0.0 \end{bmatrix}, \quad A(1) = \begin{bmatrix} -1.0 & 0.0 & 1.0 \\ 0.0 & -1.0 & 0.0 \\ 0.0 & -1.0 & -0.5 \end{bmatrix}.$$

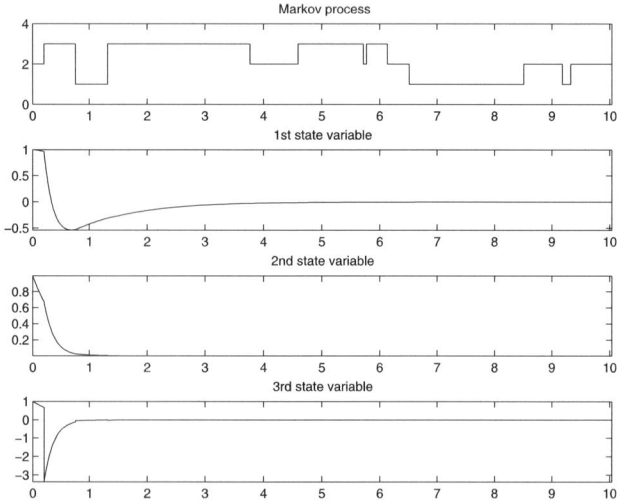

Fig. 2.1. The behavior of the system states as a function of time t

- *mode # 2:*

$$E(2) = \begin{bmatrix} 1.0 \ 0.0 \ 0.0 \\ 0.0 \ 1.0 \ 0.0 \\ 0.0 \ 0.0 \ 0.0 \end{bmatrix}, \ A(2) = \begin{bmatrix} -1.0 & 0.0 & 1.0 \\ 0.0 & -1.0 & -1.0 \\ 0.0 & 1.0 & -1.0 \end{bmatrix}.$$

- *mode # 3:*

$$E(3) = \begin{bmatrix} 1.0 \ 0.0 \ 0.0 \\ 0.0 \ 1.0 \ 0.0 \\ 0.0 \ 0.0 \ 0.0 \end{bmatrix}, \ A(3) = \begin{bmatrix} -1.0 & 0.0 & 3.0 \\ 0.0 & -1.0 & 1.0 \\ 0.0 & -1.0 & -0.2 \end{bmatrix}.$$

The switching between the three modes is supposed to be described by the following transition rates:

$$\Lambda = \begin{bmatrix} -1 & 1 & 0 \\ 0 & -2 & 2 \\ 1 & 2 & -3 \end{bmatrix}.$$

Solving LMIs (2.27)-(2.28) gives the following solution:

$$\varepsilon(1) = 0.4854, \ \varepsilon(2) = 0.1545, \ \varepsilon(3) = 0.2519$$

$$P(1) = \begin{bmatrix} 0.6300 & -0.0211 & 0.0 \\ -0.0211 & 0.8972 & 0.0 \\ 0.0785 & -0.3744 & 0.7713 \end{bmatrix},$$

$$P(2) = \begin{bmatrix} 0.4880 & 0.0210 & 0.0 \\ 0.0210 & 0.5912 & 0.0 \\ 0.1087 & 0.0595 & 0.6083 \end{bmatrix},$$

$$P(3) = \begin{bmatrix} 0.3703 & -0.1248 & 0.0 \\ -0.1248 & 0.7061 & 0.0 \\ 0.3562 & -0.5094 & 1.1098 \end{bmatrix}.$$

The three matrices are nonsingular matrices, which implies that conditions of the first theorem are satisfied. Therefore, according to Theorem 2.2.7, the system under study is piecewise regular, impulse-free and stochastically stable.

A simulation of this system using MATLAB gives the behavior of the different states in function of time t as illustrated in Fig. 2.2.

Example 2.4.3 *To show the usefulness of the results of the theorem on robust stability, let us consider the same system of Example 2.4.2 with the following data:*

- *mode # 1:*

$$D_A(1) = \begin{bmatrix} 0.1 \\ 0.0 \\ 0.0 \end{bmatrix}, \ E_A(1) = \begin{bmatrix} 0.1 \ 0.0 \ 0.0 \end{bmatrix}.$$

- *mode # 2:*

$$D_A(2) = \begin{bmatrix} 0.2 \\ 0.0 \\ 0.0 \end{bmatrix}, \ E_A(2) = \begin{bmatrix} 0.0 \ 0.0 \ 0.2 \end{bmatrix}.$$

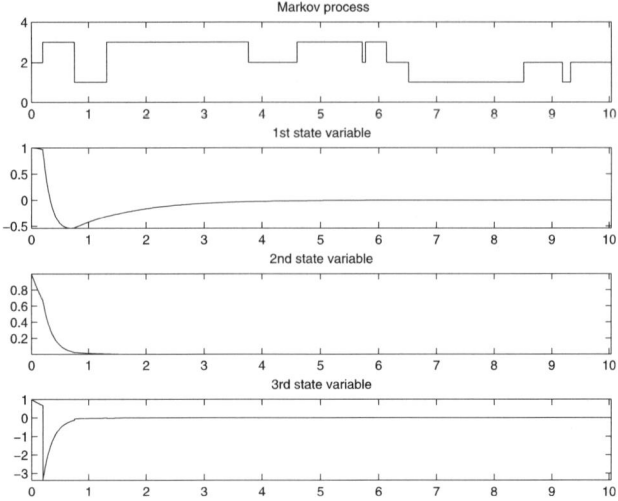

Fig. 2.2. The behavior of the system states as a function of time t

- *mode # 3:*

$$D_A(3) = \begin{bmatrix} 0.1 \\ 0.1 \\ 0.0 \end{bmatrix}, \quad E_A(3) = \begin{bmatrix} 0.0 & 0.0 & 0.2 \end{bmatrix}.$$

Solving the set of coupled LMIs (2.39)-(2.40) gives:

$$\varepsilon(1) = 0.8677, \quad \varepsilon(2) = 1.0347, \quad \varepsilon(3) = 1.3051$$
$$\varepsilon_A(1) = 1.0026, \quad \varepsilon_A(2) = 0.9922, \quad \varepsilon_A(3) = 0.9622$$

$$P(1) = \begin{bmatrix} 0.6020 & -0.1330 & 0.0 \\ -0.1330 & 1.6290 & 0.0 \\ 0.5751 & -1.6710 & 0.8840 \end{bmatrix},$$

$$P(2) = \begin{bmatrix} 0.4374 & 0.0268 & 0.0 \\ 0.0268 & 0.6407 & 0.0 \\ 0.4433 & -0.1702 & 0.7009 \end{bmatrix},$$

$$P(3) = \begin{bmatrix} 0.3333 & -0.2071 & 0.0 \\ -0.2071 & 1.0065 & 0.0 \\ 1.1929 & -1.3244 & 1.2822 \end{bmatrix},$$

which gives three nonsingular matrices and therefore the system is robustly stochastically stable.

Based on these examples, we have remarked that the last two approaches are more efficient compared to the first one and more stable. Therefore in the rest of this volume, we will continue presenting the three approaches but for the examples, we will use only the second and the third approaches if they apply.

2.5 Notes

This chapter dealt with the stability problem of the singular class of systems with random abrupt changes. The stochastic stability and the robust stochastic stability problems have been considered and LMI conditions were developed. The conditions we developed in this chapter are tractable using commercial optimization tools like LMI MATLAB toolbox, YALMIP and SeDuMi (based on MATLAB) or Scilab. The content of this chapter is mainly based on the work of the author and his coauthors [2, 21].

Part III

Stabilization

The stabilization problem is one of the most important control problems. It consists of designing a controller that will guarantee that the closed-loop state equation of the considered class of systems will be piecewise regular, impulse-free and stochastically stable and has some given specifications. Most often, the desired specifications are:

- the behavior of the transient regime;
- and the behavior at the steady state;

The stability should be the first requirement since almost all the designed systems should be stable except for some special applications. For a given dynamical systems, the stability of this system is related to the values taken by the outputs that should take finite values for bounded inputs. The controllers are in general designed to make the closed-loop dynamics stable. Regarding the transient regime, in general we are interested to control the overshoot and at the same time the settling time for all the outputs. The steady state behavior traduces the desired precision we should give to the controlled outputs.

The stabilization problem has attracted a lot of researchers from the mathematical and the control communities and a lot of results have been reported in the literature either for the deterministic and the stochastic frameworks. Many approaches have been developed to stabilize dynamical systems. Among these approaches, we quote the state feedback stabilization and the output feedback stabilization techniques.

The state feedback stabilization consists of designing a controller that assumes the complete access to the state vector at each time t. This may be too restrictive in some circumstances where the technology is not available to measure some state variables or due to the shortcut in the budget. To overcome this, the output feedback control can be used. It consists to use the measurement of the outputs to control the dynamics. Notice also, that we can still use the state feedback by designing an estimator that can estimates the state vector using the measurement of the output that will replace the real state of the system. This technique is referred to as the observer-based controller.

For the deterministic singular linear systems, the stabilization problem has been tackled by many researchers and many approaches have been reported in the literature. Among these approaches we quote the ones of state feedback and output feedback.

For the state feedback stabilization of the class of linear singular system we quote the works of [52, 58, 81, 85, 87, 92, 91, 108, 127, 122, 123, 141, 142] and the references therein. In these references, assuming the complete access to the state vector of the singular systems, the design of a controller that makes the closed-loop system regular, impulse-free and stable is established and in some references an LMI approach is used for the continuous-time case. The robust state stabilization has also been tackled and sufficient conditions have been established. To the best of our knowledge, no results on the stabilization of the discrete-time cases that uses the LMI setting exists in the literature.

In some circumstances, the state feedback stabilization may not be possible due to the lack of the appropriate sensors to measure some of the state vector or sometimes due to the limitations in the budget. To overcome this, the output feedback stabilization can be used. This technique use the output measurement to design the controller. The stabilization by output measurement has attracted a lot of researchers and many results have been reported in the literature among them we quote the works of [9, 80, 75, 86, 36, 38, 82, 83, 84, 131, 110, 119, 120, 115, 113, 114, 130] and the references therein. The stabilization by output feedback has been tackled for both the continuous-time and the discrete-time singular systems. The robust stabilization using the output measurement has also been addressed. Only the design problem for the continuous-time can be stated as LMI conditions. The discrete-time case remains an open problem.

In some circumstances, the dynamical systems may have external disturbances that can not be modeled by Gaussian process to use the linear quadratic Gaussian technique to design the desired control. Under the assumption of finite energy or power of these external disturbances, the $\mathscr{H}_\infty$ stabilization has been proposed to design controller to stabilize dynamical systems. For the last two decades, this stabilization problem has been tackled by some researchers among them we quote [40, 108, 68, 134, 89, 137, 128, 48, 124, 136, 93, 46, 106, 107] and the references therein.

This part deals with the stochastic stabilization of the class of systems with random abrupt changes. Our goal is to develop LMI conditions that can help us to design a controller that guarantees that the closed-loop state equation of the class of systems under study is piecewise regular, impulse-free and stochastically stable. Few results have been reported in the literature. Among them we quote the works of [2, 21, 22, 18, 23, 20].

This part deals with the stabilization problem of the class of singular systems with abrupt changes in the dynamics. The state feedback stabilization, the output feedback stabilization and the $\mathscr{H}_\infty$ stabilization and their robustness will be covered.

3

State Feedback Stabilization

The stabilization problem is one of the most important control problem. It consists of designing a controller which guarantees that the closed-loop state equation will be stable and has some desired specifications. This problem has attracted a lot of researchers and many results have been reported in the literature either for the deterministic systems and the stochastic ones. For more details on this subject, we refer the reader to [27] for the deterministic systems and to [27, 14] for the stochastic ones. More references can be found in the cited ones.

In the literature, we can find different stabilization techniques among them we quote the state feedback stabilization. In our setting, under the complete access to the mode and to the state vector of the system, it consists of designing a controller that will guarantee that the closed-loop state equation of the singular systems with random abrupt changes is piecewise regular, impulse-free and stochastically stable.

This chapter deals with the stabilization problem of the singular class of systems with random abrupt changes. The stochastic stabilizability and the robust stochastic stabilizability problems will be considered and LMI conditions will be developed. A state feedback controller that assures that the closed-loop state equation either for the nominal system or the uncertain one is piecewise regular, impulse-free and stochastically stable is designed in the LMI setting. The conditions we will develop in this chapter are tractable using commercial optimization tools like MATLAB LMI toolbox, YALMIP and SeDuMi (based on MATLAB) or Scilab. The content of this chapter is mainly based on the work of the author and his coauthors.

3.1 Problem Statement

Let us consider a dynamical singular linear system with random abrupt changes defined in a fundamental probability space $(\Omega, \mathcal{F}, \mathbb{P})$ and assume that its state equation is described by the following differential-algebraic systems:

$$\begin{cases} E(r_t)\dot{x}(t) = A(r_t, t)x(t) + B(r_t, t)u(t), \\ x(0) = x_0, \end{cases} \tag{3.1}$$

where $x(t) \in \mathbb{R}^n$ is the state vector, $x_0 \in \mathbb{R}^n$ is the initial state, $u(t) \in \mathbb{R}^m$ is the control input, $\{r_t, t \geq 0\}$ is the continuous-time Markov process taking values in a finite space $\mathscr{S} = \{1, 2, \cdots, N\}$ and describes the evolution of the mode at time t, $E(i)$ is a known singular matrix with rank $(E(i)) = n_r \leq n$ for all $i \in \mathscr{S}$, $A(r_t, t) \in \mathbb{R}^{n \times n}$ and $B(r_t, t) \in \mathbb{R}^{n \times m}$ are matrices with the following forms for every $r_t = i \in \mathscr{S}$:

$$A(i, t) = A(i) + D_A(i)F_A(i)E_A(i),$$
$$B(i, t) = B(i) + D_B(i)F_B(i)E_B(i),$$

with $A(i) \in \mathbb{R}^{n \times n}$, $D_A(i) \in \mathbb{R}^{n \times n_D}$, $E_A(i) \in \mathbb{R}^{n_E \times n}$, $B(i) \in \mathbb{R}^{n \times m}$, $D_B(i) \in \mathbb{R}^{n \times m_D}$ and $E_B(i) \in \mathbb{R}^{m_E \times m}$ are real known matrices with appropriate dimensions, and $F_A(i) \in \mathbb{R}^{n_D \times n_E}$ and $F_B(i) \in \mathbb{R}^{m_D \times m_E}$ are unknown real matrices that satisfy the following:

$$\begin{cases} F_A^\top(i)F_A(i) \leq \mathbb{I}, \\ F_B^\top(i)F_B(i) \leq \mathbb{I}. \end{cases} \tag{3.2}$$

The Markov process $\{r_t, t \geq 0\}$ beside taking values in the finite set $\mathscr{S}$, represents the switching between the different modes and its state equation is described by the following probability transitions:

$$\mathbb{P}\left[r_{t+h} = j | r_t = i\right] = \begin{cases} \lambda_{ij}h + o(h), & \text{when } r_t \text{ jumps from } i \text{ to } j, \\ 1 + \lambda_{ii}h + o(h), & \text{otherwise,} \end{cases} \tag{3.3}$$

where λ_{ij} is the transition rate from mode i to mode j with $\lambda_{ij} \geq 0$ when $i \neq j$ and $\lambda_{ii} = -\sum_{j=1, j \neq i}^{N} \lambda_{ij}$ and $o(h)$ is such that $\lim_{h \to 0} \frac{o(h)}{h} = 0$.

As we did previously, we will assume here when it is necessary that the transition matrix, Λ, belongs to a polytope, i. e.,

$$\Lambda = \sum_{k=1}^{\kappa} \alpha_k \Lambda_k, \tag{3.4}$$

where κ is a positive given integer, $0 \leq \alpha_k \leq 1$, with $\sum_{k=1}^{\kappa} \alpha_k = 1$ and Λ_k is a known transition matrix and its expression is given by:

$$\Lambda_k = \begin{bmatrix} \lambda_{11}^k & \cdots & \lambda_{1N}^k \\ \vdots & \ddots & \vdots \\ \lambda_{N1}^k & \cdots & \lambda_{NN}^k \end{bmatrix}, \tag{3.5}$$

where λ_{ij}^k keeps the same meaning as before.

Remark 3.1.1 *The uncertainties satisfying the condition (3.2), (3.4) are referred to as admissible. The uncertainty term, in (3.2), is supposed to depend on the system's mode, r_t.*

Remark 3.1.2 *The matrix $E(i)$, for all $i \in \mathscr{S}$, is supposed to be singular which makes the state equation (3.1) different from the one usually used to describe the* *behavior of the time-invariant dynamical systems as it is the normal practice. Notice that when $E(i)$, for all $i \in \mathscr{S}$, is not singular, (3.1) can be transformed easily to the class of Markov jump linear systems and the results developed in the literature (see Mariton [95], Boukas and Liu [27], Boukas [14] and the references therein), can be used to check the stochastic stability, of this class of systems.*

Throughout this chapter, we assume that the system state $x(t)$ and the system mode r_t are completely accessible when necessary.

Definition 3.1.1 *Nominal system (3.1) is said to be stochastically stabilizable if there exists a control*

$$u(t) = K(r_t)x(t),\qquad\qquad(3.6)$$

with $K(i) \in \mathbb{R}^{m \times n}$, $i \in \mathscr{S}$, a constant matrix such that the closed-loop system is piecewise regular, impulse-free and stochastically stable.

Definition 3.1.2 *Uncertain system (3.1) is said to be robust stochastically stabilizable if there exists a control of the form (3.6) such that the closed-loop system is piecewise regular, impulse-free and robust stochastically stable.*

The goal of this chapter is to develop conditions that allow the design of state feedback controller in the form (3.6) that makes the closed-loop dynamics piecewise regular, impulse-free robust and stochastically stable. Both nominal and uncertain systems are considered and the conditions in the LMI form are established.

3.2 Design of State Feedback Stabilization

Let us first of all concentrate on the design of a state feedback controller of the form (3.6) that will guarantee that the closed-loop state equation of the system (3.1) will be piecewise regular, impulse-free and stochastically stable. The block diagram of the closed-loop system under the state feedback controller is represented by Fig. 3.1.

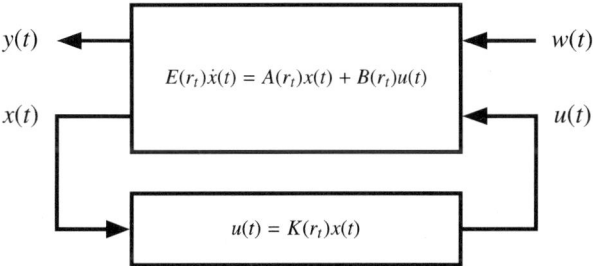

Fig. 3.1. State feedback stabilization block diagram

For the stabilization problem of the nominal system, combining the system state equation (3.1) and the controller expression (3.6) gives the following closed-loop dynamics:

$$E(r_t)\dot{x}(t) = A(r_t)x(t) + B(r_t)K(r_t)x(t)$$
$$= [A(r_t) + B(r_t)K(r_t)]\, x(t)$$
$$= A_{cl}(r_t)x(t)\,,$$

with $A_{cl}(r_t) = A(r_t) + B(r_t)K(r_t)$.

Based on the stochastic stability results (Corollary 2.2.1), this closed-loop system is piecewise regular, impulse-free and stochastically stable if there exists a set of nonsingular matrices $P = (P(1), \cdots, P(N))$, with $P(i) \in \mathbb{R}^{n \times n}$ such that the following hold for a given $\varepsilon_P > 0$:

$$\begin{cases} \varepsilon_P \left[P(i) + P^{\top}(i)\right] \geq E^{\top}(i)P(i) = P^{\top}(i)E(i) \geq 0 \\ A_{cl}^{\top}(i)P(i) + P^{\top}(i)A_{cl}(i) + \lambda_{ii}E^{\top}(i)P(i) \\ \quad + \sum_{j=1, j\neq i}^{N} \varepsilon_P \lambda_{ij} \left[P(j) + P^{\top}(j)\right] < 0\,. \end{cases}$$

Replacing $A_{cl}(i)$ by its expression in the second inequality gives:

$$A^{\top}(i)P(i) + P^{\top}(i)A(i) + K^{\top}(i)B^{\top}(i)P(i) + \lambda_{ii}E^{\top}(i)P(i)$$

$$+ P^{\top}(i)B(i)K(i) + \sum_{j=1, j\neq i}^{N} \varepsilon_P \lambda_{ij} \left[P(j) + P^{\top}(j)\right] < 0\,.$$

Notice that the left hand side of this matrix inequality is nonlinear in the design parameters $P(i)$ and $K(i)$. To transform it into an LMI, let $X(i) = P^{-1}(i)$ for each $i \in \mathscr{S}$. Let us now pre- and post-multiply the left hand side term by $X^{\top}(i)$ and $X(i)$ respectively. This gives:

$$X^{\top}(i)A^{\top}(i) + A(i)X(i) + X^{\top}(i)K^{\top}(i)B^{\top}(i) + B(i)K(i)X(i)$$

$$+ \lambda_{ii}X^{\top}(i)E^{\top}(i) + \sum_{j=1, j\neq i}^{N} \varepsilon_P \lambda_{ij} X^{\top}(i) \left[X^{-1}(j) + X^{-\top}(j)\right] X(i) < 0\,.$$

Based on Lemma 1.5.2, we get:

$$X^{-1}(j) + X^{-\top}(j) \leq \mathbb{I} + X^{-1}(j)X^{-\top}(j)$$
$$= \mathbb{I} + \left[X^{\top}(j)X(j)\right]^{-1}\,,$$

that gives in turn:

$$X^{\top}(i)A^{\top}(i) + A(i)X(i) + X^{\top}(i)K^{\top}(i)B^{\top}(i) + B(i)K(i)X(i)$$

$$+ \lambda_{ii}X^{\top}(i)E^{\top}(i) + \sum_{j=1, j\neq i}^{N} \varepsilon_P \lambda_{ij} X^{\top}(i) \left[\mathbb{I} + \left(X^{\top}(j)X(j)\right)^{-1}\right] X(i) < 0\,.$$

Notice that

$$\sum_{j=1,j\neq i}^{N} \varepsilon_P \lambda_{ij} X^\top(i) X(i) = \mathcal{Z}_i(X) \mathcal{Z}_i^\top(X),$$

$$\sum_{j=1,j\neq i}^{N} \varepsilon_P \lambda_{ij} X^\top(i) \left(X^\top(j)X(j)\right)^{-1} X(i) = \mathcal{S}_i(X) V^{-1}(X) \mathcal{S}_i^\top(X),$$

with

$$V(X) = \text{diag}\left[\varepsilon_P^{-1} X^\top(1) X(1), \cdots, \varepsilon_P^{-1} X^\top(i-1)X(i-1),\right.$$
$$\left. \varepsilon_P^{-1} X^\top(i+1) X(i+1), \cdots, \varepsilon_P^{-1} X^\top(N)X(N)\right],$$
$$\mathcal{S}_i(X) = \left[\sqrt{\lambda_{i1}} X^\top(i), \cdots, \sqrt{\lambda_{ii-1}} X^\top(i), \sqrt{\lambda_{ii+1}} X^\top(i),\right.$$
$$\left. \cdots, \sqrt{\lambda_{iN}} X^\top(i)\right],$$
$$\mathcal{Z}_i(X) = \left[\sqrt{\varepsilon_P \lambda_{i1}} X^\top(i), \cdots, \sqrt{\varepsilon_P \lambda_{ii-1}} X^\top(i), \sqrt{\varepsilon_P \lambda_{ii+1}} X^\top(i),\right.$$
$$\left. \cdots, \sqrt{\varepsilon_P \lambda_{iN}} X^\top(i)\right].$$

Using Schur complement we get:

$$\begin{bmatrix} J_0(i) & \mathcal{Z}_i(X) & \mathcal{S}_i(X) \\ \mathcal{Z}_i^\top(X) & -\mathbb{I} & 0 \\ \mathcal{S}_i^\top(X) & 0 & -V(X) \end{bmatrix},$$

with

$$J_0(i) = X^\top(i)A^\top(i) + A(i)X(i) + X^\top(i)K^\top(i)B^\top(i) + B(i)K(i)X(i)$$
$$+ \lambda_{ii} X^\top(i) E^\top(i).$$

Using again Lemma 1.5.2, we have:

$$\varepsilon_P^{-1} X^\top(j) X(j) \geq -\varepsilon_P \mathbb{I} + X^\top(j) + X(j).$$

Let $Y(i)$ and $\mathcal{X}_i(X)$ be defined as follows:

$$Y(i) = K(i)X(i),$$
$$\mathcal{X}_i(X) = \text{diag}\left[-\varepsilon_P \mathbb{I} + X^\top(1) + X(1), \cdots, -\varepsilon_P \mathbb{I} + X^\top(i-1) + X(i-1),\right.$$
$$\left. -\varepsilon_P \mathbb{I} + X^\top(i+1) + X(i+1), \cdots, -\varepsilon_P \mathbb{I} + X^\top(N) + X(N)\right].$$

The previous inequality matrix will hold if the following is satisfied:

$$\begin{bmatrix} J_0(i) & \mathcal{Z}_i(X) & \mathcal{S}_i(X) \\ \mathcal{Z}_i^\top(X) & -\mathbb{I} & 0 \\ \mathcal{S}_i^\top(X) & 0 & -\mathcal{X}_i(X) \end{bmatrix} < 0,$$

with

$$J_0(i) = X^\top(i)A^\top(i) + A(i)X(i) + Y^\top(i)B^\top(i) + B(i)Y(i)$$
$$+ \lambda_{ii}X^\top(i)E^\top(i).$$

For the condition $\varepsilon_P[P(i) + P^\top(i)] \geq E^\top(i)P(i) = P^\top(i)E(i) \geq 0$, we can transform it in a similar way to $\varepsilon_P[X(i) + X^\top(i)] \geq X^\top(i)E^\top(i) = E(i)X(i) \geq 0$.

If these conditions are satisfied for some set of nonsingular matrices $X = (X(1), \cdots, X(N)) > 0$ and a set of matrices $Y = (Y(1), \cdots, Y(N))$ for a fixed $\varepsilon_P > 0$, the closed-loop system will be piecewise regular, impulse-free and stochastically stable under the state feedback controller with a gain given by $K(i) = Y(i)X^{-1}(i), i \in \mathscr{S}$.

The results of this development are summarized by the following theorem.

Theorem 3.2.1 *Let ε_P be a given positive scalar. There exists a state feedback controller of the form (3.6) such that the closed-loop state equation of the nominal system (3.1) is piecewise regular, impulse-free and stochastically stable if there exist a set of nonsingular matrices $X = (X(1), \cdots, X(N))$, with $X(i) \in \mathbb{R}^{n \times n}$, and a set of matrices $Y = (Y(1), \cdots, Y(N))$, with $Y(i) \in \mathbb{R}^{m \times n}$, such that the following set of coupled LMIs holds for each $i \in \mathscr{S}$:*

$$\begin{bmatrix} J_0(i) & \mathcal{Z}_i(X) & \mathcal{S}_i(X) \\ \mathcal{Z}_i^\top(X) & -\mathbb{I} & 0 \\ \mathcal{S}_i^\top(X) & 0 & -\mathcal{X}_i(X) \end{bmatrix} < 0, \tag{3.7}$$

where

$$J_0(i) = A(i)X(i) + X^\top(i)A^\top(i) + B(i)Y(i) + Y^\top(i)B^\top(i)$$
$$+ \lambda_{ii}X^\top(i)E^\top(i),$$

$$\mathcal{X}_i(X) = \mathrm{diag}\Big[-\varepsilon_P\mathbb{I} + X^\top(1) + X(1), \cdots, -\varepsilon_P\mathbb{I} + X^\top(i-1) + X(i-1),$$
$$-\varepsilon_P\mathbb{I} + X^\top(i+1) + X(i+1), \cdots, -\varepsilon_P\mathbb{I} + X^\top(N) + X(N)\Big],$$

$$\mathcal{S}_i(X) = \Big[\sqrt{\lambda_{i1}}X^\top(i), \cdots, \sqrt{\lambda_{ii-1}}X^\top(i), \sqrt{\lambda_{ii+1}}X^\top(i),$$
$$\cdots, \sqrt{\lambda_{iN}}X^\top(i)\Big],$$

$$\mathcal{Z}_i(X) = \Big[\sqrt{\varepsilon_P\lambda_{i1}}X^\top(i), \cdots, \sqrt{\varepsilon_P\lambda_{ii-1}}X^\top(i), \sqrt{\varepsilon_P\lambda_{ii+1}}X^\top(i),$$
$$\cdots, \sqrt{\varepsilon_P\lambda_{iN}}X^\top(i)\Big],$$

with the following constraints:

$$\varepsilon_P\Big[X(i) + X^\top(i)\Big] \geq X^\top(i)E^\top(i) = E(i)X(i) \geq 0. \tag{3.8}$$

The stabilizing controller gain is given by $K(i) = Y(i)X^{-1}(i)$, $K(i) \in \mathbb{R}^{m \times n}$, $i \in \mathscr{S}$.

Remark 3.2.1 *The results we developed in the previous theorem can be extended easily to the mode-dependent $\varepsilon_P(j), \forall j \in \mathscr{S}$ case.*

If we consider that the parameter ε_P is mode-dependent, we get the following results.

Corollary 3.2.1 *Let $\varepsilon_P = (\varepsilon_P(i), \cdots, \varepsilon_P(N))$ be a given set of positive scalars. There exists a state feedback controller of the form (3.6) such that the closed-loop state equation of the nominal system (3.1) is piecewise regular, impulse-free and stochastically stable if there exist a set of nonsingular matrices $X = (X(1), \cdots, X(N))$, with $X(i) \in \mathbb{R}^{n \times n}$, and a set of matrices $Y = (Y(1), \cdots, Y(N))$, with $Y(i) \in \mathbb{R}^{m \times n}$, such that the following set of coupled LMIs holds for each $i \in \mathscr{S}$:*

$$\begin{bmatrix} J_0(i) & \mathcal{Z}_i(X) & S_i(X) \\ \mathcal{Z}_i^\top(X) & -\mathbb{I} & 0 \\ S_i^\top(X) & 0 & -\mathcal{X}_i(X) \end{bmatrix} < 0, \tag{3.9}$$

where

$$J_0(i) = A(i)X(i) + X^\top(i)A^\top(i) + B(i)Y(i) + Y^\top(i)B^\top(i) \\ + \lambda_{ii}X^\top(i)E^\top(i),$$

$$\mathcal{X}_i(X) = \text{diag}\Big[-\varepsilon_P(1)\mathbb{I} + X^\top(1) + X(1), \cdots, -\varepsilon_P(i-1)\mathbb{I} + X^\top(i-1) + X(i-1), \\ -\varepsilon_P(i+1)\mathbb{I} + X^\top(i+1) + X(i+1), \cdots, -\varepsilon_P(N)\mathbb{I} + X^\top(N) + X(N) \Big],$$

$$S_i(X) = \Big[\sqrt{\lambda_{i1}}X^\top(i), \cdots, \sqrt{\lambda_{ii-1}}X^\top(i), \sqrt{\lambda_{ii+1}}X^\top(i), \\ \cdots, \sqrt{\lambda_{iN}}X^\top(i) \Big],$$

$$\mathcal{Z}_i(X) = \Big[\sqrt{\varepsilon_P(1)\lambda_{i1}}X^\top(i), \cdots, \sqrt{\varepsilon_P(i-1)\lambda_{ii-1}}X^\top(i), \sqrt{\varepsilon_P(i+1)\lambda_{ii+1}}X^\top(i), \\ \cdots, \sqrt{\varepsilon_P(N)\lambda_{iN}}X^\top(i) \Big],$$

with the following constraints:

$$\varepsilon_P(i)\Big[X(i) + X^\top(i)\Big] \geq X^\top(i)E^\top(i) = E(i)X(i) \geq 0. \tag{3.10}$$

The stabilizing controller gain is given by $K(i) = Y(i)X^{-1}(i)$, $K(i) \in \mathbb{R}^{m \times n}$, $i \in \mathscr{S}$.

Let us establish another way of stabilizing the class of systems we are considering. For this purpose, based on the stability results of the Corollary 2.2.5 the closed-loop state equation when $A_{cl}(i)$ is replaced by its expression will be piecewise regular, impulse-free and stochastically stable if there exists a set of nonsingular matrices $P = (P(1), \cdots, P(N))$, with $P(i) \in \mathbb{R}^{n \times n}$ such that the following conditions hold for each $i \in S$:

$$\begin{cases} \varepsilon_P[P(i) + P^\top(i)] \geq E^\top(i)P(i) = P^\top(i)E(i) \geq 0, \\ A^\top(i)P(i) + P^\top(i)A(i) + K^\top(i)B^\top(i)P(i) + P^\top(i)B(i)K(i) \\ \quad + \lambda_{ii}E^\top(i)P(i) + \sum_{j=1, j\neq i}^N \varepsilon_P \lambda_{ij}[P(j) + P^\top(j)] < 0. \end{cases}$$

Let us now assume this time that $P^\top(i)B(i) = B(i)L(i)$ holds for every $i \in \mathscr{S}$ with $L(i) \in \mathbb{R}^{m \times m}$ a nonsingular matrix. This requires that the matrix $B(i)$ is full column rank. Using this condition, the previous conditions become:

$$\begin{cases} \varepsilon_P[P(i) + P^\top(i)] \geq E^\top(i)P(i) = P^\top(i)E(i) \geq 0, \\ A^\top(i)P(i) + P^\top(i)A(i) + K^\top(i)L^\top(i)B^\top(i) + \lambda_{ii}E^\top(i)P(i) \\ \quad + B(i)L(i)K(i) + \sum_{j=1, j\neq i}^N \varepsilon_P \lambda_{ij}[P(j) + P^\top(j)] < 0. \end{cases}$$

Letting $F(i) = L(i)K(i)$, with $F(i) \in \mathbb{R}^{m \times n}$ we get:

$$\begin{cases} \varepsilon_P \left[P(i) + P^\top(i) \right] \geq E^\top(i)P(i) = P^\top(i)E(i) \geq 0, \\ A^\top(i)P(i) + P(i)A(i) + F^\top(i)B^\top(i) + B(i)F(i) \\ \quad + \lambda_{ii}E^\top(i)P(i) + \sum_{j=1, j \neq i}^{N} \varepsilon_P \lambda_{ij} \left[P(j) + P^\top(j) \right] < 0. \end{cases}$$

The following theorem gives the results that can be used to determine the stabilizing state feedback controller.

Theorem 3.2.2 *Let ε_P be a given positive scalar. There exists a state feedback controller of the form (3.6) such that the closed-loop state equation of the nominal system (3.1) is piecewise regular, impulse-free and stochastically stable if there exist sets of nonsingular matrices $P = (P(1), \cdots, P(N))$, with $P(i) \in \mathbb{R}^{n \times n}$ and $L = (L(1), \cdots, L(N))$, with $L(i) \in \mathbb{R}^{m \times m}$ a set of matrices $F = (F(1), \cdots, F(N))$, with $F(i) \in \mathbb{R}^{m \times n}$ such that the following set of coupled LMIs holds for each $i \in \mathscr{S}$:*

$$A^\top(i)P(i) + P(i)A(i) + F^\top(i)B^\top(i) + B(i)F(i)$$
$$+ \lambda_{ii}E^\top(i)P(i) + \sum_{j=1, j \neq i}^{N} \varepsilon_P \lambda_{ij} \left[P(j) + P^\top(j) \right] < 0. \qquad (3.11)$$

with the following constraints:

$$\begin{cases} \varepsilon_P \left[P(i) + P^\top(i) \right] \geq E^\top(i)P(i) = P^\top(i)E(i) \geq 0, \\ P^\top(i)B(i) = B(i)L(i). \end{cases} \qquad (3.12)$$

The stabilizing controller gain is given by $K(i) = L^{-1}(i)F(i)$, $K(i) \in \mathbb{R}^{m \times n}$ $i \in \mathscr{S}$.

The results of the previous theorems may be conservative. To avoid this, we will establish other results with less conservatism. These results are based on the bounds we gave in Chap. 2 for the term $E^\top(i)P(i)$.

If we use the results of Corollary 2.2.5, the closed-loop system requires to be piecewise regular, impulse-free and stochastically stable that there exist a set of nonsingular matrices $P = (P(1), \cdots, P(N))$, with $P(i) \in \mathbb{R}^{n \times n}$ and a set of positive scalars $\varepsilon = (\varepsilon(1), \cdots, \varepsilon(N))$ with $(\varepsilon(i) > 0)$ such that the following hold:

$$P^\top(i)A_{cl}(i) + A_{cl}^\top(i)P(i) + \lambda_{ii}E^\top(i)P(i)$$
$$+ \sum_{j=1, j \neq i}^{N} \varepsilon(j)\lambda_{ij} \left[P^\top(j)P(j) \right] < 0,$$

with the following constraints:

$$\varepsilon(i) \left[P^\top(i)P(i) \right] \geq E^\top(i)P(i) = P^\top(i)E(i) \geq 0.$$

Replacing $A_{cl}(i)$ by its expression in the first inequality gives:

$$A^\top(i)P(i) + P^\top(i)A(i) + K^\top(i)B^\top(i)P(i) + \lambda_{ii}E^\top(i)P(i)$$
$$+ P^\top(i)B(i)K(i) + \sum_{j=1, j \neq i}^{N} \varepsilon(j)\lambda_{ij} \left[P^\top(j)P(j) \right] < 0.$$

Notice that the left hand side of this matrix inequality is nonlinear in the design parameters $P(i)$ and $K(i)$. To transform it into an LMI, let $X(i) = P^{-1}(i)$ for each $i \in \mathcal{S}$. Let us now pre- and post-multiply the left hand side term by $X^\top(i)$ and $X(i)$ respectively. This gives:

$$X^\top(i)A^\top(i) + A(i)X(i) + X^\top(i)K^\top(i)B^\top(i) + B(i)K(i)X(i)$$
$$+ \lambda_{ii}X^\top(i)E^\top(i) + \sum_{j=1,j\neq i}^{N} \varepsilon(j)\lambda_{ij}X^\top(i)\left[X^{-\top}(j)X^{-1}(j)\right]X(i) < 0.$$

Notice that:

$$\sum_{j=1,j\neq i}^{N} \lambda_{ij}X^\top(i)\left[\varepsilon^{-1}(j)X(j)X^\top(j)\right]^{-1}X(i) = S_i(X)V^{-1}(X)S_i^\top(X),$$

with

$$V(X) = \mathrm{diag}\left[\varepsilon^{-1}(1)X(1)X^\top(1), \cdots, \varepsilon^{-1}(i-1)X(i-1)X^\top(i-1),\right.$$
$$\left.\varepsilon^{-1}(i+1)X(i+1)X^\top(i+1), \cdots, \varepsilon^{-1}(N)X(N)X^\top(N)\right],$$
$$S_i(X) = \left[\sqrt{\lambda_{i1}}X^\top(i), \cdots, \sqrt{\lambda_{ii-1}}X^\top(i), \sqrt{\lambda_{ii+1}}X^\top(i), \cdots, \sqrt{\lambda_{iN}}X^\top(i)\right].$$

Using again Lemma 1.5.2, we have:

$$\varepsilon^{-1}(j)X(j)X^\top(j) \geq -\varepsilon(j)\mathbb{I} + X^\top(j) + X(j).$$

Let $Y(i)$ and $\mathcal{X}_i(X)$ be defined as follows:

$$Y(i) = K(i)X(i),$$
$$\mathcal{X}_i(X) = \mathrm{diag}\left[-\varepsilon(1)\mathbb{I} + X^\top(1) + X(1), \cdots, -\varepsilon(i-1)\mathbb{I} + X^\top(i-1) + X(i-1),\right.$$
$$\left.-\varepsilon(i+1)\mathbb{I} + X^\top(i+1) + X(i+1), \cdots, -\varepsilon(N)\mathbb{I} + X^\top(N) + X(N)\right],$$

and by Schur complement, the previous inequality matrix becomes:

$$\begin{bmatrix} J_0(i) & S_i(X) \\ S_i^\top(X) & -\mathcal{X}_i(X) \end{bmatrix} < 0,$$

with

$$J_0(i) = X^\top(i)A^\top(i) + A(i)X(i) + Y^\top(i)B^\top(i) + B(i)Y(i)$$
$$+ \lambda_{ii}X^\top(i)E^\top(i).$$

For the condition $\varepsilon(i)\left[P^\top(i)P(i)\right] \geq E^\top(i)P(i) = P^\top(i)E(i) \geq 0$, we can transform it in a similar way to $\varepsilon(i)\mathbb{I} \geq X^\top(i)E^\top(i) = E(i)X(i) \geq 0$.

If these inequalities are satisfied for a set of nonsingular matrices $X = (X(1), \cdots, X(N)) > 0$, a set of matrices $Y = (Y(1), \cdots, Y(N))$ and a set of positive scalars $\varepsilon = (\varepsilon(1), \cdots, \varepsilon(N))$, the closed-loop system will be piecewise regular, impulse-free and stochastically stable under the state feedback controller with a gain given by $K(i) = Y(i)X^{-1}(i), i \in \mathcal{S}$.

The results of this development are summarized by the following theorem.

Theorem 3.2.3 *There exists a state feedback controller of the form (3.6) such that the closed-loop state equation of the nominal system (3.1) is piecewise regular, impulse-free and stochastically stable if there exist a set of nonsingular matrices $X = (X(1), \cdots, X(N))$, with $X(i) \in \mathbb{R}^{n \times n}$, a set of matrices $Y = (Y(1), \cdots, Y(N))$, with $Y(i) \in \mathbb{R}^{m \times n}$, and a set of positive scalars $\varepsilon = (\varepsilon(1), \cdots, \varepsilon(N))$ such that the following set of coupled LMIs holds for each $i \in \mathscr{S}$:*

$$\begin{bmatrix} \widehat{J_0}(i) & S_i(X) \\ S_i^\top(X) & -X_i(X) \end{bmatrix} < 0, \tag{3.13}$$

where

$$\widehat{J_0}(i) = A(i)X(i) + X^\top(i)A^\top(i) + B(i)Y(i) + Y^\top(i)B^\top(i)$$
$$+ \lambda_{ii} X^\top(i)E^\top(i),$$
$$S_i(X) = \left[\sqrt{\lambda_{i1}} X^\top(i), \cdots, \sqrt{\lambda_{ii-1}} X^\top(i), \sqrt{\lambda_{ii+1}} X^\top(i), \cdots, \sqrt{\lambda_{iN}} X^\top(i) \right],$$
$$X_i(X) = \text{diag}\left[-\varepsilon(1)\mathbb{I} + X^\top(1) + X(1), \cdots, -\varepsilon(i-1)\mathbb{I} + X^\top(i-1) + X(i-1), \right.$$
$$\left. -\varepsilon(i+1)\mathbb{I} + X^\top(i+1) + X(i+1), \cdots, -\varepsilon(N)\mathbb{I} + X^\top(N) + X(N) \right],$$

with the following constraints:

$$\varepsilon(i)\mathbb{I} \geq X^\top(i)E^\top(i) = E(i)X(i) \geq 0. \tag{3.14}$$

The stabilizing controller gain is given by $K(i) = Y(i)X^{-1}(i)$, $K(i) \in \mathbb{R}^{m \times n}$, $i \in \mathscr{S}$.

Now if we refer to the third stability conditions we developed in Chap. 2, the results of Corollary 2.2.7 require for the closed-loop system to be piecewise regular, impulse-free and stochastically stable that there exist a set of nonsingular matrices $P = (P(1), \cdots, P(N))$, with $P(i) \in \mathbb{R}^{n \times n}$ and a set of positive scalars $\varepsilon = (\varepsilon(1), \cdots, \varepsilon(N))$ such that the following hold:

$$P^\top(i)A_{cl}(i) + A_{cl}^\top(i)P(i) + \lambda_{ii}E^\top(i)P(i)$$
$$+ \sum_{j=1, j \neq i}^{N} \lambda_{ij} \left[\frac{1}{4}\varepsilon^{-1}(j)\mathbb{I} + \varepsilon(j)E^\top(j)P(j)P^\top(j)E(j) \right] < 0,$$

with the following constraints:

$$E^\top(i)P(i) = P^\top(i)E(i) \geq 0.$$

Replacing $A_{cl}(i)$ by its expression in the second inequality gives:

$$A^\top(i)P(i) + P^\top(i)A(i) + K^\top(i)B^\top(i)P(i) + \lambda_{ii}E^\top(i)P(i)$$
$$+ P^\top(i)B(i)K(i) + \sum_{j=1, j \neq i}^{N} \lambda_{ij} \left[\frac{1}{4}\varepsilon^{-1}(j)\mathbb{I} + \varepsilon(j)E^\top(j)P(j)P^\top(j)E(j) \right] < 0.$$

Notice that the left hand side of this matrix inequality is nonlinear in the design parameters $P(i)$ and $K(i)$. To transform it into an LMI, let $X(i) = P^{-1}(i)$ for each $i \in \mathscr{S}$. Let us now pre- and post-multiply the left hand side term by $X^\top(i)$ and $X(i)$ respectively. This gives:

$$X^\top(i)A^\top(i) + A(i)X(i) + X^\top(i)K^\top(i)B^\top(i) + B(i)K(i)X(i)$$

$$+ \lambda_{ii}X^\top(i)E^\top(i) + \sum_{j=1,j\neq i}^{N} \varepsilon^{-1}(j)\frac{1}{4}\lambda_{ij}X^\top(i)X(i)$$

$$+ \sum_{j=1,j\neq i}^{N} \lambda_{ij}X^\top(i)E^\top(j)\left[\varepsilon^{-1}(j)X^\top(j)X(j)\right]^{-1}E(j)X(i) < 0 \,.$$

Now if we define $\mathcal{X}_i(\varepsilon)$, $\mathcal{V}_i(X)$, $\mathcal{Z}_i(X)$ and $\mathcal{S}_i(X)$ as follows:

$$\mathcal{Z}_i(X) = \left[\sqrt{\lambda_{i1}}X^\top(i), \cdots, \sqrt{\lambda_{ii-1}}X^\top(i), \right.$$

$$\left. \sqrt{\lambda_{ii+1}}X^\top(i), \cdots, \sqrt{\lambda_{iN}}X^\top(i) \right],$$

$$\mathcal{S}_i(X) = \left[\sqrt{\lambda_{i1}}X^\top(i)E^\top(1), \cdots, \sqrt{\lambda_{ii-1}}X^\top(i)E^\top(i-1), \right.$$

$$\left. \sqrt{\lambda_{ii+1}}X^\top(i)E^\top(i+1), \cdots, \sqrt{\lambda_{iN}}X^\top(i)E^\top(N) \right],$$

$$\mathcal{X}_i(\varepsilon) = \mathrm{diag}\left[4\varepsilon(1)\mathbb{I}, \cdots, 4\varepsilon(i-1)\mathbb{I}, 4\varepsilon(i+1)\mathbb{I}, \cdots, 4\varepsilon(N)\mathbb{I}\right],$$

$$\mathcal{V}_i(X) = \mathrm{diag}\left[\varepsilon^{-1}(1)X^\top(1)X(1), \cdots, \varepsilon^{-1}(i-1)X^\top(i-1)X(i-1), \right.$$

$$\left. \varepsilon^{-1}(i+1)X^\top(i+1)X(i+1), \cdots, \varepsilon^{-1}(N)X^\top(N)X(N) \right],$$

we get:

$$\sum_{j=1,j\neq i}^{N} \varepsilon^{-1}(j)\frac{1}{4}\lambda_{ij}X^\top(i)X(i) = \mathcal{Z}_i(X)\mathcal{X}_i^{-1}(\varepsilon)\mathcal{Z}_i^\top(X),$$

$$\sum_{j=1,j\neq i}^{N} \lambda_{ij}X^\top(i)E^\top(j)\left[\varepsilon^{-1}(j)X^\top(j)X(j)\right]^{-1}E(j)X(i) = \mathcal{S}_i(X)\mathcal{V}_i^{-1}(\varepsilon)\mathcal{S}_i^\top(X).$$

Using again Lemma 1.5.2, we have:

$$\varepsilon^{-1}(j)X^\top(j)X(j) \geq -\varepsilon(j)\mathbb{I} + X^\top(j) + X(j).$$

Let $Y(i)$ and $\mathcal{X}_i(X)$ be defined as follows:

$$Y(i) = K(i)X(i),$$

$$\mathcal{X}_i(X) = \mathrm{diag}\left[-\varepsilon(1)\mathbb{I} + X^\top(1) + X(1), \cdots, -\varepsilon(i-1)\mathbb{I} + X^\top(i-1) + X(i-1), \right.$$

$$\left. -\varepsilon(i+1)\mathbb{I} + X^\top(i+1) + X(i+1), \cdots, -\varepsilon(N)\mathbb{I} + X^\top(N) + X(N) \right].$$

and by Schur complement, the previous inequality matrix becomes:

$$\begin{bmatrix} J_0(i) & \mathcal{Z}_i(X) & \mathcal{S}_i(X) \\ \mathcal{Z}_i^\top(X) & -\mathcal{X}_i(\varepsilon) & 0 \\ \mathcal{S}_i^\top(X) & 0 & -\mathcal{X}_i(X) \end{bmatrix} < 0,$$

with

$$J_0(i) = X^\top(i)A^\top(i) + A(i)X(i) + Y^\top(i)B^\top(i) + B(i)Y(i)$$
$$+ \lambda_{ii}X^\top(i)E^\top(i).$$

For the condition $E^\top(i)P(i) = P^\top(i)E(i) \geq 0$, we can transform it in a similar way to $X^\top(i)E^\top(i) = E(i)X(i) \geq 0$.

If these inequalities are satisfied for a set of nonsingular matrices $X = (X(1), \cdots, X(N)) > 0$, a set of matrices $Y = (Y(1), \cdots, Y(N))$ and a set of positive scalars $\varepsilon = (\varepsilon(1), \cdots, \varepsilon(N))$, the closed-loop system will be piecewise regular, impulse-free and stochastically stable under the state feedback controller with a gain given by $K(i) = Y(i)X^{-1}(i), i \in \mathscr{S}$.

The results of this development are summarized by the following theorem.

Theorem 3.2.4 *There exists a state feedback controller of the form (3.6) such that the closed-loop state equation of the nominal system (3.1) is piecewise regular, impulse-free and stochastically stable if there exist a set of nonsingular matrices $X = (X(1), \cdots, X(N))$, with $X(i) \in \mathbb{R}^{n \times n}$, a set of matrices $Y = (Y(1), \cdots, Y(N))$, with $Y(i) \in \mathbb{R}^{m \times n}$, and a set of positive scalars $\varepsilon = (\varepsilon(1), \cdots, \varepsilon(N))$ such that the following set of coupled LMIs holds for each $i \in \mathscr{S}$:*

$$\begin{bmatrix} J_0(i) & \mathcal{Z}_i(X) & \mathcal{S}_i(X) \\ \mathcal{Z}_i^\top(X) & -\mathcal{X}_i(\varepsilon) & 0 \\ \mathcal{S}_i^\top(X) & 0 & -\mathcal{X}_i(X) \end{bmatrix} < 0, \tag{3.15}$$

where

$$J_0(i) = A(i)X(i) + X^\top(i)A^\top(i) + B(i)Y(i) + B^\top(i)Y^\top(i) + \lambda_{ii}X^\top(i)E^\top(i),$$
$$\mathcal{Z}_i(X) = \left[\sqrt{\lambda_{i1}}X^\top(i), \cdots, \sqrt{\lambda_{ii-1}}X^\top(i), \sqrt{\lambda_{ii+1}}X^\top(i), \cdots, \sqrt{\lambda_{iN}}X^\top(i) \right],$$
$$\mathcal{S}_i(X) = \left[\sqrt{\lambda_{i1}}X^\top(i)E^\top(1), \cdots, \sqrt{\lambda_{ii-1}}X^\top(i)E^\top(i-1), \right.$$
$$\left. \sqrt{\lambda_{ii+1}}X^\top(i)E^\top(i+1), \cdots, \sqrt{\lambda_{iN}}X^\top(i)E^\top(N) \right],$$
$$\mathcal{X}_i(\varepsilon) = \text{diag}\left[4\varepsilon(1)\mathbb{I}, \cdots, 4\varepsilon(i-1)\mathbb{I}, 4\varepsilon(i+1)\mathbb{I}, \cdots, 4\varepsilon(N)\mathbb{I}\right],$$
$$\mathcal{X}_i(X) = \text{diag}\left[-\varepsilon(1)\mathbb{I} + X^\top(1) + X(1), \cdots, -\varepsilon(i-1)\mathbb{I} + X^\top(i-1) + X(i-1), \right.$$
$$\left. -\varepsilon(i+1)\mathbb{I} + X^\top(i+1) + X(i+1), \cdots, -\varepsilon(N)\mathbb{I} + X^\top(N) + X(N)\right],$$

with the following constraints:

$$X^\top(i)E^\top(i) = E(i)X(i) \geq 0. \tag{3.16}$$

The stabilizing controller gain is given by $K(i) = Y(i)X^{-1}(i)$, $K(i) \in \mathbb{R}^{m \times n}$, $i \in \mathscr{S}$.

Let us now consider the robust stabilization using a state feedback controller. As we did for the nominal case, let us see how we can extend the previous results of the theorems on stabilization. Combining the expression of the controller and the state equation of the system, we get the following closed-loop one:

$$
\begin{aligned}
E(r_t)\dot{x}(t) &= A(r_t, t)x(t) + B(r_t, t)K(r_t)x(t) \\
&= [A(r_t) + B(r_t)K(r_t) + D_A(r_t)F_A(r_t, t)E_A(r_t) \\
&\quad + D_B(r_t)F_B(r_t, t)E_B(r_t)K(r_t)]\, x(t) \\
&= A_{cl}(r_t, t)x(t)\,,
\end{aligned}
$$

with

$$
\begin{aligned}
A_{cl}(r_t, t) &= A(r_t) + B(r_t)K(r_t) + D_A(r_t)F_A(r_t, t)E_A(r_t) \\
&\quad + D_B(r_t)F_B(r_t, t)E_B(r_t)K(r_t)\,.
\end{aligned}
$$

Based on the stochastic stability results (Corollary 2.2.1), the closed-loop uncertain system is piecewise regular, impulse-free and stochastically stable if there exists a set of nonsingular matrices $P = (P(1), \cdots, P(N))$, with $P(i) \in \mathbb{R}^{n \times n}$ such that the following hold:

$$
\begin{cases}
\varepsilon_P\left[P(i) + P^\top(i)\right] \geq E^\top(i)P(i) = P^\top(i)E(i) \geq 0\,, \\
A_{cl}^\top(i, t)P(i) + P^\top(i)A_{cl}(i, t) + \lambda_{ii}E^\top(i)P(i) \\
\quad + \sum_{j=1, j\neq i}^{N} \varepsilon_P\lambda_{ij}\left[P(j) + P^\top(j)\right] < 0\,.
\end{cases}
$$

Replacing $A_{cl}(i, t)$ by its expression we get:

$$
\begin{aligned}
&A^\top(i)P(i) + P^\top(i)A(i) + K^\top(i)B^\top(i)P(i) + P^\top(i)B(i)K(i) \\
&\quad + E_A^\top(i)F_A^\top(i)D_A^\top(i)P(i) + P^\top(i)D_A(i)F_A(i)E_A(i) \\
&\quad + K^\top(i)E_B^\top(i)F_B^\top(i)D_B^\top(i)P(i) + P^\top(i)D_B(i)F_B(i)E_B(i)K(i) \\
&\quad + \lambda_{ii}E^\top(i)P(i) + \sum_{j=1, j\neq i}^{N} \varepsilon_P\lambda_{ij}\left[P(j) + P^\top(j)\right] < 0\,,
\end{aligned}
$$

for all admissible uncertainties.

The left hand side of this matrix inequality is nonlinear in the design parameters $P(i)$ and $K(i)$. To transform it into an LMI, let $X(i) = P^{-1}(i)$ for each $i \in \mathscr{S}$. Pre- and post-multiplying the left hand side term by $X^\top(i)$ and $X(i)$ respectively give:

$$
\begin{aligned}
&X^\top(i)A^\top(i) + A(i)X(i) + X^\top(i)K^\top(i)B^\top(i) + B(i)K(i)X(i) \\
&\quad + X^\top(i)E_A^\top(i)F_A^\top(i)D_A^\top(i) + D_A(i)F_A(i)E_A(i)X(i) \\
&\quad + X^\top(i)K^\top(i)E_B^\top(i)F_B^\top(i)D_B^\top(i) + D_B(i)F_B(i)E_B(i)K(i)X(i) \\
&\quad + \lambda_{ii}X^\top(i)E^\top(i) + \sum_{j=1, j\neq i}^{N} \varepsilon_P\lambda_{ij}X^\top(i)\left[X^{-1}(j) + X^{-\top}(j)\right]X(i) < 0\,.
\end{aligned}
$$

Using Lemma 1.5.2, we get:

$$
\begin{aligned}
X^{-1}(j) + X^{-\top}(j) &\leq \mathbb{I} + X^{-1}(j)X^{-\top}(j) \\
&= \mathbb{I} + \left[X^\top(j)X(j)\right]^{-1}\,.
\end{aligned}
$$

The previous inequality matrix will hold if the following one is satisfied for each $i \in \mathscr{S}$:

$$
\begin{aligned}
& X^\top(i)A^\top(i) + A(i)X(i) + X^\top(i)K^\top(i)B^\top(i) + B(i)K(i)X(i) \\
& \quad + X^\top(i)E_A^\top(i)F_A^\top(i)D_A^\top(i) + D_A(i)F_A(i)E_A(i)X(i) \\
& \quad + X^\top(i)K^\top(i)E_B^\top(i)F_B^\top(i)D_B^\top(i) + D_B(i)F_B(i)E_B(i)K(i)X(i) \\
& \quad + \lambda_{ii}X^\top(i)E^\top(i) + \sum_{j=1, j\neq i}^{N} \varepsilon_P \lambda_{ij} X^\top(i)\left[\mathbb{I} + \left(X^\top(j)X(j)\right)^{-1}\right]X(i) < 0.
\end{aligned}
$$

Based on Lemma (1.5.1) we have:

$$
\begin{aligned}
& X^\top(i)E_A^\top(i)F_A^\top(i)D_A^\top(i) + D_A(i)F_A(i)E_A(i)X(i) \\
& \quad \leq \varepsilon_A(i)D_A(i)D_A^\top(i) + \varepsilon_A^{-1}(i)X^\top(i)E_A^\top(i)E_A(i)X(i), \\
& X^\top(i)K^\top(i)E_B^\top(i)F_B^\top(i)D_B^\top(i) + D_B(i)F_B(i)E_B(i)K(i)X(i) \\
& \quad \leq \varepsilon_B(i)D_B(i)D_B^\top(i) + \varepsilon_B^{-1}(i)X^\top(i)K^\top(i)E_B^\top(i)E_B(i)K(i)X(i),
\end{aligned}
$$

for any $\varepsilon_A > 0$ and $\varepsilon_A > 0$.

Using this, we get in turn:

$$
\begin{aligned}
& X^\top(i)A^\top(i) + A(i)X(i) + X^\top(i)K^\top(i)B^\top(i) + B(i)K(i)X(i) \\
& \quad + \varepsilon_A(i)D_A(i)D_A^\top(i) + \varepsilon_A^{-1}(i)X^\top(i)E_A^\top(i)E_A(i)X(i) \\
& \quad + \varepsilon_B(i)D_B(i)D_B^\top(i) + \varepsilon_B^{-1}(i)X^\top(i)K^\top(i)E_B^\top(i)E_B(i)K(i)X(i) \\
& \quad + \lambda_{ii}X^\top(i)E^\top(i) + \sum_{j=1, j\neq i}^{N} \varepsilon_P \lambda_{ij} X^\top(i)\left[\mathbb{I} + \left(X^\top(j)X(j)\right)^{-1}\right]X(i) < 0,
\end{aligned}
$$

that can be transformed in the LMI setting following the same steps as before.

Similarly, $\varepsilon_P\left[P(i) + P^\top(i)\right] \geq E^\top P(i) = P^\top(i)E \geq 0$ can be transformed to $\varepsilon_P\left[X(i) + X^\top(i)\right] \geq X^\top(i)E^\top = EX(i) \geq 0$.

If we define $Y(i)$, $\mathcal{S}_i(X)$ and $\mathcal{X}_i(X)$ as follows:

$$
\begin{aligned}
Y(i) &= K(i)X(i) \\
\mathcal{S}_i(X) &= \left[\sqrt{\varepsilon_P \lambda_{i1}}X^\top(i), \cdots, \sqrt{\varepsilon_P \lambda_{ii-1}}X^\top(i), \right. \\
& \qquad \left. \sqrt{\varepsilon_P \lambda_{ii+1}}X^\top(i), \cdots, \sqrt{\varepsilon_P \lambda_{iN}}X^\top(i)\right], \\
\mathcal{X}_i(X) &= \operatorname{diag}\left[-\mathbb{I} + X^\top(1) + X(1), \cdots, -\mathbb{I} + X^\top(i-1) + X(i-1), \right. \\
& \qquad \left. -\mathbb{I} + X^\top(i+1) + X(i+1), \cdots, -\mathbb{I} + X^\top(N) + X(N)\right],
\end{aligned}
$$

and proceeding similarly as before, the previous inequality matrix gives the following equivalent ones to guarantee that the closed state equation is piecewise regular, impulse-free and robust stochastically stable:

$$\varepsilon_P \left[X(i) + X^\top(i) \right] \geq X^\top(i)E^\top(i) = E(i)X(i) \geq 0$$

$$\begin{bmatrix} J(i) & X^\top(i)E_A^\top(i) & Y^\top(i)E_B^\top(i) & S_i(X) & S_i(X) \\ E_A(i)X(i) & -\varepsilon_A(i)\mathbb{I} & 0 & 0 & 0 \\ E_B(i)Y(i) & 0 & -\varepsilon_B(i)\mathbb{I} & 0 & 0 \\ S_i^\top(X) & 0 & 0 & -\mathbb{I} & 0 \\ S_i^\top(X) & 0 & 0 & 0 & -X_i(X) \end{bmatrix} < 0,$$

where

$$J(i) = X^\top(i)A^\top(i) + A(i)X(i) + Y^\top(i)B^\top(i) + B(i)Y(i)$$
$$+ \lambda_{ii}X^\top(i)E^\top(i) + \varepsilon_A(i)D_A(i)D_A^\top(i) + \varepsilon_B(i)D_B(i)D_B^\top(i).$$

The following theorem summarizes the results on robust stochastic stabilization:

Theorem 3.2.5 *Let ε_P be a given positive scalar. There exists a state feedback controller of the form (3.6) such that the closed-loop state equation of the system (3.1) is piecewise regular, impulse-free and stochastically stable if there exist set of nonsingular matrices $X = (X(1), \cdots, X(N))$, $X(i) \in \mathbb{R}^{n \times n}$, a set of matrices $Y = (Y(1), \cdots, Y(N))$, $Y(i) \in \mathbb{R}^{m \times n}$, and sets of positive scalars $\varepsilon_A = (\varepsilon_A(1), \cdots, \varepsilon_A(N))$ and $\varepsilon_B = (\varepsilon_B(1), \cdots, \varepsilon_B(N))$, such that the following holds for each $i \in \mathscr{S}$:*

$$\begin{cases} \varepsilon_P \left[X(i) + X^\top(i) \right] \geq X^\top(i)E(i) = E(i)X(i) \geq 0, \\ \begin{bmatrix} J(i) & X^\top(i)E_A^\top(i) & Y^\top(i)E_B^\top(i) & S_i(X) & S_i(X) \\ E_A(i)X(i) & -\varepsilon_A(i)\mathbb{I} & 0 & 0 & 0 \\ E_B(i)Y(i) & 0 & -\varepsilon_B(i)\mathbb{I} & 0 & 0 \\ S_i^\top(X) & 0 & 0 & -\mathbb{I} & 0 \\ S_i^\top(X) & 0 & 0 & 0 & -X_i(X) \end{bmatrix} < 0, \end{cases}$$

where

$$J(i) = X^\top(i)A^\top(i) + A(i)X(i) + Y^\top(i)B^\top(i) + B(i)Y(i)$$
$$+ \lambda_{ii}X^\top(i)E^\top(i) + \varepsilon_A(i)D_A(i)D_A^\top(i) + \varepsilon_B(i)D_B(i)D_B^\top(i),$$
$$S_i(X) = \left[\sqrt{\varepsilon_P \lambda_{i1}}X^\top(i), \cdots, \sqrt{\varepsilon_P \lambda_{ii-1}}X^\top(i), \right.$$
$$\left. \sqrt{\varepsilon_P \lambda_{ii+1}}X^\top(i), \cdots, \sqrt{\varepsilon_P \lambda_{iN}}X^\top(i) \right],$$
$$X_i(X) = \text{diag}\left[-\mathbb{I} + X^\top(1) + X(1), \cdots, -\mathbb{I} + X^\top(i-1) + X(i-1), \right.$$
$$\left. -\mathbb{I} + X^\top(i+1) + X(i+1), \cdots, -\mathbb{I} + X^\top(N) + X(N) \right].$$

The stabilizing controller gain is given by $K(i) = Y(i)X^{-1}(i)$, $K(i) \in \mathbb{R}^{m \times n}$, $i \in \mathscr{S}$.

Let us extend the results of the other stabilization design approach that we used for nominal system and adapt them for uncertain systems. For this purpose, applying the results of Theorem 3.2.4, we know that there exists a state feedback controller of the form (3.6), with gain $K(i) = Y(i)X^{-1}(i)$, $i \in \mathcal{S}$, that makes the closed-loop state equation of the uncertain system (3.1) is piecewise regular, impulse-free and stochastically stable if there exist a set of nonsingular matrices $X = (X(1), \cdots, X(N))$, with $X(i) \in \mathbb{R}^{n \times n}$, a set of matrices $Y = (Y(1), \cdots, Y(N))$, with $Y(i) \in \mathbb{R}^{m \times n}$, and a set of positive scalars $\varepsilon = (\varepsilon(1), \cdots, \varepsilon(N))$ such that the following set of coupled LMIs holds for each $i \in \mathcal{S}$:

$$\begin{bmatrix} \widehat{J_0}(i) & S_i(X) \\ S_i^{\top}(X) & -X_i(X) \end{bmatrix} < 0, \tag{3.17}$$

where

$$\begin{aligned}
\widehat{J_0}(i) &= [A(i) + D_A(i)F_A(i)E_A(i)]X(i) + D_B(i)F_B(i)E_B(i)Y(i) \\
&\quad + X^{\top}(i)[A(i) + D_A(i)F_A(i)E_A(i)]^{\top} + [D_B(i)F_B(i)E_B(i)Y(i)]^{\top} \\
&\quad + B(i)Y(i) + Y^{\top}(i)B^{\top}(i) + \lambda_{ii}X^{\top}(i)E^{\top}(i),
\end{aligned}$$

$$S_i(X) = \left[\sqrt{\lambda_{i1}}X^{\top}(i), \cdots, \sqrt{\lambda_{ii-1}}X^{\top}(i), \sqrt{\lambda_{ii+1}}X^{\top}(i), \cdots, \sqrt{\lambda_{iN}}X^{\top}(i) \right],$$

$$\begin{aligned}
X_i(X) = \operatorname{diag}\big[&-\varepsilon(1)\mathbb{I} + X^{\top}(1) + X(1), \cdots, -\varepsilon(i-1)\mathbb{I} + X^{\top}(i-1) + X(i-1), \\
&-\varepsilon(i+1)\mathbb{I} + X^{\top}(i+1) + X(i+1), \cdots, -\varepsilon(N)\mathbb{I} + X^{\top}(N) + X(N) \big],
\end{aligned}$$

with the following constraints:

$$\varepsilon(i)\mathbb{I} \geq X^{\top}(i)E^{\top}(i) = E(i)X(i) \geq 0. \tag{3.18}$$

Notice that we have:

$$\begin{bmatrix} D_A(i)F_A(i)E_A(i)X(i) & 0 \\ 0 & 0 \end{bmatrix} = \begin{bmatrix} D_A(i) \\ 0 \end{bmatrix} F_A(i) \begin{bmatrix} E_A(i)X(i) & 0 \end{bmatrix},$$

$$\begin{bmatrix} D_B(i)F_B(i)E_B(i)Y(i) & 0 \\ 0 & 0 \end{bmatrix} = \begin{bmatrix} D_B(i) \\ 0 \end{bmatrix} F_B(i) \begin{bmatrix} E_B(i)Y(i) & 0 \end{bmatrix},$$

using Lemma 1.5.1, we get:

$$\begin{bmatrix} D_A(i)F_A(i)E_A(i)X(i) & 0 \\ 0 & 0 \end{bmatrix} + \begin{bmatrix} D_A(i)F_A(i)E_A(i)X(i) & 0 \\ 0 & 0 \end{bmatrix}^{\top}$$
$$\leq \begin{bmatrix} \varepsilon_A(i)D_A(i)D_A^{\top}(i) & 0 \\ 0 & 0 \end{bmatrix} + \begin{bmatrix} \varepsilon_A^{-1}(i)X^{\top}(i)E_A^{\top}(i)E_A(i)X(i) & 0 \\ 0 & 0 \end{bmatrix},$$

$$\begin{bmatrix} D_B(i)F_B(i)E_B(i)Y(i) & 0 \\ 0 & 0 \end{bmatrix} + \begin{bmatrix} D_B(i)F_B(i)E_B(i)Y(i) & 0 \\ 0 & 0 \end{bmatrix}^{\top}$$
$$\leq \begin{bmatrix} \varepsilon_B(i)D_B(i)D_B^{\top}(i) & 0 \\ 0 & 0 \end{bmatrix} + \begin{bmatrix} \varepsilon_B^{-1}(i)Y^{\top}(i)E_B^{\top}(i)E_A(i)Y(i) & 0 \\ 0 & 0 \end{bmatrix},$$

for $\varepsilon_A(i) > 0$ and $\varepsilon_B(i) > 0$, $i \in \mathcal{S}$.

Using now Schur complement, we get:

$$\begin{bmatrix} J(i) & X^{\top}(i)E_A^{\top}(i) & Y^{\top}(i)E_B^{\top}(i) & S_i(X) \\ E_A(i)X(i) & -\varepsilon_A(i)\mathbb{I} & 0 & 0 \\ E_B(i)Y(i) & 0 & -\varepsilon_B(i)\mathbb{I} & 0 \\ S_i^{\top}(X) & 0 & 0 & -X_i(X) \end{bmatrix} < 0,$$

with

$$J(i) = A(i)X(i) + X^\top(i)A^\top(i) + Y^\top(i)B^\top(i) + B(i)Y(i)$$
$$+ \lambda_{ii}X^\top(i)E^\top(i) + \varepsilon_A(i)D_A(i)D_A^\top(i) + \varepsilon_B(i)D_B(i)D_B^\top(i)$$

The following theorem gives the results that can be used to determine the stabilizing state feedback controller.

Theorem 3.2.6 *There exists a state feedback controller of the form (3.6) such that the closed-loop state equation of the system (3.1) is piecewise regular, impulse-free and stochastically stable if there exist sets of nonsingular matrices $X = (X(1), \cdots, X(N))$, $X(i) \in \mathbb{R}^{n \times n}$, and $Y = (Y(1), \cdots, Y(N))$, $Y(i) \in \mathbb{R}^{m \times n}$, and sets of positive scalars $\varepsilon_A = (\varepsilon_A(1), \cdots, \varepsilon_A(N))$, $\varepsilon_B = (\varepsilon_B(1), \cdots, \varepsilon_B(N))$ and $\varepsilon = (\varepsilon(1), \cdots, \varepsilon(N))$, such that the following holds for each $i \in \mathscr{S}$:*

$$\begin{bmatrix} J(i) & X^\top(i)E_A^\top(i) & Y^\top(i)E_B^\top(i) & S_i(X) \\ E_A(i)X(i) & -\varepsilon_A(i)\mathbb{I} & 0 & 0 \\ E_B(i)Y(i) & 0 & -\varepsilon_B(i)\mathbb{I} & 0 \\ S_i^\top(X) & 0 & 0 & -\mathcal{X}_i(X) \end{bmatrix} < 0, \tag{3.19}$$

where

$$J(i) = A(i)X(i) + X^\top(i)A^\top(i) + Y^\top(i)B^\top(i) + B(i)Y(i)$$
$$+ \lambda_{ii}X^\top(i)E^\top(i) + \varepsilon_A(i)D_A(i)D_A^\top(i) + \varepsilon_B(i)D_B(i)D_B^\top(i),$$
$$S_i(X) = \left[\sqrt{\lambda_{i1}}X^\top(i), \cdots, \sqrt{\lambda_{ii-1}}X^\top(i), \sqrt{\lambda_{ii+1}}X^\top(i), \cdots, \sqrt{\lambda_{iN}}X^\top(i) \right],$$
$$\mathcal{X}_i(X) = \text{diag}\left[-\varepsilon(1)\mathbb{I} + X^\top(1) + X(1), \cdots, -\varepsilon(i-1)\mathbb{I} + X^\top(i-1) + X(i-1), \right.$$
$$\left. -\varepsilon(i+1)\mathbb{I} + X^\top(i+1) + X(i+1), \cdots, -\varepsilon(N)\mathbb{I} + X^\top(N) + X(N) \right],$$

with the following constraints:

$$\varepsilon(i)\mathbb{I} \geq X^\top(i)E^\top(i) = E(i)X(i) \geq 0. \tag{3.20}$$

The controller gain is given by $K(i) = Y(i)X^{-1}(i)$, $K(i) \in \mathbb{R}^{m \times n}$.

Now if we consider the third approach, we know that based on Theorem 3.2.5, that there exists a state feedback controller of the form (3.6), with gain $K(i) = Y(i)X^{-1}(i)$, $i \in \mathscr{S}$ such that the closed-loop state equation of the uncertain system (3.1) is piecewise regular, impulse-free and stochastically stable if there exist a set of nonsingular matrices $X = (X(1), \cdots, X(N))$, with $X(i) \in \mathbb{R}^{n \times n}$, a set of matrices $Y = (Y(1), \cdots, Y(N))$, with $Y(i) \in \mathbb{R}^{m \times n}$, and a set of positive scalars $\varepsilon = (\varepsilon(1), \cdots, \varepsilon(N))$ such that the following set of coupled LMIs holds for each $i \in \mathscr{S}$:

$$\begin{bmatrix} J_0(i) & \mathcal{Z}_i(X) & S_i(X) \\ \mathcal{Z}_i^\top(X) & -\mathcal{X}_i(\varepsilon) & 0 \\ S_i^\top(X) & 0 & -\mathcal{X}_i(X) \end{bmatrix} < 0,$$

where

$$
\begin{aligned}
J_0(i) &= [A(i) + D_A(i)F_A(i)E_A(i)]\,X(i) + D_A(i)F_A(i)E_A(i)Y(i) \\
&\quad + X^\top(i)\,[A(i) + D_A(i)F_A(i)E_A(i)]^\top + [D_A(i)F_A(i)E_A(i)Y(i)]^\top \\
&\quad + B(i)Y(i) + B^\top(i)Y^\top(i) + \lambda_{ii}X^\top(i)E^\top(i)\,,
\end{aligned}
$$

$$
\begin{aligned}
S_i(X) &= \Big[\, \sqrt{\lambda_{i1}}X^\top(i)E^\top(1), \cdots, \sqrt{\lambda_{ii-1}}X^\top(i)E^\top(i-1), \\
&\qquad \sqrt{\lambda_{ii+1}}X^\top(i)E^\top(i+1), \cdots, \sqrt{\lambda_{iN}}X^\top(i)E^\top(N)\,\Big],
\end{aligned}
$$

$$
\begin{aligned}
Z_i(X) &= \Big[\, \sqrt{\lambda_{i1}}X^\top(i), \cdots, \sqrt{\lambda_{ii-1}}X^\top(i), \\
&\qquad \sqrt{\lambda_{ii+1}}X^\top(i), \cdots, \sqrt{\lambda_{iN}}X^\top(i)\,\Big],
\end{aligned}
$$

$$
\begin{aligned}
\mathcal{X}_i(X) &= \operatorname{diag}\Big[-\varepsilon(1)\mathbb{I} + X^\top(1) + X(1), \cdots, -\varepsilon(i-1)\mathbb{I} + X^\top(i-1) + X(i-1), \\
&\qquad -\varepsilon(i+1)\mathbb{I} + X^\top(i+1) + X(i+1), \cdots, -\varepsilon(N)\mathbb{I} + X^\top(N) + X(N)\Big],
\end{aligned}
$$

$$
\mathcal{X}_i(\varepsilon) = \operatorname{diag}\,[4\varepsilon(1)\mathbb{I}, \cdots, 4\varepsilon(i-1)\mathbb{I}, 4\varepsilon(i+1)\mathbb{I}, \cdots, 4\varepsilon(N)\mathbb{I}]\,,
$$

with the following constraints:

$$
X^\top(i)E^\top(i) = E(i)X(i) \geq 0\,.
$$

Using Lemma 1.5.1 and following the same steps as before we get the following theorem that gives the results that can be used to determine the stabilizing state feedback controller.

Theorem 3.2.7 *There exists a state feedback controller of the form (3.6) such that the closed-loop state equation of the system (3.1) is piecewise regular, impulse-free and stochastically stable if there exist sets of nonsingular matrices $X = (X(1), \cdots, X(N))$, $X(i) \in \mathbb{R}^{n \times n}$ and $Y = (Y(1), \cdots, Y(N))$, $Y(i) \in \mathbb{R}^{m \times n}$, and sets of positive scalars $\varepsilon_A = (\varepsilon_A(1), \cdots, \varepsilon_A(N))$, $\varepsilon_B = (\varepsilon_B(1), \cdots, \varepsilon_B(N))$ and $\varepsilon = (\varepsilon(1), \cdots, \varepsilon(N))$ such that the following holds for each $i \in \mathscr{S}$:*

$$
\begin{bmatrix}
J(i) & X^\top(i)E_A^\top(i) & Y^\top(i)E_B^\top(i) & Z_i(X) & S_i(X) \\
E_A(i)X(i) & -\varepsilon_A(i)\mathbb{I} & 0 & 0 & 0 \\
E_B(i)Y(i) & 0 & -\varepsilon_B(i)\mathbb{I} & 0 & 0 \\
Z_i^\top(X) & 0 & 0 & -\mathcal{X}_i(\varepsilon) & 0 \\
S_i^\top(X) & 0 & 0 & 0 & -\mathcal{X}_i(X)
\end{bmatrix} < 0\,, \tag{3.21}
$$

where

$$
\begin{aligned}
J(i) &= A(i)X(i) + X^\top(i)A^\top(i) + Y^\top(i)B^\top(i) + B(i)Y(i) \\
&\quad + \lambda_{ii}X^\top(i)E^\top(i) + \varepsilon_A(i)D_A(i)D_A^\top(i) + \varepsilon_B(i)D_B(i)D_B^\top(i)
\end{aligned}
$$

$$
\begin{aligned}
S_i(X) &= \Big[\, \sqrt{\lambda_{i1}}X^\top(i)E^\top(1), \cdots, \sqrt{\lambda_{ii-1}}X^\top(i)E^\top(i-1), \\
&\qquad \sqrt{\lambda_{ii+1}}X^\top(i)E^\top(i+1), \cdots, \sqrt{\lambda_{iN}}X^\top(i)E^\top(N)\,\Big],
\end{aligned}
$$

$$
\begin{aligned}
Z_i(X) &= \Big[\, \sqrt{\lambda_{i1}}X^\top(i), \cdots, \sqrt{\lambda_{ii-1}}X^\top(i), \\
&\qquad \sqrt{\lambda_{ii+1}}X^\top(i), \cdots, \sqrt{\lambda_{iN}}X^\top(i)\,\Big],
\end{aligned}
$$

$$X_i(X) = \text{diag}\left[-\varepsilon(1)\mathbb{I} + X^\top(1) + X(1), \cdots, -\varepsilon(i-1)\mathbb{I} + X^\top(i-1) + X(i-1),\right.$$
$$-\varepsilon(i+1)\mathbb{I} + X^\top(i+1) + X(i+1), \cdots, -\varepsilon(N)\mathbb{I} + X^\top(N) + X(N)\Big],$$
$$X_i(\varepsilon) = \text{diag}\left[4\varepsilon(1)\mathbb{I}, \cdots, 4\varepsilon(i-1)\mathbb{I}, 4\varepsilon(i+1)\mathbb{I}, \cdots, 4\varepsilon(N)\mathbb{I}\right],$$

with the following constraints:

$$X^\top(i)E^\top(i) = E(i)X(i) \geq 0. \tag{3.22}$$

The controller gain is given by $K(i) = Y(i)X^{-1}(i)$, $K(i) \in \mathbb{R}^{m \times n}$.

Remark 3.2.2 *The constraints* $X^\top(i)E^\top(i) = E(i)X(i)$ *may be difficult to solve with some commercial LMI toolboxes like LMI-Toolbox of MATLAB. To overcome this we can use the following LMI condition that may approximate this constraint:*

$$\left[X^\top(i)E^\top(i) - E(i)X(i)\right]^\top \left[X^\top(i)E^\top(i) - E(i)X(i)\right] \leq \beta\mathbb{I}, \tag{3.23}$$

that gives the following LMI:

$$\begin{bmatrix} -\beta\mathbb{I} & \left[X^\top(i)E^\top(i) - E(i)X(i)\right]^\top \\ \left[X^\top(i)E^\top(i) - E(i)X(i)\right] & -\mathbb{I} \end{bmatrix} \leq 0. \tag{3.24}$$

Therefore, the solution of our problem is brought to the minimization of β subject to the appropriate LMIs of the ones we developed with (3.24) and $\beta \geq 0$ that we should minimize.

3.3 Constant Gain State Feedback Stabilization

The fact that all the stabilization results we developed earlier have controller gains that are mode dependent, this limits sometimes the use of such controller when the mode is not available for feedback. Notice that even in this case, the mode can be estimated and the state feedback control can be used. This way of doing may require more time in computation which may limit the use of such approach. To avoid this, an alternative consists of using the state feedback controller with constant gain. This will be our goal in the rest of this chapter.

Let us prove that by choosing a constant nonsingular matrix instead of a set of nonsingular matrices as it was done in Chap. 2, we can get a simple approach that may be restrictive, which shows that the system is piecewise regular, impulse-free and stochastically stable. To show that the free system is piecewise regular, impulse-free, the steps are the same as in Chap. 2, for this reason the details is omitted. To show the stochastic stability, let denote by (x, i) the values of the state vector $x(t)$ and the mode r_t at time t and consider the following Lyapunov candidate function:

$$V(x(t), r(t)) = x^\top(t)E^\top(r(t))Px(t),$$

where P is a nonsingular matrix.

The weak infinitesimal generator of the Markov Process, $\{(x(t), i), t \geq 0\}$, emanating from the point (x, i) at time t is given by:

$$\mathscr{L}V(x(t), i) = x^{\top}(t)P^{\top}A(i)x(t) + x^{\top}(t)A^{\top}(i)Px(t)$$

$$+ x^{\top}(t) \sum_{j=1}^{N} \lambda_{ij}E^{\top}(j)Px(t),$$

that can be rewritten as follows:

$$\mathscr{L}V(x(t), i) = x^{\top}(t)\Gamma(i)x(t),$$

with $\Gamma(i) = A^{\top}(i)P + P^{\top}A(i) + \sum_{j=1}^{N} \lambda_{ij}E^{\top}(j)P$.

If there exists a nonsingular matrix P such that $\Gamma(i) < 0$ holds for every mode $i \in \mathscr{S}$, then system (3.1) is piecewise regular, impulse-free and stochastically stable. In fact if the condition

$$A^{\top}(i)P + P^{\top}A(i) + \sum_{j=1}^{N} \lambda_{ij}E^{\top}(j)P < 0, \qquad (3.25)$$

holds, then it results that:

$$\mathscr{L}V(x(t), i) \leq - \min_{i \in \mathscr{S}} \lambda_{min}(-\Gamma(i)) x^{\top}(t)x(t),$$

where $\lambda_{min}(D)$ is the minimum eigenvalue of the matrix D.

Now this together with Dynkin's formula yields:

$$\mathbb{E}\left[V(x(t), i)\right] - V(x(0), r_0) = \mathbb{E}\left[\int_0^t \mathscr{L}V(x(s), r_s)ds|(x_0, r_0)\right]$$

$$\leq - \min_{i \in \mathscr{S}}\{\lambda_{min}(-\Gamma(i))\}\mathbb{E}\left[\int_0^t x^{\top}(s)x(s)ds|(x_0, r_0)\right],$$

implying, in turn,

$$\min_{i \in \mathscr{S}}\{\lambda_{min}(-\Gamma(i))\}\mathbb{E}\left[\int_0^t x^{\top}(s)x(s)ds|(x_0, r_0)\right]$$

$$\leq \mathbb{E}\left[V(x(0), r_0)\right] - \mathbb{E}\left[V(x(t), i)\right]$$

$$\leq V(x(0), r_0).$$

This yields that

$$\mathbb{E}\left[\int_0^t x^{\top}(s)x(s)ds|(x_0, r_0)\right] \leq \frac{V(x(0), r_0)}{\min_{i \in \mathscr{S}}\{\lambda_{min}(-\Gamma(i))\}},$$

holds for any $t > 0$. This means that the system (3.1) is stochastically stable. The following corollary gives the results of this development.

Corollary 3.3.1 *If there exists a nonsingular matrix $P > 0$, $P \in \mathbb{R}^{n \times n}$ such that the following set of LMIs hold for every $i \in \mathscr{S}$:*

$$A^\top(i)P + P^\top A(i) + \sum_{j=1}^{N} \lambda_{ij} E^\top(j)P < 0, \tag{3.26}$$

with the following constraints:

$$E^\top(i)P = P^\top E(i) \geq 0,$$

then system (3.1) is piecewise regular, impulse-free and stochastically stable.

As we did in Chap. 2, we can establish the following results when using the different bounds for the term $E^\top(i)P$. The corresponding results are given.

When $E^\top(i)P \leq \varepsilon(i)[P^\top + P]$, the corresponding results are given by the following corollary.

Corollary 3.3.2 *Let $\varepsilon = (\varepsilon(1), \cdots, \varepsilon(N))$ be a given set of positive scalars. If there exists a nonsingular matrix $P > 0$, $P \in \mathbb{R}^{n \times n}$ such that the following set of the matrix inequalities hold for every $i \in \mathscr{S}$:*

$$A^\top(i)P + P^\top A(i) + \lambda_{ii} E^\top(i)P + \sum_{j=1, j \neq i}^{N} \varepsilon(j)\lambda_{ij}\left[P^\top + P\right] < 0, \tag{3.27}$$

with the following constraints:

$$\varepsilon(j)\left[P^\top + P\right] \geq E^\top(i)P = P^\top E(i) \geq 0,$$

then system (3.1) is piecewise regular, impulse-free and stochastically stable.

When $E^\top(i)P \leq \varepsilon(i)[P^\top P]$, the corresponding results are given by the following corollary.

Corollary 3.3.3 *Let $\varepsilon = (\varepsilon(1), \cdots, \varepsilon(N))$ be a given set of positive scalars. If there exists a nonsingular matrix $P > 0$, $P \in \mathbb{R}^{n \times n}$ such that the following set of the matrix inequalities hold for every $i \in \mathscr{S}$:*

$$A^\top(i)P + P^\top A(i) + \lambda_{ii} E^\top(i)P + \sum_{j=1, j \neq i}^{N} \varepsilon(j)\lambda_{ij}\left[P^\top P\right] < 0, \tag{3.28}$$

with the following constraints:

$$\varepsilon(i)\left[P^\top P\right] \geq E^\top(i)P = P^\top E(i) \geq 0,$$

then system (3.1) is piecewise regular, impulse-free and stochastically stable.

When $E^\top(i)P \leq \frac{1}{4}\varepsilon(i)\mathbb{I} + \varepsilon(i)E^\top(i)PP^\top E(i)$, the corresponding results are given by the following corollary.

Corollary 3.3.4 *Let* $\varepsilon = (\varepsilon(1), \cdots, \varepsilon(N))$ *be a given set of positive scalars. If there exists a nonsingular matrix* $P > 0$, $P \in \mathbb{R}^{n \times n}$ *such that the following set of the matrix inequalities holds for every* $i \in \mathscr{S}$:

$$A^\top(i)P + P^\top A(i) + \lambda_{ii} E^\top(i)P + \sum_{j=1,j\neq i}^{N} \lambda_{ij}\left[\frac{1}{4}\varepsilon(j)\mathbb{I} + \varepsilon(j)E^\top(j)PP^\top E(j)\right] < 0, \quad (3.29)$$

with the following constraints:

$$E^\top(i)P = P^\top E(i) \geq 0,$$

then system (3.1) is piecewise regular, impulse-free and stochastically stable.

The state feedback controller with constant gain is given by:

$$u(t) = \mathscr{K} x(t), \quad (3.30)$$

with $\mathscr{K}$ a constant gain to be determined.

Plugging the controller expression (3.30) in the state equation (3.1) we get the following closed-loop dynamics:

$$E\dot{x}(t) = A_{cl}(i)x(t),$$

with $A_{cl}(i) = A(i) + B(i)\mathscr{K}$.

To design the stabilizing controller with constant gain (3.30) that makes the closed-loop dynamics piecewise regular, impulse-free and stochastically stable, we can use the different bounds we use in Chap. 2.

If we assume that there exists a set of positive scalars $\varepsilon_P = (\varepsilon_P(1), \cdots, \varepsilon_P(N))$ such that the following holds for every mode $i \in \mathscr{S}$:

$$E^\top(i)P \leq \varepsilon_P(i)\left[P + P^\top\right].$$

Using this, the closed-loop system is piecewise regular, impulse-free and stochastically stable, if there exists a nonsingular P such that the following hold:

$$\begin{cases} \varepsilon_P(i)\left[P + P^\top\right] \geq E^\top(i)P = P^\top E(i) \geq 0, \\ A^\top(i)P + P^\top A(i) + \mathscr{K}^\top B^\top(i)P + P^\top B(i)\mathscr{K} + \lambda_{ii}E^\top(i)P \\ \quad + \sum_{j=1,j\neq i}^{N} \lambda_{ij}\varepsilon_P(j)\left[P + P^\top\right] < 0. \end{cases}$$

The second condition is nonlinear in the design parameters $\mathscr{K}$ and P. To transform it into an LMI, let $X = P^{-1}$ and pre- and post-multiply this inequality respectively by $X^\top$ and X implies:

$$X^\top A^\top(i) + A(i)X + X^\top \mathscr{K}^\top B^\top(i) + B(i)\mathscr{K}X + \lambda_{ii}X^\top E^\top(i)$$

$$+ \sum_{j=1,j\neq i}^{N} \lambda_{ij}\varepsilon_P(j)\left[X + X^\top\right] < 0.$$

Letting $K = \mathcal{K} X$ gives:

$$X^{\top} A^{\top}(i) + A(i)X + K^{\top} B^{\top}(i) + B(i)K + \lambda_{ii} X^{\top} E^{\top}(i)$$

$$+ \sum_{j=1, j \neq i}^{N} \lambda_{ij} \varepsilon_P(j) \left[X + X^{\top} \right] < 0 \,.$$

Also notice that the condition, $\varepsilon_P(i) \left[P + P^{\top} \right] \geq E^{\top}(i)P = P^{\top} E(i) \geq 0$, can be transformed to $\varepsilon_P(i) \left[X + X^{\top} \right] \geq X^{\top} E^{\top}(i) = E(i)X \geq 0$.

The state feedback controller gain can be designed by the results of the following theorem.

Theorem 3.3.1 *Let $\varepsilon_P = (\varepsilon_P(1), \cdots, \varepsilon_P(N))$ be a given set of positive scalars. There exists a state feedback controller with constant gain that guarantees that the closed-loop state equation will be piecewise regular, impulse-free and stochastically stable if there exist a nonsingular matrix $X > 0$, $X \in \mathbb{R}^{n \times n}$ and a matrix K, $K \in \mathbb{R}^{m \times n}$ such that the following set of LMI holds for every $i \in \mathscr{S}$:*

$$X^{\top} A^{\top}(i) + A(i)X + K^{\top} B^{\top}(i) + B(i)K + \lambda_{ii} X^{\top} E^{\top}(i)$$

$$+ \sum_{j=1, j \neq i}^{N} \lambda_{ij} \varepsilon_P(j) \left[X + X^{\top} \right] < 0 \,, \qquad (3.31)$$

with the following constraints:

$$\varepsilon_P(i) \left[X + X^{\top} \right] \geq X^{\top} E^{\top} = EX \geq 0 \,. \qquad (3.32)$$

The controller gain is given by $\mathcal{K} = KX^{-1}$, $\mathcal{K} \in \mathbb{R}^{m \times n}$.

Let us now consider that the following hold:

$$E^{\top}(i)P \leq \varepsilon(i)P^{\top}P \,.$$

Using this, the closed-loop state equation will be piecewise regular, impulse-free and stochastically stable if there exists a nonsingular matrix P such that the following holds for every $i \in \mathscr{S}$:

$$\begin{cases} \varepsilon(i)P^{\top}P \geq E^{\top}(i)P = P^{\top}E(i) \geq 0 \,, \\ A^{\top}(i)P + P^{\top}A(i) + \mathcal{K}^{\top}B^{\top}(i)P + P^{\top}B(i)\mathcal{K} + \lambda_{ii}E^{\top}(i)P \\ \quad + \sum_{j=1, j \neq i}^{N} \lambda_{ij} \varepsilon(j) \left[P^{\top}P \right] < 0 \,. \end{cases}$$

Let $X = P^{-1}$ and pre- and post-multiply the previous conditions, we get:

$$\begin{cases} \varepsilon(i)\mathbb{I} \geq X^{\top}E^{\top}(i) = E(i)X \geq 0 \,, \\ X^{\top}A^{\top}(i) + A(i)X + X^{\top}\mathcal{K}^{\top}B^{\top}(i) + B(i)\mathcal{K}X + \lambda_{ii}X^{\top}E^{\top}(i) \\ \quad + \sum_{j=1, j \neq i}^{N} \lambda_{ij} \varepsilon(j)\mathbb{I} < 0 \,. \end{cases}$$

Letting $K = \mathcal{K} X$, we get the following theorem which gives another design procedure for a state feedback controller with constant gain.

Theorem 3.3.2 *There exists a state feedback controller with constant gain that guarantees that the closed-loop state equation will be piecewise regular, impulse-free and stochastically stable if there exist a nonsingular matrix X, $X \in \mathbb{R}^{n \times n}$ a matrix K, $K \in \mathbb{R}^{m \times n}$ and a set of positive scalars $\varepsilon = (\varepsilon(1), \cdots, \varepsilon(N))$ such that the following set of LMI holds for every $i \in \mathscr{S}$:*

$$X^\top A^\top(i) + A(i)X + K^\top B^\top(i) + B(i)K + \lambda_{ii} X^\top E^\top(i)$$

$$+ \sum_{j=1, j\neq i}^{N} \lambda_{ij} \varepsilon(j) \mathbb{I} < 0, \tag{3.33}$$

with the following constraints:

$$\varepsilon(i)\mathbb{I} \geq X^\top E^\top(i) = E(i)X \geq 0, \forall i \in \mathscr{S}. \tag{3.34}$$

The controller gain is given by $\mathscr{K} = KX^{-1}$, $\mathscr{K} \in \mathbb{R}^{m \times n}$.

If we consider now that the following holds

$$E^\top(i)P \leq \frac{1}{4}\varepsilon^{-1}(i)\mathbb{I} + \varepsilon(i)E^\top(i)PP^\top E(i),$$

with $\varepsilon(i) > 0$.

The closed-loop system will be in this case piecewise regular, impulse-free and stochastically stable if there exists a nonsingular matrix P such that the following hold for each $i \in \mathscr{S}$:

$$\begin{cases} E^\top(i)P = P^\top E(i) \geq 0, \\ A^\top(i)P + P^\top A(i) + \mathscr{K}^\top B^\top(i)P + P^\top B(i)\mathscr{K} + \lambda_{ii} E^\top(i)P \\ \quad + \sum_{j=1, j\neq i}^{N} \lambda_{ij}\left[\frac{1}{4}\varepsilon^{-1}(j)\mathbb{I} + \varepsilon(j)E^\top(j)PP^\top E(j)\right] < 0. \end{cases}$$

Let $X = P^{-1}$ and pre- and post-multiply the previous conditions we get:

$$\begin{cases} X^\top E^\top(i) = E(i)X \geq 0, \\ X^\top A^\top(i) + A(i)X + X^\top \mathscr{K}^\top B^\top(i) + B(i)\mathscr{K}X + \lambda_{ii} X^\top E^\top(i) \\ \quad + \sum_{j=1, j\neq i}^{N} \lambda_{ij}\frac{1}{4}\varepsilon^{-1}(j)X^\top \mathbb{I}X \\ \quad + \sum_{j=1, j\neq i}^{N} \lambda_{ij}\varepsilon(j)X^\top E^\top(j)X^{-1}X^{-\top} E(j)X < 0. \end{cases}$$

Letting $K = \mathscr{K}X$ and proceeding in a similar way as we did previously, we get the following theorem gives another design procedure for a state feedback controller with constant gain.

Theorem 3.3.3 *There exists a state feedback controller with constant gain that guarantees that the closed-loop state equation will be piecewise regular, impulse-free and stochastically stable if there exist a nonsingular matrix X, $X \in \mathbb{R}^{n \times n}$ and a matrix K, $K \in \mathbb{R}^{m \times n}$ such that the following set of LMI holds for every $i \in \mathscr{S}$:*

$$\begin{bmatrix} J(i) & \mathscr{Z}_i(X) & \mathscr{S}_i(X) \\ \mathscr{Z}_i^\top(X) & -\mathscr{X}_i(\varepsilon) & 0 \\ \mathscr{S}_i^\top(X) & 0 & -\mathscr{X}_i(X) \end{bmatrix} < 0, \tag{3.35}$$

where

$$J(i) = X^\top A^\top(i) + A(i)X + K^\top B^\top(i) + B(i)K + \lambda_{ii} X^\top E^\top(i)$$

$$S_i(X) = \left[\sqrt{\lambda_{i1}} X^\top E^\top(1), \cdots, \sqrt{\lambda_{ii-1}} X^\top E^\top(i-1), \right.$$

$$\left. \sqrt{\lambda_{ii+1}} X^\top E^\top(i+1), \cdots, \sqrt{\lambda_{iN}} X^\top E^\top(N) \right],$$

$$Z_i(X) = \left[\sqrt{\lambda_{i1}} X^\top, \cdots, \sqrt{\lambda_{ii-1}} X^\top, \sqrt{\lambda_{ii+1}} X^\top, \cdots, \sqrt{\lambda_{iN}} X^\top \right],$$

$$X_i(X) = \mathrm{diag}\left[-\varepsilon(1)\mathbb{I} + X^\top + X, \cdots, -\varepsilon(i-1)\mathbb{I} + X^\top + X, \right.$$

$$\left. -\varepsilon(i+1)\mathbb{I} + X^\top + X, \cdots, -\varepsilon(N)\mathbb{I} + X^\top + X \right],$$

$$X_i(\varepsilon) = \mathrm{diag}\left[4\varepsilon(1)\mathbb{I}, \cdots, 4\varepsilon(i-1)\mathbb{I}, 4\varepsilon(i+1)\mathbb{I}, \cdots, 4\varepsilon(N)\mathbb{I} \right],$$

with the following constraints:

$$X^\top E^\top(i) = E(i)X \geq 0. \tag{3.36}$$

The controller gain is given by $\mathscr{K} = KX^{-1}$, $\mathscr{K} \in \mathbb{R}^{m \times n}$.

Let us now consider that the state equation has uncertainties and see how the previous results can be extended. In a similar manner, the closed-loop uncertain state equation will be piecewise regular, impulse-free and stochastically stable if there exists a nonsingular matrix P such that the following hold for every $i \in \mathscr{S}$

$$\begin{cases} E^\top P(i) = P^\top E(i) \geq 0, \\ A_{cl}^\top(i,t)P + P^\top A_{cl}(i,t) + \mathscr{K}^\top B^\top(i,t)P + P^\top B(i,t)\mathscr{K} + \lambda_{ii} E^\top(i)P \\ + \sum_{j=1,j\neq i}^{N} \lambda_{ij} E^\top(j)P < 0. \end{cases}$$

Using the expression of $A_{cl}(i,t)$ and $B(i,t)$, the second matrix inequality becomes:

$$A^\top(i)P + P^\top A(i) + \mathscr{K}^\top B^\top(i)P + P^\top B(i)\mathscr{K}$$

$$+ E_A^\top(i)F_A^\top(i)D_A^\top(i)P + P^\top D_A(i)F_A(i)E_A(i)$$

$$+ \mathscr{K}^\top E_B^\top(i)F_B^\top(i)D_B^\top(i)P + P^\top D_B(i)F_B(i)E_B(i)\mathscr{K}$$

$$+ \lambda_{ii} E^\top(i)P + \sum_{j=1,j\neq i}^{N} \lambda_{ij} E^\top(j)P < 0.$$

Based now on Lemma 1.5.1, we obtain:

$$A^\top(i)P + P^\top A(i) + \mathscr{K}^\top B^\top(i)P + P^\top B(i)\mathscr{K}$$

$$+ \varepsilon_A(i)P^\top D_A(i)D_A^\top(i)P + \varepsilon_A^{-1}(i)E_A^\top(i)E_A(i)$$

$$+ \varepsilon_B(i)P^\top D_B(i)D_B^\top(i)P + \varepsilon_B^{-1}(i)\mathscr{K}^\top E_B^\top(i)E_B(i)\mathscr{K}$$

$$+ \lambda_{ii} E^\top(i)P + \sum_{j=1,j\neq i}^{N} \lambda_{ij} E^\top(j)P < 0.$$

Let $X = P^{-1}$. Pre- and post-multiplying this inequality respectively by $X^\top$ and X, we get:

$$X^\top A^\top(i) + A(i)X + X^\top \mathcal{K}^\top B^\top(i) + B(i)\mathcal{K}X + \varepsilon_A(i)D_A(i)D_A^\top(i)$$
$$+ \varepsilon_A^{-1}(i)X^\top E_A^\top(i)E_A(i)X + \varepsilon_B(i)D_B(i)D_B^\top(i)$$
$$+ \varepsilon_B^{-1}(i)X^\top \mathcal{K}^\top E_B^\top(i)E_B(i)\mathcal{K}X$$
$$+ \lambda_{ii}X^\top E^\top(i) + \sum_{j=1,j\neq i}^{N} \lambda_{ij}X^\top E^\top(j) < 0.$$

Letting $K = \mathcal{K}X$, we get:

$$X^\top A^\top(i) + A(i)X + K^\top B^\top(i) + B(i)K + \varepsilon_A(i)D_A(i)D_A^\top(i)$$
$$+ \varepsilon_A^{-1}(i)X^\top E_A^\top(i)E_A(i)X + \varepsilon_B(i)D_B(i)D_B^\top(i)$$
$$+ \varepsilon_B^{-1}(i)K^\top E_B^\top(i)E_B(i)K$$
$$+ \lambda_{ii}X^\top E^\top(i) + \sum_{j=1,j\neq i}^{N} \lambda_{ij}X^\top E^\top(j) < 0.$$

that can be rewritten as follows:

$$\begin{bmatrix} J_0(i) & X^\top E_A^\top(i) & Y^\top E_B^\top(i) \\ E_A(i)X & -\varepsilon_A(i)\mathbb{I} & 0 \\ E_B(i)Y & 0 & -\varepsilon_B(i)\mathbb{I} \end{bmatrix},$$

where

$$J_0(i) = X^\top A^\top(i) + A(i)X + K^\top B^\top(i) + B(i)K + \varepsilon_A(i)D_A(i)D_A^\top(i)$$
$$+ \varepsilon_B(i)D_B(i)D_B^\top(i) + \lambda_{ii}X^\top E^\top(i) + \sum_{j=1,j\neq i}^{N} \lambda_{ij}X^\top E^\top(j).$$

with the following constraints:

$$X^\top E^\top(i) = E(i)X \geq 0.$$

We can also use the upper bound we gave before to establish other conditions. Using the first upper bound for the term $X^\top E^\top(i)$ for a set of positive scalars $\varepsilon_P = (\varepsilon_P(1), \cdots, \varepsilon_P(N)), \varepsilon_P(i) > 0$, we obtain after using the Schur complement:

$$\begin{bmatrix} J(i) & X^\top E_A^\top(i) & K^\top E_B^\top(i) \\ E_A(i)X & -\varepsilon_A(i)\mathbb{I} & 0 \\ E_B(i)K & 0 & -\varepsilon_B(i)\mathbb{I} \end{bmatrix} < 0,$$

with

$$J(i) = X^\top A^\top(i) + A(i)X + K^\top B^\top(i) + B(i)K + \lambda_{ii}X^\top E^\top(i)$$
$$+ \varepsilon_A(i)D_A(i)D_A^\top(i) + \varepsilon_B(i)D_B(i)D_B^\top(i)$$
$$+ \sum_{j=1,j\neq i}^{N} \varepsilon_P(j)\lambda_{ij}\left[X + X^\top\right].$$

The following theorem gives a design procedure for a constant gain state feedback controller.

Theorem 3.3.4 *Let $\varepsilon_P = (\varepsilon_P(1), \cdots, \varepsilon_P(N))$ be a given set of positive scalars. There exists a state feedback controller with constant gain that guarantees that the closed-loop system is piecewise regular, impulse-free and stochastically stable if there exist a nonsingular matrix X, $X \in \mathbb{R}^{n \times n}$ a matrix K, $K \in \mathbb{R}^{m \times n}$ and sets of positive scalars $\varepsilon_A = (\varepsilon_A(1), \cdots, \varepsilon_A(N)) > 0$ and $\varepsilon_B = (\varepsilon_B(1), \cdots, \varepsilon_B(N))$, such that the following set of LMIs holds for every $i \in \mathscr{S}$ and all admissible uncertainties:*

$$\begin{bmatrix} J(i) & X^{\top} E_A^{\top}(i) & K^{\top} E_B^{\top}(i) \\ E_A(i)X & -\varepsilon_A(i)\mathbb{I} & 0 \\ E_B(i)K & 0 & -\varepsilon_B(i)\mathbb{I} \end{bmatrix} < 0, \tag{3.37}$$

where

$$\begin{aligned} J(i) = X^{\top} A^{\top}(i) + A(i)X + K^{\top} B^{\top}(i) + B(i)K + \lambda_{ii} X^{\top} E^{\top}(i) \\ + \varepsilon_A(i) D_A(i) D_A^{\top}(i) + \varepsilon_B(i) D_B(i) D_B^{\top}(i) \\ + \sum_{j=1, j \neq i}^{N} \varepsilon_P(j) \lambda_{ij} \left[X + X^{\top} \right]. \end{aligned}$$

with the following constraints:

$$\varepsilon_P(i) \left[X + X^{\top} \right] \geq X^{\top} E^{\top}(i) = E(i)X \geq 0. \tag{3.38}$$

The controller gain is given by $\mathscr{K} = KX^{-1}$, $\mathscr{K} \in \mathbb{R}^{m \times n}$.

For the second upper bound for the term $E^{\top}(i)P$, the closed-loop system will be stable if there exist a nonsingular matrix X and a set of positive scalars $\varepsilon = (\varepsilon(1), \cdots, \varepsilon(N))$ such that the following hold:

$$\begin{cases} \varepsilon(i)\mathbb{I} \geq X^{\top} E^{\top} = EX \geq 0, \\ X^{\top} A^{\top}(i) + A(i)X + X^{\top} \mathscr{K}^{\top} B^{\top}(i) + B(i)\mathscr{K} X \\ + \varepsilon_A(i) D_A(i) D_A^{\top}(i) + \varepsilon_A^{-1}(i) X^{\top} E_A^{\top}(i) E_A(i) X \\ + \varepsilon_B(i) D_B(i) D_B^{\top}(i) + \varepsilon_B^{-1}(i) X^{\top} \mathscr{K}^{\top} E_B^{\top}(i) E_B(i) \mathscr{K} X \\ + \lambda_{ii} X^{\top} E^{\top}(i) + \sum_{j=i, j \neq i}^{N} \lambda_{ij} \varepsilon(j)\mathbb{I} < 0, \end{cases}$$

Letting $K = \mathscr{K} X$ and using Schur complement, we get:

$$\begin{bmatrix} \tilde{J}(i) & X^{\top} E_A^{\top}(i) & K^{\top} E_B^{\top}(i) \\ E_A(i)X & -\varepsilon_A(i)\mathbb{I} & 0 \\ E_B(i)K & 0 & -\varepsilon_B(i)\mathbb{I} \end{bmatrix} < 0,$$

with

$$\begin{aligned} \tilde{J}(i) = X^{\top} A^{\top}(i) + A(i)X + K^{\top} B^{\top}(i) + B(i)K + \varepsilon_A(i) D_A(i) D_A^{\top}(i) \\ + \varepsilon_B(i) D_B(i) D_B^{\top}(i) + \lambda_{ii} X^{\top} E^{\top}(i) + \sum_{j=i, j \neq i}^{N} \lambda_{ij} \varepsilon(j)\mathbb{I}. \end{aligned}$$

The following theorem summarizes the results in this case.

Theorem 3.3.5 *There exists a state feedback controller with constant gain that guarantees that the closed-loop state equation will be piecewise regular, impulse-free and stochastically stable if there exist a nonsingular matrix X, $X \in \mathbb{R}^{n \times n}$ a matrix K, $K \in \mathbb{R}^{m \times n}$ and sets of positive scalars $\varepsilon = (\varepsilon(1), \cdots, \varepsilon(N))$, $\varepsilon_A = (\varepsilon_A(1), \cdots, \varepsilon_A(N))$ and $\varepsilon_B = (\varepsilon_B(1), \cdots, \varepsilon_B(N))$ such that the following set of LMI holds for every $i \in \mathscr{S}$:*

$$\begin{bmatrix} \tilde{J}(i) & X^\top E_A^\top(i) & K^\top E_B^\top(i) \\ E_A(i)X & -\varepsilon_A(i)\mathbb{I} & 0 \\ E_B(i)K & 0 & -\varepsilon_B(i)\mathbb{I} \end{bmatrix} < 0, \tag{3.39}$$

where

$$\tilde{J}(i) = X^\top A^\top(i) + A(i)X + K^\top B^\top(i) + B(i)K + \varepsilon_A(i)D_A(i)D_A^\top(i)$$
$$+ \varepsilon_B(i)D_B(i)D_B^\top(i) + \lambda_{ii}X^\top E^\top(i) + \sum_{j=i, j\neq i}^{N} \lambda_{ij}\varepsilon(j)\mathbb{I},$$

with the following constraints:

$$\varepsilon(i)\mathbb{I} \geq X^\top E^\top(i) = E(i)X \geq 0. \tag{3.40}$$

The controller gain is given by $\mathscr{K} = KX^{-1}$, $\mathscr{K} \in \mathbb{R}^{m \times n}$.

For the third upper bound, we can proceed similarly and the following theorem gives the design procedure for a constant gain controller.

Theorem 3.3.6 *There exists a state feedback controller with constant gain that guarantees that the closed-loop state equation will be piecewise regular, impulse-free and stochastically stable if there exist a nonsingular matrix X, $X \in \mathbb{R}^{n \times n}$ a matrix K, $K \in \mathbb{R}^{m \times n}$ and sets of positive scalars $\varepsilon = (\varepsilon(1), \cdots, \varepsilon(N))$, $\varepsilon_A = (\varepsilon_A(1), \cdots, \varepsilon_A(N))$ and $\varepsilon_B = (\varepsilon_B(1), \cdots, \varepsilon_B(N))$ such that the following set of LMI holds for every $i \in \mathscr{S}$:*

$$\begin{bmatrix} J(i) & \mathcal{Z}_i(X) & S_i(X) & X^\top E_A^\top(i) & K^\top E_B^\top(i) \\ \mathcal{Z}_i^\top(X) & -X_i(\varepsilon) & 0 & 0 & 0 \\ S_i^\top(X) & 0 & -X_i(X) & 0 & 0 \\ E_A(i)X & 0 & 0 & -\varepsilon_A(i)\mathbb{I} & 0 \\ E_B(i)K & 0 & 0 & 0 & -\varepsilon_B(i)\mathbb{I} \end{bmatrix} < 0, \tag{3.41}$$

where

$$J(i) = X^\top A^\top(i) + A(i)X + K^\top B^\top(i) + B(i)K + \lambda_{ii}X^\top E^\top(i)$$
$$+ \varepsilon_A(i)D_A(i)D_A^\top(i) + \varepsilon_B(i)D_B(i)D_B^\top(i)$$
$$S_i(X) = \left[\sqrt{\lambda_{i1}}X^\top E^\top(1), \cdots, \sqrt{\lambda_{ii-1}}X^\top E^\top(i-1), \right.$$
$$\left. \sqrt{\lambda_{ii+1}}X^\top E^\top(i+1), \cdots, \sqrt{\lambda_{iN}}X^\top E^\top(N) \right],$$

$$\mathcal{Z}_i(X) = \left[\sqrt{\lambda_{i1}} X^\top, \cdots, \sqrt{\lambda_{ii-1}} X^\top, \right.$$
$$\left. \sqrt{\lambda_{ii+1}} X^\top, \cdots, \sqrt{\lambda_{iN}} X^\top \right],$$
$$\mathcal{X}_i(X) = \text{diag} \left[-\varepsilon(1)\mathbb{I} + X^\top + X, \cdots, -\varepsilon(i-1)\mathbb{I} + X^\top + X, \right.$$
$$\left. -\varepsilon(i+1)\mathbb{I} + X^\top + X, \cdots, -\varepsilon(N)\mathbb{I} + X^\top + X \right],$$
$$\mathcal{X}_i(\varepsilon) = \text{diag} \left[4\varepsilon(1)\mathbb{I}, \cdots, 4\varepsilon(i-1)\mathbb{I}, 4\varepsilon(i+1)\mathbb{I}, \cdots, 4\varepsilon(N)\mathbb{I} \right],$$

with the following constraints:

$$X^\top E^\top(i) = E(i)X \geq 0. \tag{3.42}$$

The controller gain is given by $\mathcal{K} = KX^{-1}$, $\mathcal{K} \in \mathbb{R}^{m \times n}$.

3.4 Numerical Examples

The goal of this section is to show the effectiveness the developed results. For this purpose we give some numerical examples and using the existing tools to solve our LMIs.

Example 3.4.1 *Let us consider a two modes dynamical singular system with a dynamics described by (3.1) and assume that the data is given by:*

- *mode # 1:*

$$E1 = \begin{bmatrix} 1 & 0 & 0 \\ 0 & 1 & 0 \\ 0 & 0 & 0 \end{bmatrix},$$

$$A(1) = \begin{bmatrix} 0.0 & 1.0 & 1.0 \\ -1.0 & 3.0 & 0.0 \\ -1.0 & -1.0 & 0.0 \end{bmatrix}, \ B(1) = \begin{bmatrix} 0.0 & 0.2 \\ 1.0 & 0.0 \\ -0.1 & 1.0 \end{bmatrix}.$$

- *mode # 2:*

$$E2 = \begin{bmatrix} 1 & 0 & 0 \\ 0 & 1 & 0 \\ 0 & 0 & 0 \end{bmatrix},$$

$$A(2) = \begin{bmatrix} -1.0 & 0.0 & 1.0 \\ 0.0 & 0.0 & 1.0 \\ 0.0 & 1.0 & -1.0 \end{bmatrix}, \ B(2) = \begin{bmatrix} 0.0 & 0.2 \\ 1.2 & 0.0 \\ -0.1 & 1.2 \end{bmatrix}.$$

The switching between the two modes is described by the following transition matrix:

$$\Lambda = \begin{bmatrix} -1 & 1 \\ 1.1 & -1.1 \end{bmatrix}.$$

Solving (3.7)-(3.8) with $\varepsilon_P = 1.1$, we get the following solution:

$$X(1) = \begin{bmatrix} 0.9142 & -0.1288 & 0.0 \\ -0.1288 & 1.1204 & 0.0 \\ -0.2589 & 0.0612 & 0.9919 \end{bmatrix},$$

$$X(2) = \begin{bmatrix} 1.1020 & -0.0658 & 0.0 \\ -0.0658 & 1.0665 & 0.0 \\ -0.2710 & 0.0345 & 0.9671 \end{bmatrix},$$

$$Y(1) = \begin{bmatrix} 0.2568 & -5.0733 & -0.0000 \\ -1.1325 & 0.5783 & -0.9649 \end{bmatrix},$$

$$Y(2) = \begin{bmatrix} 0.3937 & -1.2776 & 0.0000 \\ -0.6820 & -1.8362 & -0.8733 \end{bmatrix}.$$

In view of Theorem 3.2.1, we conclude that the system under study is stochastically stabilizable and a set of stabilizing gains are given by

$$K(1) = \begin{bmatrix} -0.3629 & -4.5698 & -0.0000 \\ -1.4578 & 0.4017 & -0.9728 \end{bmatrix},$$

$$K(2) = \begin{bmatrix} 0.2868 & -1.1803 & 0.0000 \\ -0.9455 & -1.7509 & -0.9030 \end{bmatrix}.$$

Solving (3.13)-(3.14), we get the following solution:

$$\varepsilon(1) = 0.9567, \ \varepsilon(2) = 0.9614$$

$$X(1) = \begin{bmatrix} 0.7309 & -0.0165 & 0.0 \\ -0.0165 & 0.7615 & 0.0 \\ -0.1153 & 0.0119 & 1.0618 \end{bmatrix},$$

$$X(2) = \begin{bmatrix} 0.7914 & -0.0040 & 0.0 \\ -0.0040 & 0.7579 & 0.0 \\ -0.0402 & 0.0038 & 1.0165 \end{bmatrix},$$

$$Y(1) = \begin{bmatrix} -0.1443 & -2.9763 & -0.2262 \\ -1.2424 & 0.6992 & 0.0000 \end{bmatrix},$$

$$Y(2) = \begin{bmatrix} 0.2533 & -0.5652 & 0.0000 \\ -0.8881 & -1.5191 & -0.0058 \end{bmatrix}.$$

In view of Theorem 3.2.3, we conclude that the system under study is stochastically stabilizable and a set of stabilizing gains are given by

$$K(1) = \begin{bmatrix} -0.3191 & -3.9118 & -0.2130 \\ -1.6800 & 0.8819 & 0.0000 \end{bmatrix},$$

$$K(2) = \begin{bmatrix} 0.3162 & -0.7441 & 0.0000 \\ -1.1326 & -2.0103 & -0.0057 \end{bmatrix}.$$

Solving (3.15)-(3.16), we get the following solution:

$$\varepsilon(1) = 0.3305, \quad \varepsilon(2) = 0.3387$$

$$X(1) = \begin{bmatrix} 0.5956 & -0.0363 & 0.0 \\ -0.0363 & 0.6265 & 0.0 \\ -0.1839 & 0.0168 & 0.5727 \end{bmatrix},$$

$$X(2) = \begin{bmatrix} 0.6259 & -0.0149 & 0.0 \\ -0.0149 & 0.6130 & 0.0 \\ 0.0446 & -0.0011 & 0.5130 \end{bmatrix},$$

$$Y(1) = \begin{bmatrix} 0.0920 & -2.4734 & 0.4279 \\ -0.7818 & -0.0367 & 0.0000 \end{bmatrix},$$

$$Y(2) = \begin{bmatrix} 0.1021 & -0.4498 & -0.0000 \\ -0.3507 & -0.9766 & -0.0827 \end{bmatrix}.$$

In view of Theorem 3.2.4, we conclude that the system under study is stochastically stabilizable and a set of stabilizing gains are given by

$$K(1) = \begin{bmatrix} 0.1438 & -3.9597 & 0.7471 \\ -1.3209 & -0.1351 & 0.0000 \end{bmatrix},$$

$$K(2) = \begin{bmatrix} 0.1457 & -0.7303 & -0.0000 \\ -0.5872 & -1.6078 & -0.1612 \end{bmatrix}.$$

Example 3.4.2 *As a second example to show the results on robust stabilization, let us consider the same system of the previous example with the following extra data:*

- *mode # 1:*

$$D_A(1) = \begin{bmatrix} 0.0 \\ 0.1 \\ 0.0 \end{bmatrix}, \quad E_A(1) = \begin{bmatrix} 0.0 & 0.1 & 0.0 \end{bmatrix}$$

$$D_B(1) = \begin{bmatrix} 0.0 \\ 0.1 \\ 0.2 \end{bmatrix}, \quad E_B(1) = \begin{bmatrix} 0.1 & 0.2 \end{bmatrix}$$

- *mode # 2:*

$$D_A(2) = \begin{bmatrix} 0.1 \\ 0.0 \\ 0.0 \end{bmatrix}, \quad E_A(2) = \begin{bmatrix} 0.0 & 0.0 & 0.2 \end{bmatrix}$$

$$D_B(2) = \begin{bmatrix} 0.0 \\ 0.2 \\ 0.1 \end{bmatrix}, \quad E_B(2) = \begin{bmatrix} 0.2 & 0.1 \end{bmatrix}$$

To solve the robust stabilization problem we can chose any approach we developed earlier for this purpose. In this example we will use the third approach. Solving (3.21)-(3.22), we get the following solution:

$$\varepsilon(1) = 0.3296,\ \varepsilon(2) = 0.3379$$
$$\varepsilon_A(1) = 1.0019,\ \varepsilon_A(2) = 1.0056$$
$$\varepsilon_B(1) = 1.0220,\ \varepsilon_B(2) = 1.0152$$

$$X(1) = \begin{bmatrix} 0.5970 & -0.0414 & 0.0 \\ -0.0414 & 0.6215 & 0.0 \\ -0.1861 & 0.0123 & 0.5652 \end{bmatrix},$$

$$X(2) = \begin{bmatrix} 0.6283 & -0.0161 & 0.0 \\ -0.0161 & 0.6068 & 0.0 \\ 0.0418 & -0.0086 & 0.5068 \end{bmatrix},$$

$$Y(1) = \begin{bmatrix} -0.0614 & -2.4726 & -0.5601 \\ -0.7584 & 0.9271 & -0.1218 \end{bmatrix},$$

$$Y(2) = \begin{bmatrix} 0.0553 & -0.4595 & -0.3176 \\ -0.3443 & -0.6690 & -0.1197 \end{bmatrix}.$$

In view of Theorem 3.2.7, we conclude that the system under study is robustly stochastically stabilizable and a set of stabilizing gains are given by

$$K(1) = \begin{bmatrix} -0.6894 & -4.0048 & -0.9910 \\ -1.2396 & 1.4134 & -0.2155 \end{bmatrix},$$

$$K(2) = \begin{bmatrix} 0.1101 & -0.7632 & -0.6266 \\ -0.5611 & -1.1207 & -0.2362 \end{bmatrix}.$$

3.5 Notes

This chapter dealt with the stabilizability problem of the class of singular systems with random abrupt changes. The stochastic stabilizability and the robust stochastic stabilizability problems have been considered and LMI conditions were developed. A state feedback controller that assures that the closed-loop state equation either for the nominal system or the uncertain is piecewise regular, impulse-free and stochastically stable is designed using the LMI framework. Three approaches for designing a state feedback controller were developed. The conditions we developed in this chapter are tractable using commercial optimization tools. The content of this chapter is mainly based on the work of the author and his coauthors [2, 22].

4

$\mathcal{H}_\infty$ Stabilization

The $\mathcal{H}_\infty$ stabilization is one of the popular approach that has been proposed to stabilize dynamical systems with external disturbance. This technique is an alternative to the linear quadratic control with Gaussian disturbance. It requires reasonable assumptions compared to the linear quadratic control with Gaussian disturbance that needs some statistical properties that are very difficult to satisfy in practice. The $\mathcal{H}_\infty$ stabilization requires only that the external disturbance has finite energy or finite power. This control problem has attracted a lot of researchers for more details, we refer the reader to [27] and the references therein either for the deterministic or the stochastic systems. For the class of singular systems in the deterministic framework, this problem has also been tackled by some authors and for more details we refer the reader to the works of [40, 46, 48, 68, 89, 93, 106, 107, 108, 128, 124, 126, 137, 136] and the references therein.

For the class of systems we are treating in this volume only few papers have been reported in the literature among them we quote the works of [2, 14]. The goal of this chapter consists of presenting the foundations of the $\mathcal{H}_\infty$ stabilization and its robustness. Some LMI conditions for designing a state feedback controller that makes the closed-loop system piecewise regular, impulse-free and stochastically stable and at the same time guarantees a prescribed disturbance rejection of a given level. The chapter is organized as follows. In Sect. 2, the problem is stated. In Sect. 3, the $\mathcal{H}_\infty$ stabilization with a state feedback controller is studied and LMI conditions are established either for the nominal and the uncertain systems. In Sect. 4, a constant gain state feedback controller is designed to make the closed-loop system piecewise regular, impulse-free and stochastically stable and at the same time assures the disturbance rejection.

4.1 Problem Statement

Let us consider a dynamical system of the class of linear singular systems with random abrupt changes we are dealing with and suppose that it is described by the following differential-algebraic equations:

$$\begin{cases} E(r_t)\dot{x}(t) = A(r_t,t)x(t) + B(r_t,t)u(t) + B_\omega(r_t)w(t), \ x(0) = x_0, \\ z(t) = C_z(r_t,t)x(t) + D_z(r_t,t)u(t) + B_z(r_t)w(t), \end{cases} \tag{4.1}$$

where $x(t) \in \mathbb{R}^n$ is the state vector, $u(t) \in \mathbb{R}^m$ is the control vector, $z(t) \in \mathbb{R}^q$ is the controlled output and $w(t) \in \mathbb{R}^l$ is the system external disturbance, $E(i)$ is a known singular matrix with rank $(E(i)) = n_r \leq n$ for all $i \in \mathscr{S}$, the matrices $A(r_t,t)$, $B(r_t,t)$, $C_z(r_t,t)$ and $D_z(r_t,t)$ are given by when $r_t - i \in \mathscr{S}$:

$$\begin{cases} A(i,t) = A(i) + D_A(i)F_A(i)E_A(i), \\ B(i,t) = B(i) + D_B(i)F_B(i)E_B(i), \\ C_z(i,t) = C_z(i) + D_{C_z}(i)F_{C_z}(i)E_{C_z}(i), \\ D_z(i,t) = D_z(i) + D_{D_z}(i)F_{D_z}(i)E_{D_z}(i), \end{cases}$$

with $A(i) \in \mathbb{R}^{n\times n}$, $B(i) \in \mathbb{R}^{n\times m}$, $B_\omega(i) \in \mathbb{R}^{n\times l}$, $C_z(i) \in \mathbb{R}^{q\times n}$, $D_z(i) \in \mathbb{R}^{q\times m}$, $B_z(i) \in \mathbb{R}^{q\times l}$, $D_A(i) \in \mathbb{R}^{n\times n_D}$, $E_A(i) \in \mathbb{R}^{n_E\times n}$, $D_B(i) \in \mathbb{R}^{n\times m_D}$, $E_B(i) \in \mathbb{R}^{m_E\times m}$, $D_{C_z}(i) \in \mathbb{R}^{q\times q_D}$, $E_{C_z}(i) \in \mathbb{R}^{q_E\times n}$, $D_{D_z}(i) \in \mathbb{R}^{q\times l_D}$, and $E_{D_z}(i) \in \mathbb{R}^{l_E\times m}$ are known real matrices with appropriate dimensions, and the matrices $F_A(i) \in \mathbb{R}^{n_D\times n_E}$, $F_B(i) \in \mathbb{R}^{m_D\times m_E}$, $F_{C_z}(i) \in \mathbb{R}^{q_D\times q_E}$, $F_{D_z}(i) \in \mathbb{R}^{l_D\times l_E}$ are time-varying unknown matrices satisfying the following:

$$\begin{cases} F_A^\top(i)F_A(i) \leq \mathbb{I}, \\ F_B^\top(i)F_B(i) \leq \mathbb{I}, \\ F_{C_z}^\top(i)F_{C_z}(i) \leq \mathbb{I}, \\ F_{D_z}^\top(i)F_{D_z}(i) \leq \mathbb{I}. \end{cases}$$

The system disturbance, $w(t)$, is assumed to belong to $\mathscr{L}_2[0,\infty)$ which means that the following holds:

$$\mathbb{E}\left[\int_0^\infty \omega^\top(t)\omega(t)dt\right] < \infty. \tag{4.2}$$

This implies that the disturbance has finite energy.

The objective of this chapter is to develop LMI conditions that can be used to design a state feedback controller that makes the closed-loop dynamics piecewise regular, impulse-free and stochastically stable and at the same time guarantees the disturbance rejection of a certain desired level of the external signal.

4.2 $\mathscr{H}_\infty$ State Feedback Stabilization

In the rest of this chapter, we will deal with the design of state feedback controller that renders the closed-loop state equation of the system (4.1) piecewise regular, impulse-free and stochastically stable, and guarantees the disturbance rejection with a certain level $\gamma > 0$. We are also interested by the design of robust controllers that guarantee the same goal. Mathematically, we are concerned with the design of a controller that guarantees the following for all $\omega \in \mathscr{L}_2[0, \infty)$:

$$\|z(t)\|_2 < \gamma \left[\|\omega(t)\|_2^2 + M(x_0, r_0)\right]^{\frac{1}{2}},$$

where $\gamma > 0$ is a prescribed level of disturbance rejection to be achieved, x_0 and r_0 are the initial conditions of the state vector and the mode respectively at time $t = 0$, and $M(x_0, r_0)$ is a constant that depends on the initial conditions (x_0, r_0).

Let us start by developing results for the nominal system, i. e., all the uncertainties are equal to zero. Before proceeding let us define the different concepts we will use for this purpose.

Definition 4.2.1 *Let $\gamma > 0$ be a given positive constant. System (4.1) with $u(t) \equiv 0$ and uncertainties equal to zero is said to be stochastically stable with γ-disturbance attenuation if there exists a constant $M(x_0, r_0)$ with $M(0, r_0) = 0$, for all $r_0 \in \mathscr{S}$, such that the following holds:*

$$\|z\|_2 \triangleq \left[\mathbb{E}\int_0^\infty z^\top(t)z(t)dt|(x_0, r_0)\right]^{1/2} \leq \gamma \left[\|w\|_2^2 + M(x_0, r_0)\right]^{\frac{1}{2}}. \tag{4.3}$$

Definition 4.2.2 *System (4.1) with $u(t) \equiv 0$ and uncertainties equal to zero is said to be internally stochastically stable if there exists a set of nonsingular matrices $P = (P(1), \cdots, P(N))$, satisfying the following for every $i \in \mathscr{S}$:*

$$E^\top(i)P(i) = P^\top(i)E(i) \geq 0, \tag{4.4}$$

$$P^\top(i)A(i) + A^\top(i)P(i) + \lambda_{ii}E^\top(i)P(i)$$

$$+ \sum_{j=1, j\neq i}^N \lambda_{ij}E^\top(j)P(j) < 0. \tag{4.5}$$

Remark 4.2.1 *Based on this definition and the different bounds on the term $E^\top(j)P(j)$, the conditions (4.4)-(4.5) become respectively for the three bounds we developed earlier:*

$$\varepsilon_P(i)\left[P(i) + P^\top(i)\right] \geq E^\top(i)P(i) = P^\top(i)E(i) \geq 0, \tag{4.6}$$

$$P^\top(i)A(i) + A^\top(i)P(i) + \lambda_{ii}E^\top(i)P(i)$$

$$+ \sum_{j=1, j\neq i}^N \varepsilon_P(j)\lambda_{ij}\left[P(j) + P^\top(j)\right] < 0, \tag{4.7}$$

for $\varepsilon_P(i) > 0$.

$$\varepsilon(i)\left[P^\top(i)P(i)\right] \geq E^\top(i)P(i) = P^\top(i)E(i) \geq 0, \tag{4.8}$$

$$P^\top(i)A(i) + A^\top(i)P(i) + \lambda_{ii}E^\top(i)P(i)$$

$$+ \sum_{j=1, j\neq i}^N \varepsilon(j)\lambda_{ij}\left[P^\top(j)P(j)\right] < 0, \tag{4.9}$$

for $\varepsilon(i) > 0$.

$$E^\top(i)P(i) = P^\top(i)E(i) \geq 0, \tag{4.10}$$

$$P^\top(i)A(i) + A^\top(i)P(i) + \lambda_{ii}E^\top(i)P(i)$$

$$+ \sum_{j=1,j\neq i}^{N} \lambda_{ij}\left[\frac{1}{4}\varepsilon^{-1}(j)\mathbb{I} + \varepsilon(j)E^\top(j)P(j)P^\top(j)E(j)\right] < 0, \tag{4.11}$$

for $\varepsilon(i) > 0$.

By virtue of Definition 2.1.1, it is obvious that internally stochastically stable means that system (4.1) is stochastically stable in case of $w(t) \equiv 0$, i. e., system (4.1) being free of input disturbance. Likewise, we can give the following definitions:

Definition 4.2.3 *System (4.1) with $u(t) \equiv 0$ and uncertainties equal to zero is said to be internally stochastically stable if it is stochastically stable in case of $w(t) \equiv 0$.*

Definition 4.2.4 *The nominal system (4.1) is said to be stabilizable with γ-disturbance in the stochastic sense if there exists a control law such that the closed-loop system under this control law is piecewise regular, impulse-free and stochastically stable and satisfies (4.3).*

For the uncertain system, we will have similar definitions that we can summarize as follows:

Definition 4.2.5 *Let $\gamma > 0$ be a given positive constant. System (4.1) with $u(t) \equiv 0$ is said to be robust stochastically stable with γ-disturbance attenuation if there exists a constant $M(x_0, r_0)$ with $M(0, r_0) = 0$, for all $r_0 \in \mathscr{S}$, such that the following holds for all admissible uncertainties:*

$$\|z\|_2 \overset{\triangle}{=} \left[\mathbb{E}\int_0^\infty z^\top(t)z(t)dt|(x_0, r_0)\right]^{\frac{1}{2}} \leq \gamma\left[\|w\|_2^2 + M(x_0, r_0)\right]^{\frac{1}{2}}. \tag{4.12}$$

Definition 4.2.6 *System (4.1) with $u(t) \equiv 0$ is said to be internally robust stochastically stable if there exists a set of nonsingular matrices $P = (P(1), \cdots, P(N))$, satisfying the following for every $i \in \mathscr{S}$ and for all admissible uncertainties:*

$$E^\top(i)P(i) = P^\top(i)E(i) \geq 0, \tag{4.13}$$

$$P^\top(i)A(i,t) + A^\top(i,t)P(i) + \lambda_{ii}E^\top(i)P(i)$$

$$+ \sum_{j=1,j\neq i}^{N} \lambda_{ij}E^\top(j)P(j) < 0. \tag{4.14}$$

Remark 4.2.2 *Based on this definition and the different bounds on the term $E^\top(j)P(j)$, similarly to the nominal system, the conditions (4.13)-(4.14) become respectively for the three bounds we developed earlier:*

$$\varepsilon_P(i)\left[P(i) + P^\top(i)\right] \geq E^\top(i)P(i) = P^\top(i)E(i) \geq 0, \tag{4.15}$$

$$P^\top(i)A(i,t) + A^\top(i,t)P(i) + \lambda_{ii}E^\top(i)P(i)$$
$$+ \sum_{j=1,j\neq i}^{N} \varepsilon_P(j)\lambda_{ij}\left[P(j) + P^\top(j)\right] < 0, \qquad (4.16)$$

for $\varepsilon_P(i) > 0$.

$$\varepsilon(i)\left[P^\top(i)P(i)\right] \geq E^\top(i)P(i) = P^\top(i)E(i) \geq 0, \qquad (4.17)$$
$$P^\top(i)A(i,t) + A^\top(i,t)P(i) + \lambda_{ii}E^\top(i)P(i)$$
$$+ \sum_{j=1,j\neq i}^{N} \varepsilon(j)\lambda_{ij}\left[P^\top(j)P(j)\right] < 0, \qquad (4.18)$$

for $\varepsilon(i) > 0$.

$$E^\top(i)P(i) = P^\top(i)E(i) \geq 0, \qquad (4.19)$$
$$P^\top(i)A(i,t) + A^\top(i,t)P(i) + \lambda_{ii}E^\top(i)P(i)$$
$$+ \sum_{j=1,j\neq i}^{N} \lambda_{ij}\left[\frac{1}{4}\varepsilon^{-1}(j)\mathbb{I} + \varepsilon(j)E^\top(j)P(j) + P^\top(j)E(j)\right] < 0, \qquad (4.20)$$

for $\varepsilon(i) > 0$.

By virtue of Definition 2.1.2, it is obvious that internally robust stochastically stable means that system (4.1) is robust stochastically stable in case of $w(t) \equiv 0$, i. e., system (4.1) being free of input disturbance. Likewise, we can give the following definitions:

Definition 4.2.7 *System (4.1) with $u(t) \equiv 0$ is said to be robust internally stochastically stable if it is robust stochastically stable in case of $w(t) \equiv 0$.*

Definition 4.2.8 *System (4.1) is said to be robust stabilizable with γ-disturbance in the robust stochastically stable sense if there exists a control law such that the closed-loop system under this control law is robust stochastically stable and satisfies (4.3).*

The following theorem shows that in case of $w(t) \not\equiv 0$, internally stochastically stable implies stochastic stability.

Theorem 4.2.1 *If system (4.1) with $u(t) \equiv 0$ is internally stochastically stable, then it is stochastically stable.*

Proof: To prove this theorem, let us consider the initial conditions at time t are respectively denoted by $x(t) = x$ and $r_t = i$ and choose a candidate Lyapunov function be defined as follows:

$$V(x(t), r(t)) = x^\top(t)E^\top(r(t))P(r(t))x(t),$$

where $P(i)$ is a nonsingular matrix for every $i \in \mathscr{S}$.

As we did previously, we can compute the infinitesimal operator $\mathscr{L}$ emanating from the point (x, i) at time t as follows:

$$\mathscr{L}V(x(t), i) = \dot{x}^{\mathsf{T}}(t)E^{\mathsf{T}}(i)P(i)x(t) + x^{\mathsf{T}}(t)E^{\mathsf{T}}(i)P(i)\dot{x}(t)$$

$$+ \lambda_{ii}x^{\mathsf{T}}(t)E^{\mathsf{T}}(i)P(i)x(t) + \sum_{j=1, j\neq i}^{N} \lambda_{ij}x^{\mathsf{T}}(t)E^{\mathsf{T}}(j)P(j)x(t)$$

$$= x^{\mathsf{T}}(t)\left[A^{\mathsf{T}}(i)P(i) + P^{\mathsf{T}}(i)A(i)\right]x(t)$$

$$+ x^{\mathsf{T}}(t)\left[\lambda_{ii}E^{\mathsf{T}}(i)P(i) + \sum_{j=1, j\neq i}^{N} \lambda_{ij}E^{\mathsf{T}}P(j)\right]x(t)$$

$$+ 2x^{\mathsf{T}}(t)P^{\mathsf{T}}(i)B_\omega(i)\omega(t). \tag{4.21}$$

Using now Lemma 1.5.1, we get the following for any $\varepsilon_w(i) > 0$

$$2x^{\mathsf{T}}(t)P^{\mathsf{T}}(i)B_\omega(i)\omega(t) \leq \varepsilon_w(i)x^{\mathsf{T}}(t)P^{\mathsf{T}}(i)B_\omega(i)B_\omega^{\mathsf{T}}(i)P(i)x(t)$$
$$+\varepsilon_w^{-1}(i)\omega^{\mathsf{T}}(t)\omega(t).$$

Combining this with (4.21) yields

$$\mathscr{L}V(x(t), i) \leq x^{\mathsf{T}}(t)\left[A^{\mathsf{T}}(i)P(i) + P^{\mathsf{T}}(i)A(i) + \lambda_{ii}E^{\mathsf{T}}(i)P(i)\right]x(t)$$

$$+ x^{\mathsf{T}}(t)\left[\sum_{j=1, j\neq i}^{N} \lambda_{ij}E^{\mathsf{T}}(j)P(j)\right]x(t)$$

$$+ \varepsilon_w(i)x^{\mathsf{T}}(t)P^{\mathsf{T}}(i)B_\omega(i)B_\omega^{\mathsf{T}}(i)P(i)x(t) + \varepsilon_w^{-1}(i)\omega^{\mathsf{T}}(t)\omega(t)$$

$$= x^{\mathsf{T}}(t)\left[A^{\mathsf{T}}(i)P(i) + P^{\mathsf{T}}(i)A(i) + \lambda_{ii}E^{\mathsf{T}}(i)P(i)\right]x(t)$$

$$+ x^{\mathsf{T}}(t)\left[\sum_{j=1, j\neq i}^{N} \lambda_{ij}E^{\mathsf{T}}(j)P(j)\right]x(t)$$

$$+ x^{\mathsf{T}}(t)\left[\varepsilon_w(i)P^{\mathsf{T}}(i)B_\omega(i)B_\omega^{\mathsf{T}}(i)P(i)\right]x(t) + \varepsilon_w^{-1}(i)\omega^{\mathsf{T}}(t)\omega(t),$$

$$= x^{\mathsf{T}}(t)\Xi(i)x(t) + \varepsilon_w^{-1}(i)\omega^{\mathsf{T}}(t)\omega(t),$$

with

$$\Xi(i) = A^{\mathsf{T}}(i)P(i) + P^{\mathsf{T}}(i)A(i) + \lambda_{ii}E^{\mathsf{T}}(i)P(i) + \sum_{j=1, j\neq i}^{N} \lambda_{ij}E^{\mathsf{T}}(j)P(j)$$

$$+ \varepsilon_w(i)P^{\mathsf{T}}(i)B_\omega(i)B_\omega^{\mathsf{T}}(i)P(i).$$

Based on Dynkin's formula, we get the following:

$$\mathbb{E}[V(x(t), i)] - V(x_0, r_0) = \mathbb{E}\left[\int_0^t \mathscr{L}V(x(s), r_s)ds | x_0, r_0\right],$$

which combined with the previous inequality yields

$$\mathbb{E}[V(x(t), i)] - V(x_0, r_0) \leq \mathbb{E}\left[\int_0^t x^\top(s)\Xi(r_s)x(s)ds|x_0, r_0\right]$$
$$+\varepsilon_w^{-1}(i)\int_0^t \omega^\top(s)\omega(s)ds. \qquad (4.22)$$

Since $V(x(t), i)$ is non-negative, (4.22) implies

$$\mathbb{E}[V(x(t), i)] + \mathbb{E}\left[\int_0^t x^\top(s)[-\Xi(r_s)]x(s)ds|x_0, r_0\right]$$
$$\leq V(x_0, r_0) + \varepsilon_w^{-1}(i)\int_0^t \omega^\top(s)\omega(s)ds,$$

which yields

$$\min_{i \in \mathscr{S}}\{\lambda_{min}(-\Xi(i))\}\mathbb{E}\left[\int_0^\infty x^\top(s)x(s)ds\right] \leq \mathbb{E}\left[\int_0^\infty x^\top(s)[-\Xi(r_s)]x(s)ds\right]$$
$$\leq V(x_0, r_0) + \varepsilon_w^{-1}(i)\int_0^\infty \omega^\top(s)\omega(s)ds.$$

This proves that system (4.1) is stochastically stable.

Let us now establish what conditions should we satisfy if we want the system (4.1) to be piecewise regular, impulse-free and stochastically stable and has γ-disturbance rejection. The following theorem gives such conditions.

Theorem 4.2.2 *Let ε_P and γ be given positive scalars. If there exists a set of non-singular matrices $P = (P(1), \cdots, P(N))$, with $P(i) \in \mathbb{R}^{n \times n}$ such that the following set of coupled LMIs holds for every $i \in \mathscr{S}$:*

$$\begin{bmatrix} J_0(i) & P^\top(i)B_\omega(i) + C_z^\top(i)B_z(i) \\ B_\omega^\top(i)P(i) + B_z^\top(i)C_z(i) & B_z^\top(i)B_z(i) - \gamma^2\mathbb{I} \end{bmatrix} < 0, \qquad (4.23)$$

where

$$J_0(i) = A^\top(i)P(i) + P^\top(i)A(i) + \lambda_{ii}E^\top(i)P(i)$$
$$+ \sum_{j=1, j\neq i}^N \varepsilon_P\lambda_{ij}\left[P(j) + P^\top(j)\right] + C_z^\top(i)C_z(i),$$

with the following constraints:

$$\varepsilon_P\left[P(i) + P^\top(i)\right] \geq E^\top(i)P(i) = P^\top(i)E(i) \geq 0, \qquad (4.24)$$

then system (4.1) with $u(t) \equiv 0$ is piecewise regular, impulse-free and stochastically stable and satisfies the following:

$$\|z\|_2 \leq \left[\gamma^2\|w\|_2^2 + x_0^\top E^\top(r_0)P(r_0)x_0\right]^{\frac{1}{2}}, \qquad (4.25)$$

which means that the system with $u(t) = 0$ for all $t \geq 0$ is piecewise regular, impulse-free and stochastically stable with γ-disturbance attenuation.

Proof: From (4.23) and using Schur complement, we get the following inequality

$$A^\top(i)P(i) + P^\top(i)A(i) + \lambda_{ii}E^\top(i)P(i) + \sum_{j=1,j\neq i}^{N} \varepsilon_P \lambda_{ij}\left[P(j) + P^\top(j)\right]$$
$$+ C_z^\top(i)C_z(i) < 0 .$$

which implies the following since $C_z^\top(i)C_z(i) \geq 0$

$$A^\top(i)P(i) + P^\top(i)A(i) + \lambda_{ii}E^\top(i)P(i) + \sum_{j=1,j\neq i}^{N} \varepsilon_P \lambda_{ij}\left[P(j) + P^\top(j)\right] < 0 .$$

Based on Definition 4.2.2, this and the conditions (4.24) proves that the system under study is piecewise regular, impulse-free and internally stochastically stable. Using now Theorem 4.2.1, we conclude that system (4.1) with $u(t) \equiv 0$ is piecewise regular, impulse-free and stochastically stable.

Let us now prove that (4.25) is satisfied. To this end, let us define the following performance function:

$$J_T = \mathbb{E}\left[\int_0^T [z^\top(t)z(t) - \gamma^2\omega^\top(t)\omega(t)]dt\right] .$$

To prove (4.25), it suffices to establish that J_∞ is bounded, i. e.,

$$J_\infty \leq V(x_0, r_0) = x_0^\top E^\top(r_0)P(r_0)x_0 .$$

First of all notice that for $V(x(t), i) = x^\top(t)E^\top(i)P(i)x(t)$, where (x, i) represent respectively the value of $x(t)$ and r_t at time t, the infinitesimal operator $\mathscr{L}$ emanating from the point (x, i) at time t is given by:

$$\mathscr{L}V(x(t), i) \leq x^\top(t)\left[A^\top(i)P(i) + P^\top(i)A(i) + \lambda_{ii}E^\top(i)P(i)\right]x(t)$$
$$+ x^\top(t)\left[\sum_{j=1,j\neq}^{N} \varepsilon_P \lambda_{ij}\left[P(j) + P^\top(j)\right]\right]x(t)$$
$$+ x^\top(t)P^\top(i)B_\omega(i)\omega(t) + \omega^\top(t)B_\omega^\top(i)P(i)x(t) ,$$

and

$$z^\top(t)z(t) - \gamma^2\omega^\top(t)\omega(t)$$
$$= [C_z(i)x(t) + B_z(i)\omega(t)]^\top [C_z(i)x(t) + B_z(i)\omega(t)] - \gamma^2\omega^\top(t)\omega(t)$$
$$= x^\top(t)C_z^\top(i)C_z(i)x(t) + x^\top(t)C_z^\top(i)B_z(i)\omega(t)$$
$$+ \omega^\top(t)B_z^\top(i)C_z(i)x(t) + \omega^\top(t)B_z^\top(i)B_z(i)\omega(t) - \gamma^2\omega^\top(t)\omega(t) ,$$

which implies the following equality:

$$z^\top(t)z(t) - \gamma^2\omega^\top(t)\omega(t) + \mathscr{L}V(x(t), i) = \eta^\top(t)\Theta(i)\eta(t) ,$$

with

$$\Theta(i) = \begin{bmatrix} J_0(i) & P^\top(i)B_\omega(i) + C_z^\top(i)B_z(i) \\ B_\omega^\top(i)P(i) + B_z^\top(i)C_z(i) & B_z^\top(i)B_z(i) - \gamma^2\mathbb{I} \end{bmatrix}$$

$$\eta^\top(t) = \begin{bmatrix} x^\top(t) & \omega^\top(t) \end{bmatrix}.$$

Therefore,

$$J_T = \mathbb{E}\left[\int_0^T [z^\top(t)z(t) - \gamma^2\omega^\top(t)\omega(t) + \mathscr{L}V(x(t), r_t)]dt \right]$$
$$- \mathbb{E}\left[\int_0^T \mathscr{L}V(x(t), r_t)]dt \right].$$

Using now Dynkin's formula, i. e.,

$$\mathbb{E}\left[\int_0^T \mathscr{L}V(x(t), r_t)]dt | x_0, r_0 \right] = \mathbb{E}[V(x(T), r_T)] - V(x_0, r_0),$$

and the fact that $\Theta(i) < 0$ for all $i \in \mathscr{S}$, we get:

$$J_T = \mathbb{E}\left[\int_0^T \eta^\top(t)\Theta(r_t)\eta(t)dt \right] - \mathbb{E}[V(x(T), r_T)] + V(x_0, r_0). \qquad (4.26)$$

Since $\Theta(i) < 0$ and $\mathbb{E}[V(x(T), r_T)] \geq 0$, (4.26) implies the following:

$$J_T \leq V(x_0, r_0),$$

which yields $J_\infty \leq V(x_0, r_0)$, i. e.,

$$\|z\|_2^2 - \gamma^2\|\omega\|_2^2 \leq x_0^\top E^\top(r_0)P(r_0)x_0.$$

This gives the desired results:

$$\|z\|_2 \leq \left[\gamma^2\|\omega\|_2^2 + x_0^\top E^\top(r_0)P(r_0)x_0 \right]^{\frac{1}{2}}$$

This ends the proof of the theorem.

For the other bounds of the term $E^\top(i)P(i)$, we can establish in a similar manner the following results:

Theorem 4.2.3 *Let $\varepsilon = (\varepsilon(1), \cdots, \varepsilon(N))$ and γ be given positive scalars. If there exists a set of nonsingular matrices $P = (P(1), \cdots, P(N))$, such that the following set of coupled matrix inequalities holds for every $i \in \mathscr{S}$*

$$\begin{bmatrix} J_0(i) & P^\top(i)B_\omega(i) + C_z^\top(i)B_z(i) \\ B_\omega^\top(i)P(i) + B_z^\top(i)C_z(i) & B_z^\top(i)B_z(i) - \gamma^2\mathbb{I} \end{bmatrix} < 0, \qquad (4.27)$$

where

$$J_0(i) = A^\top(i)P(i) + P^\top(i)A(i) + \lambda_{ii}E^\top(i)P(i)$$
$$+ \sum_{j=1, j\neq i}^N \varepsilon(j)\lambda_{ij}\left[P^\top(j)P(j) \right] + C_z^\top(i)C_z(i),$$

with the following constraints:

$$\varepsilon(i)\left[P^\top(i)P(i)\right] \ge E^\top(i)P(i) = P^\top(i)E(i) \ge 0, \tag{4.28}$$

then system (4.1) with $u(t) \equiv 0$ is piecewise regular, impulse-free and stochastically stable and satisfies the following:

$$\|z\|_2 \le \left[\gamma^2\|w\|_2^2 + x_0^\top E^\top(r_0)P(r_0)x_0\right]^{\frac{1}{2}}, \tag{4.29}$$

which means that the system with $u(t) = 0$ for all $t \ge 0$ is piecewise regular, impulse-free and stochastically stable with γ-disturbance attenuation.

When $E^\top(i)P(i) \le \frac{1}{4}\varepsilon^{-1}(j)\mathbb{I} + \varepsilon(j)E^\top(j)P(j)P^\top(j)E(j)$ is used, we get the following results.

Theorem 4.2.4 *Let $\varepsilon = (\varepsilon(1), \cdots, \varepsilon(N))$ and γ be given positive scalars. If there exists a set of nonsingular matrices $P = (P(1), \cdots, P(N))$, with $P(i) \in \mathbb{R}^{n\times n}$ such that the following set of coupled matrix inequalities holds for every $i \in \mathscr{S}$*

$$\begin{bmatrix} J_0(i) & P^\top(i)B_\omega(i) + C_z^\top(i)B_z(i) \\ B_\omega^\top(i)P(i) + B_z^\top(i)C_z(i) & B_z^\top(i)B_z(i) - \gamma^2\mathbb{I} \end{bmatrix} < 0, \tag{4.30}$$

where

$$J_0(i) = A^\top(i)P(i) + P^\top(i)A(i) + \lambda_{ii}E^\top(i)P(i)$$
$$+ \sum_{j=1, j\ne i}^N \lambda_{ij}\left[\frac{1}{4}\varepsilon^{-1}(j)\mathbb{I} + \varepsilon(j)E^\top(j)P(j)P^\top(j)E(j)\right] + C_z^\top(i)C_z(i),$$

with the following constraints:

$$E^\top(i)P(i) = P^\top(i)E(i) \ge 0, \tag{4.31}$$

then system (4.1) with $u(t) \equiv 0$ is piecewise regular, impulse-free and stochastically stable and satisfies the following:

$$\|z\|_2 \le \left[\gamma^2\|w\|_2^2 + x_0^\top E^\top(r_0)P(r_0)x_0\right]^{\frac{1}{2}}, \tag{4.32}$$

which means that the system with $u(t) = 0$ for all $t \ge 0$ is piecewise regular, impulse-free and stochastically stable with γ-disturbance attenuation.

Let us now return to the state equation (4.1) and consider this time that the uncertainties are not equal to zero. In this case the system with $u(t) = 0$ for all $t \ge 0$ is internally stochastically stable if there exits a set of nonsingular matrices, $P = (P(1), \cdots, P(N))$, with $P(i) \in \mathbb{R}^{n\times n}$ such that the following holds for all admissible uncertainties and for every $i \in \mathscr{S}$:

$$\varepsilon_P\left[P(i) + P^\top(i)\right] \ge E^\top(i)P(i) = P^\top(i)E(i) \ge 0,$$

$$P^\top(i)A(i, t) + A^\top(i, t)P(i) + \lambda_{ii}E^\top(i)P(i) + \sum_{j=1, j\ne i}^N \varepsilon_P\lambda_{ij}\left[P(j) + P^\top(j)\right] < 0,$$

for a given positive scalar ε_P.

This condition is useless since it contains the uncertainties $F_A(i)$. Let us now transform it into an useful condition that can be used to check if a given system of the class we are considering is piecewise regular, impulse-free and robust stable.

Now if we use the expression of $A(i, t)$, we get:

$$A^\top(i)P(i) + P^\top(i)A(i) + \lambda_{ii}E^\top(i)P(i) + \sum_{j=1, j\neq i}^{N} \varepsilon_P\lambda_{ij}\left[P(j) + P^\top(j)\right]$$
$$+ P^\top(i)D_A(i)F_A(i)E_A(i) + E_A^\top(i)F_A^\top(i)D_A^\top(i)P(i) < 0.$$

Using now the lemma 1.5.1, the previous inequality will be satisfied if the following holds:

$$A^\top(i)P(i) + P^\top(i)A(i) + \sum_{j=1, j\neq i}^{N} \varepsilon_P\lambda_{ij}\left[P(j) + P^\top(j)\right] + \varepsilon_A(i)E_A^\top(i)E_A(i)$$
$$+ \lambda_{ii}E^\top(i)P(i) + \varepsilon_A^{-1}(i)P^\top(i)D_A(i)D_A^\top(i)P(i) < 0,$$

with $\varepsilon_A(i) > 0$ for all $i \in \mathscr{S}$.

Using Schur complement we get the desired condition:

$$\begin{bmatrix} J_0(i) & P^\top(i)D_A(i) \\ D_A^\top(i)P(i) & -\varepsilon_A(i)\mathbb{I} \end{bmatrix} < 0,$$

with

$$J_0(i) = A^\top(i)P(i) + P^\top(i)A(i) + \sum_{j=1, j\neq i}^{N} \varepsilon_P\lambda_{ij}\left[P(j) + P^\top(j)\right]$$
$$+ \lambda_{ii}E^\top(i)P(i) + \varepsilon_A(i)F_A^\top(i)E_A(i).$$

The result of this development is summarized by the following theorem.

Theorem 4.2.5 *Let ε_P be a given positive scalar ($\varepsilon_P > 0$). If there exist a set of nonsingular matrices $P = (P(1), \cdots, P(N)) > 0$ and a set of positive scalars $\varepsilon_A = (\varepsilon_A(1), \cdots, \varepsilon_A(N))$ such that the following set of coupled LMIs holds for every $i \in \mathscr{S}$ and for all admissible uncertainties:*

$$\begin{bmatrix} J_0(i) & P^\top(i)D_A(i) \\ D_A^\top(i)P(i) & -\varepsilon_A(i)\mathbb{I} \end{bmatrix} < 0,$$

where

$$J_0(i) = A^\top(i)P(i) + P^\top(i)A(i) + \sum_{j=1, j\neq i}^{N} \varepsilon_P\lambda_{ij}\left[P(j) + P^\top(j)\right]$$
$$+ \lambda_{ii}E^\top(i)P(i) + \varepsilon_A(i)E_A^\top(i)E_A(i),$$

with the following constraints:

$$\varepsilon_P\left[P(i) + P^\top(i)\right] \geq E^\top(i)P(i) = P^\top(i)E(i) \geq 0,$$

then system (4.1) with $u(t) = 0$ for all $t \geq 0$ is piecewise regular, impulse-free and internally stochastically stable.

The following result shows that if the system (4.1) is internally stochastically stable for all admissible uncertainties, it is also robustly stochastically stable.

Theorem 4.2.6 *Let $w(.) \in \mathscr{L}_2[0, \infty)$. If the system (4.1) with $u(t) = 0$ for all $t \geq 0$, is internally stochastically stable for all admissible uncertainties, it is also stochastically stable.*

Proof: The proof of this theorem is similar to the one of the previous theorem and the details are omitted.

Theorem 4.2.7 *Let ε_P and γ be a given positive scalars. If there exists a set of non-singular matrices $P = (P(1), \cdots, P(N))$ such that the following set of coupled LMIs holds for all admissible uncertainties:*

$$\begin{bmatrix} J_u(i) & P^\top(i)B_w(i) + C_z^\top(i,t)B_z(i) \\ B_w^\top(i)P(i) + B^\top(i)C_z(i,t) & B_z^\top(i)B_z(i) - \gamma^2 \mathbb{I} \end{bmatrix} < 0, \qquad (4.33)$$

where

$$J_u(i) = A^\top(i,t)P(i) + P^\top(i)A(i,t) + \sum_{j=1,j\neq i}^{N} \varepsilon_P \lambda_{ij} \left[P(j) + P^\top(j) \right]$$

$$+ \lambda_{ii} E^\top(i)P(i) + C_z^\top(i,t)C_z(i,t),$$

with the following constraints:

$$\varepsilon_P \left[P(i) + P^\top(i) \right] \geq E^\top(i)P(i) = P^\top(i)E(i) \geq 0, \qquad (4.34)$$

then system (4.1) with $u(t) = 0$ for all $t \geq 0$ is piecewise regular, impulse-free and robustly stochastically stable and satisfies the disturbance rejection of level γ, i. e.,

$$\|z\|_2 \leq \left[\gamma^2 \|w\|_2^2 + x_0^\top E^\top(r_0)P(r_0)x_0 \right]^{\frac{1}{2}}. \qquad (4.35)$$

Proof: Let us first of all start by proving, that if the LMI (4.33) is satisfied then it implies that the system is piecewise regular, impulse-free and robustly stochastically stable. Notice that, if the LMI (4.33) is satisfied, then it implies that

$$A^\top(i,t)P(i) + P^\top(i)A(i,t) + \sum_{j=1,j\neq i}^{N} \varepsilon_P \lambda_{ij} \left[P(j) + P^\top(j) \right]$$

$$+ \lambda_{ii} E^\top(i)P(i) + C_z^\top(i,t)C_z(i,t) < 0,$$

and since $C_z^\top(i,t)C_z(i,t) \geq 0$ for all $i \in \mathscr{S}$ and for all t, then we get:

$$A^\top(i,t)P(i) + P^\top(i)A(i,t) + \lambda_{ii}E^\top(i)P(i)$$

$$+ \sum_{j=1,j\neq i}^{N} \varepsilon_P \lambda_{ij} \left[P(j) + P^\top(j) \right] < 0,$$

which implies that the system is piecewise regular, impulse-free and mean square quadratically stable.

Let us prove also that the LMIs (4.33)-(4.34) imply that the system satisfies the disturbance rejection of level γ. For this purpose, let (x, i) denote respectively the values of the state vector, $x(t)$ and the mode, r_t at time t and choose a Lyapunov function $V(x(t), i)$ defined as:

$$V(x(t), r(t)) = x^\top(t)E^\top(r(t))P(r(t))x(t),$$

where $P(i)$ is a nonsingular matrix.

The infinitesimal operator $\mathscr{L}(.)$ emanating from the point (x, i) at time t is given by:

$$\mathscr{L}V(x(t), i) \leq x^\top(t)\left[A^\top(i, t)P(i) + P^\top(i)A(i, t) + \lambda_{ii}E^\top(i)P(i)\right]x(t)$$
$$+ x^\top(t)\left[\sum_{j=1, j \neq i}^{N} \varepsilon_P \lambda_{ij}\left[P(j) + P^\top(j)\right]\right]x(t)$$
$$+ 2x^\top(t)P^\top(i)B_w(i)w(t).$$

Let us now define the following performance function:

$$J_T = \mathbb{E}\left[\int_0^T [z^\top(t)z(t) - \gamma^2\omega^\top(t)\omega(t)]dt\right].$$

To prove (4.35), it suffices to establish that J_∞ is bounded for all admissible uncertainties, i.e.,

$$J_\infty \leq V(x_0, r_0) = x_0^\top E^\top(r_0)P(r_0)x_0.$$

Notice that:

$$\mathscr{L}V(x(t), i) \leq x^\top(t)\left[A^\top(i, t)P(i) + P^\top(i)A(i, t)\right]x(t)$$
$$+ x^\top(t)\left[\lambda_{ii}E^\top(i)P(i) + \sum_{j=1, j \neq i}^{N} \varepsilon_P \lambda_{ij}\left[P(j) + P^\top(j)\right]\right]x(t)$$
$$+ x^\top(t)P^\top(i)B_\omega(i)\omega(t) + \omega^\top(t)B_\omega^\top(i)P(i)x(t),$$

and

$$z^\top(t)z(t) - \gamma^2\omega(t)\omega(t)$$
$$= [C_z(i, t)x(t) + B_z(i)\omega(t)]^\top[C_z(i, t)x(t) + B_z(i)\omega(t)]$$
$$- \gamma^2\omega(t)\omega(t)$$
$$= x^\top(t)C_z^\top(i, t)C_z(i, t)x(t) + x^\top(t)C_z^\top(i, t)B_z(i)\omega(t)$$
$$+ \omega^\top(t)B_z^\top(i)C_z(i, t)x(t)$$
$$+ \omega^\top(t)B_z^\top(i)B_z(i)\omega(t) - \gamma^2\omega^\top(t)\omega(t),$$

which implies:

$$z^\top(t)z(t) - \gamma^2\omega^\top(t)\omega(t) + \mathscr{L}V(x(t), i) = \eta^\top(t)\Theta_u(i)\eta(t),$$

with

$$\Theta_u(i) = \begin{bmatrix} J_u(i) & P^\top(i)B_\omega(i) + C_z^\top(i,t)B_z(i) \\ B_\omega^\top(i)P(i) + B_z^\top(i)C_z(i,t) & B_z^\top(i)B_z(i) - \gamma^2\mathbb{I} \end{bmatrix},$$

$$\eta^\top(t) = \begin{bmatrix} x^\top(t) & \omega^\top(t) \end{bmatrix}.$$

Therefore,

$$J_T = \mathbb{E}\left[\int_0^T \left[z^\top(t)z(t) - \gamma^2\omega^\top(t)\omega(t) + \mathscr{L}V(x(t),r_t)\right]dt\right]$$
$$- \mathbb{E}\left[\int_0^T \mathscr{L}V(x(t),r_t)dt\right].$$

From Dynkin's formula, we get

$$\mathbb{E}\left[\int_0^T \mathscr{L}V(x(t),r_t)]dt\big|x_0,r_0\right] = \mathbb{E}[V(x(T),r_T)] - V(x_0,r_0).$$

which implies:

$$J_T = \mathbb{E}\left[\int_0^T \eta^\top(t)\Theta_u(r_t)\eta(t)dt\right] - \mathbb{E}[V(x(T),r_T)] + V(x_0,r_0). \qquad (4.36)$$

Since $\Theta_u(i) < 0$ and $\mathbb{E}[V(x(T),r_T)] \geq 0$, (4.36) implies the following:

$$J_T \leq V(x_0,r_0),$$

which yields $J_\infty \leq V(x_0,r_0)$, i. e.,

$$\|z\|_2^2 - \gamma^2\|\omega\|_2^2 \leq x_0^\top E^\top(r_0)P(r_0)x_0.$$

This gives the desired results:

$$\|z\|_2 \leq \left[\gamma^2\|\omega\|_2^2 + x_0^\top E^\top(r_0)P(r_0)x_0\right]^{\frac{1}{2}},$$

which gives (4.35). This ends the proof of the theorem.

The LMI of this theorem is useless since it depends on the uncertainties. Let us transform it to get an equivalent LMI condition that doesn't depend on the system uncertainties that we can use easily to check if a given system is piecewise regular, impulse-free and robustly stochastically stable. For this purpose, notice that:

$$\begin{bmatrix} \begin{bmatrix} A^\top(i,t)P(i) \\ +\lambda_{ii}E^\top(i)P(i) + P^\top(i)A(i,t) \\ +\sum_{j=1,j\neq i}^N \varepsilon_P\lambda_{ij}[P(j) + P^\top(j)] \\ +C_z^\top(i,t)C_z(i,t) \end{bmatrix} & \begin{bmatrix} C_z^\top(i,t)B_z(i) \\ +P^\top(i)B_\omega(i) \end{bmatrix} \\ \begin{bmatrix} B_z^\top(i)C_z(i) \\ +B_\omega^\top(i)P(i) \end{bmatrix} & B_z^\top(i)B_z(i) - \gamma^2\mathbb{I} \end{bmatrix}$$
$$= \begin{bmatrix} J_1(i,t) & P^\top(i)B_\omega(i) \\ B_\omega^\top(i)P(i) & -\gamma^2\mathbb{I} \end{bmatrix} + \begin{bmatrix} C_z^\top(i,t) \\ B_z^\top(i) \end{bmatrix}\begin{bmatrix} C_z(i,t) & B_z(i) \end{bmatrix},$$

with

$$J_1(i, t) = A^\top(i, t)P(i) + P^\top(i)A(i, t) + \lambda_{ii}E^\top(i)P(i)$$

$$+ \sum_{j=1, j\neq i}^{N} \varepsilon_P \lambda_{ij} \left[P(j) + P^\top(j)\right].$$

Using now Schur complement we show that this is equivalent to the following inequality:

$$\begin{bmatrix} J_1(i, t) & P^\top(i)B_\omega(i) & C_z^\top(i, t) \\ B_\omega^\top(i)P(i) & -\gamma^2\mathbb{I} & B_z^\top(i) \\ C_z(i, t) & B_z(i) & -\mathbb{I} \end{bmatrix} < 0.$$

Using now the expressions of $A(i, t)$ and $C(i, t)$ we get:

$$\begin{bmatrix} J_1(i) & P^\top(i)B_\omega(i) & C_z^\top(i) \\ B_\omega^\top(i)P(i) & -\gamma^2\mathbb{I} & B_z^\top(i) \\ C_z(i) & B_z(i) & -\mathbb{I} \end{bmatrix}$$

$$+ \begin{bmatrix} E_A^\top(i)F_A^\top(i)D_A^\top(i)P(i) & 0 & 0 \\ 0 & 0 & 0 \\ 0 & 0 & 0 \end{bmatrix}$$

$$+ \begin{bmatrix} P^\top(i)D_A(i)F_A(i)E_A(i) & 0 & 0 \\ 0 & 0 & 0 \\ 0 & 0 & 0 \end{bmatrix}$$

$$+ \begin{bmatrix} 0 & 0 & F_{C_z}^\top(i)F_{C_z}^\top(i)D_{C_z}^\top(i) \\ 0 & 0 & 0 \\ 0 & 0 & 0 \end{bmatrix}$$

$$+ \begin{bmatrix} 0 & 0 & 0 \\ 0 & 0 & 0 \\ D_{C_z}(i)F_{C_z}(i)E_{C_z}(i) & 0 & 0 \end{bmatrix} < 0,$$

with

$$J_1(i) = A^\top(i)P(i) + P^\top(i)A(i) + \lambda_{ii}E^\top(i)P(i) + \sum_{j=1, j\neq i}^{N} \varepsilon_P \lambda_{ij} \left[P(j) + P^\top(j)\right].$$

Notice that:

$$\begin{bmatrix} E_A^\top(i)F_A^\top(i)D_A^\top(i)P(i) & 0 & 0 \\ 0 & 0 & 0 \\ 0 & 0 & 0 \end{bmatrix}$$

$$= \begin{bmatrix} E_A^\top(i) \\ 0 \\ 0 \end{bmatrix} F_A^\top(i) \left[D_A^\top(i)P(i) \ 0 \ 0\right],$$

$$\begin{bmatrix} P^\top(i)D_A(i)F_A(i)E_A(i) \ 0 \ 0 \\ 0 \qquad\qquad 0 \ 0 \\ 0 \qquad\qquad 0 \ 0 \end{bmatrix}$$

$$= \begin{bmatrix} P^\top(i)D_A(i) \\ 0 \\ 0 \end{bmatrix} F_A(i) \begin{bmatrix} E_A(i) \ 0 \ 0 \end{bmatrix},$$

$$\begin{bmatrix} 0 \ 0 \ E_{C_z}^\top(i)F_{C_z}^\top(i)D_{C_z}^\top(i) \\ 0 \ 0 \qquad 0 \\ 0 \ 0 \qquad 0 \end{bmatrix}$$

$$= \begin{bmatrix} E_{C_z}^\top(i) \\ 0 \\ 0 \end{bmatrix} F_{C_z}^\top(i) \begin{bmatrix} 0 \ 0 \ D_{C_z}^\top(i) \end{bmatrix},$$

and

$$\begin{bmatrix} 0 \qquad\qquad 0 \ 0 \\ 0 \qquad\qquad 0 \ 0 \\ D_{C_z}(i)F_{C_z}(i)E_{C_z}(i) \ 0 \ 0 \end{bmatrix}$$

$$= \begin{bmatrix} 0 \\ 0 \\ D_{C_z}(i) \end{bmatrix} F_{C_z}(i) \begin{bmatrix} E_{C_z}(i) \ 0 \ 0 \end{bmatrix}.$$

Using now Lemma 1.5.1, we get:

$$\begin{bmatrix} P^\top(i)D_A(i)F_A(i)E_A(i) \ 0 \ 0 \\ 0 \qquad\qquad 0 \ 0 \\ 0 \qquad\qquad 0 \ 0 \end{bmatrix}$$

$$+ \begin{bmatrix} E_A^\top(i)F_A^\top(i)D_A^\top(i)P(i) \ 0 \ 0 \\ 0 \qquad\qquad 0 \ 0 \\ 0 \qquad\qquad 0 \ 0 \end{bmatrix}$$

$$\leq \varepsilon_A^{-1}(i) \begin{bmatrix} P^\top(i)D_A(i) \\ 0 \\ 0 \end{bmatrix} \begin{bmatrix} D_A^\top(i)P(i) \ 0 \ 0 \end{bmatrix}$$

$$+ \varepsilon_A(i) \begin{bmatrix} E_A^\top(i) \\ 0 \\ 0 \end{bmatrix} \begin{bmatrix} E_A(i) \ 0 \ 0 \end{bmatrix}$$

$$= \begin{bmatrix} \begin{bmatrix} \varepsilon_A^{-1}(i)P^\top(i)D_A(i)D_A^\top(i)P(i) \\ +\varepsilon_A(i)E_A^\top(i)E_A(i) \end{bmatrix} \ 0 \ 0 \\ 0 \qquad\qquad\qquad 0 \ 0 \\ 0 \qquad\qquad\qquad 0 \ 0 \end{bmatrix},$$

for any $\varepsilon_A(i) > 0$,

$$\begin{bmatrix} 0 & 0 & E_{C_z}^\top(i)F_{C_z}^\top(i)D_{C_z}^\top(i) \\ 0 & 0 & 0 \\ 0 & 0 & 0 \end{bmatrix}$$

$$+ \begin{bmatrix} 0 & 0 & 0 \\ 0 & 0 & 0 \\ D_{C_z}(i)F_{C_z}(i)E_{C_z}(i) & 0 & 0 \end{bmatrix}$$

$$\leq \varepsilon_{C_z}^{-1}(i) \begin{bmatrix} E_{C_z}^\top(i) \\ 0 \\ 0 \end{bmatrix} \begin{bmatrix} E_{C_z}(i) & 0 & 0 \end{bmatrix}$$

$$+ \varepsilon_{C_z}(i) \begin{bmatrix} 0 \\ 0 \\ D_{C_z}(i) \end{bmatrix} \begin{bmatrix} 0 & 0 & D_{C_z}^\top(i) \end{bmatrix}$$

$$= \begin{bmatrix} \varepsilon_{C_z}^{-1}(i)E_{C_z}^\top(i)E_{C_z}(i) & 0 & 0 \\ 0 & 0 & 0 \\ 0 & 0 & \varepsilon_{C_z}(i)D_{C_z}(i)D_{C_z}^\top(i) \end{bmatrix},$$

for any $\varepsilon_{C_z}(i) > 0$.

Based on these inequalities, the previous one gives:

$$\begin{bmatrix} J_n(i) & P^\top(i)B_\omega(i) & C_z^\top(i) \\ B_\omega^\top(i)P(i) & -\gamma^2\mathbb{I} & B_z^\top(i) \\ C_z(i) & B_z(i) & -\mathbb{I} \end{bmatrix}$$

$$+ \begin{bmatrix} \begin{bmatrix} \varepsilon_A^{-1}(i)P^\top(i)D_A(i)D_A^\top(i)P(i) \\ +\varepsilon_A(i)E_A^\top(i)E_A(i) \end{bmatrix} & 0 & 0 \\ 0 & 0 & 0 \\ 0 & 0 & 0 \end{bmatrix}$$

$$+ \begin{bmatrix} \varepsilon_{C_z}^{-1}(i)E_{C_z}^\top(i)E_{C_z}(i) & 0 & 0 \\ 0 & 0 & 0 \\ 0 & 0 & \varepsilon_{C_z}(i)D_{C_z}(i)D_{C_z}^\top(i) \end{bmatrix} < 0,$$

with

$$J_n(i) = A^\top(i)P(i) + P^\top(i)A(i) + \lambda_{ii}E^\top(i)P(i)$$

$$+ \sum_{j=1, j\neq i}^N \varepsilon_P\lambda_{ij}\left[P(j) + P^\top(j)\right].$$

Let $J_m(i)$, $\mathcal{W}_n(i)$ and $\mathcal{T}_m(i)$ be defined as:

$$J_m(i) = J_n(i) + \varepsilon_A(i)E_A^\top(i)E_A(i),$$

$$\mathcal{W}_m(i) = \text{diag}\left[\varepsilon_A(i)\mathbb{I}, \varepsilon_{C_z}(i)\mathbb{I}\right],$$

$$\mathcal{T}_m(i) = \left[P^\top(i)D_A(i), E_{C_z}^\top(i)\right].$$

and using Schur complement we get the equivalent inequality:

$$
\begin{bmatrix}
J_m(i) & P^\top(i)B_\omega(i) & C_z^\top(i) & \mathcal{T}_m(i) \\
B_\omega^\top(i)P(i) & -\gamma^2\mathbb{I} & B_z^\top(i) & 0 \\
C_z(i) & B_z(i) & -\mathbb{I} + \varepsilon_{C_z}(i)D_{C_z}(i)D_{C_z}^\top(i) & 0 \\
\mathcal{T}_m^\top(i) & 0 & 0 & -\mathcal{W}_m(i)
\end{bmatrix} < 0 .
$$

The following theorem summarizes the results of this development.

Theorem 4.2.8 *Let ε_P and γ be a given positive scalars. If there exist a set of non-singular matrices $P = (P(1), \cdots, P(N))$ and sets of positive scalars $\varepsilon_A = (\varepsilon_A(1), \cdots, \varepsilon_A(N))$, and $\varepsilon_{C_z} = (\varepsilon_{C_z}(1), \cdots, \varepsilon_{C_z}(N))$ such that the following set of coupled LMIs holds for every $i \in \mathscr{S}$ and for all admissible uncertainties:*

$$
\begin{bmatrix}
J_m(i) & P^\top(i)B_\omega(i) & C_z^\top(i) & \mathcal{T}_m(i) \\
B_\omega^\top(i)P(i) & -\gamma^2\mathbb{I} & B_z^\top(i) & 0 \\
C_z(i) & B_z(i) & -\mathbb{I} + \varepsilon_{C_z}(i)D_{C_z}(i)D_{C_z}^\top(i) & 0 \\
\mathcal{T}_m^\top(i) & 0 & 0 & -\mathcal{W}_m(i)
\end{bmatrix} < 0 , \tag{4.37}
$$

where

$$
J_m(i) = J_n(i) + \varepsilon_A(i)E_A^\top(i)E_A(i) ,
$$

$$
\mathcal{W}_m(i) = \mathrm{diag}\left[\varepsilon_A(i)\mathbb{I}, \varepsilon_{C_z}(i)\mathbb{I}\right] ,
$$

$$
\mathcal{T}_m(i) = \left[P^\top(i)D_A(i), E_{C_z}^\top(i)\right] ,
$$

with the following constraints:

$$
\varepsilon_P\left[P(i) + P^\top(i)\right] \geq E^\top(i)P(i) = P^\top(i)E(i) \geq 0 , \tag{4.38}
$$

then, the system (4.1) is piecewise regular, impulse-free and robustly stochastically stable and moreover the system satisfies the disturbance rejection of level γ.

Let us now focus on the design of a state feedback controller. The structure of this controller is given by the following form:

$$
u(t) = K(r_t)x(t) , \tag{4.39}
$$

where $x(t)$ is the state vector and $K(i)$, $i \in \mathscr{S}$ is a design parameter with appropriate dimension that has to be chosen. We will assume the complete access to the state vector and to the mode at each time t when it is necessary.

let us first of all drop the uncertainties from the state equation and see how we can design a controller of the form (4.39). Plugging the expression of the controller in the state equation (4.1), we get:

$$
\begin{cases}
E(r_t)\dot{x}(t) = \bar{A}(r_t)x(t) + B_w(r_t)w(t) , \\
z(t) = \bar{C}_z(r_t)x(t) + B_z(r_t)w(t) ,
\end{cases} \tag{4.40}
$$

where $\bar{A}(r_t) = A(r_t) + B(r_t)K(r_t)$ and $\bar{C}_z(r_t) = C_z(r_t) + D_z(r_t)K(r_t)$.

Using now the results of Theorem 4.2.2, we get the following one for the stochastic stability and the disturbance rejection of level $\gamma > 0$ for the dynamics of the closed-loop system.

Theorem 4.2.9 *Let ε_P and γ be given positive scalars and $K = (K(1), \cdots, K(N))$ a given set of given gains. If there exists a set of nonsingular matrices $P = (P(1), \cdots, P(N))$ with $P(i) \in \mathbb{P}^{n \times n}$, such that the following set of coupled LMIs holds for every $i \in \mathscr{S}$*

$$\begin{bmatrix} \bar{J}_0(i) & \bar{C}_z^\top(i)B_z(i) + P^\top(i)B_\omega(i) \\ B_z^\top(i)\bar{C}_z(i) + B_\omega^\top(i)P(i) & B_z^\top(i)B_z(i) - \gamma^2\mathbb{I} \end{bmatrix} < 0, \qquad (4.41)$$

where

$$\bar{J}_0(i) = \bar{A}^\top(i)P(i) + P^\top(i)\bar{A}(i) + \lambda_{ii}E^\top(i)P(i)$$
$$+ \sum_{j=1, j\neq i}^{N} \varepsilon_P\lambda_{ij}\left[P(j) + P^\top(j)\right] + \bar{C}_z^\top(i)\bar{C}_z(i),$$

with the following constraints:

$$\varepsilon_P\left[P(i) + P^\top(i)\right] \geq E^\top(i)P(i) = P^\top(i)E(i) \geq 0, \qquad (4.42)$$

then system (4.1) is stochastically stable under the controller (4.39) and satisfies the following

$$\|z\|_2 \leq \left[\gamma^2\|w\|_2^2 + x_0^\top E^\top(r_0)P(r_0)x_0\right]^{\frac{1}{2}}, \qquad (4.43)$$

which means that the system is stochastically stable with γ-disturbance attenuation.

To synthesize the controller gain, let us transform the matrix inequality (4.41) into a form that can be used easily to compute this gain for every mode $i \in \mathscr{S}$. For this purpose, notice that:

$$\begin{bmatrix} \bar{J}_0(i) & \begin{bmatrix} \bar{C}_z^\top(i)B_z(i) \\ +P^\top(i)B_\omega(i) \end{bmatrix} \\ \begin{bmatrix} B_z^\top(i)\bar{C}_z(i) \\ +B_\omega^\top(i)P(i) \end{bmatrix} & B_z^\top(i)B_z(i) - \gamma^2\mathbb{I} \end{bmatrix} = \begin{bmatrix} \bar{J}_1(i) & P^\top(i)B_\omega(i) \\ B_\omega^\top(i)P(i) & -\gamma^2\mathbb{I} \end{bmatrix} + \begin{bmatrix} \bar{C}_z^\top(i) \\ B_z^\top(i) \end{bmatrix}\begin{bmatrix} \bar{C}_z(i) & B_z(i) \end{bmatrix},$$

with

$$\bar{J}_0(i) = \bar{A}^\top(i)P(i) + P^\top(i)\bar{A}(i) + \lambda_{ii}E^\top(i)P(i)$$
$$+ \sum_{j=1, j\neq i}^{N} \varepsilon_P\lambda_{ij}\left[P(j) + P^\top(j)\right] + \bar{C}_z^\top(i)\bar{C}_z(i),$$
$$\bar{J}_1(i) = \bar{A}^\top(i)P(i) + P^\top(i)\bar{A}(i) + \lambda_{ii}E^\top(i)P(i)$$
$$+ \sum_{j=1, j\neq i}^{N} \varepsilon_P\lambda_{ij}\left[P(j) + P^\top(j)\right].$$

Using now Schur complement, we show that (4.41) is equivalent to the following inequality:

$$\begin{bmatrix} \bar{J}_1(i) & P^\top(i)B_\omega(i) & \bar{C}_z^\top(i) \\ B_\omega^\top(i)P(i) & -\gamma^2\mathbb{I} & B_z^\top(i) \\ \bar{C}_z(i) & B_z(i) & -\mathbb{I} \end{bmatrix} < 0.$$

Since $\bar{A}(i)$ is nonlinear in $K(i)$ and $P(i)$, the previous matrix inequality is then nonlinear and therefore it can not solved using existing linear algorithms. To transform it to an LMI, let $X(i) = P^{-1}(i)$. As we did many times previously, let us pre- and post-multiply this inequality by $\text{diag}[X^\top(i), \mathbb{I}, \mathbb{I}]$ and its transpose, where $\mathbb{I}$ is an appropriate identity matrix, which gives:

$$\begin{bmatrix} \bar{J}_X(i) & B_\omega(i) & X^\top(i)\bar{C}_z^\top(i) \\ B_\omega^\top(i) & -\gamma^2\mathbb{I} & B_z^\top(i) \\ \bar{C}_z(i)X(i) & B_z(i) & -\mathbb{I} \end{bmatrix} < 0,$$

with

$$\bar{J}_X(i) = X^\top(i)\bar{A}^\top(i) + \bar{A}(i)X(i) + \lambda_{ii}E^\top(i)P(i)$$

$$+ \sum_{j=1, j\neq i}^{N} \varepsilon_P\lambda_{ij}X^\top(i)\left[X^{-1}(j) + X^{-\top}(j)\right]X(i).$$

Notice that the following hold:

$$X^{-1}(i) + X^{-\top}(i) - \mathbb{I} \le \left[X^\top(i)X(i)\right]^{-1}.$$

Let $Z_i(X)$ be defined as:

$$Z_i(X) = \text{diag}\left[X^\top(1)X(1), \cdots, X^\top(i-1)X(i-1), X^\top(i+1)X(i+1),\right.$$
$$\left. \cdots, X^\top(N)X(N)\right].$$

Notice that

$$X^\top(i)\bar{A}^\top(i) + \bar{A}(i)X(i) = X^\top(i)A^\top(i) + A(i)X(i) + Y^\top(i)B^\top(i) + B(i)Y(i),$$

$$\sum_{j=1, j\neq i}^{N} \varepsilon_P\lambda_{ij}X^\top(i)\left[X^{-1}(j) + X^{-\top}(j)\right]X(i) \le \sum_{j=1, j\neq i}^{N} \varepsilon_P\lambda_{ij}X^\top(i)X(i)$$

$$+ S_i(X)Z_i^{-1}(X)S_i^\top(X),$$
$$X^\top(i)\left[C_z(i) + D_z(i)K(i)\right]^\top = X^\top(i)C_z^\top(i) + Y^\top(i)D_z^\top(i),$$

where $Y(i) = K(i)X(i)$, and $S_i(X)$ is defined as follows:

$$S_i(X) = \left[\sqrt{\varepsilon_P\lambda_{i1}}X^\top(i), \cdots, \sqrt{\varepsilon_P\lambda_{ii-1}}X^\top(i), \sqrt{\varepsilon_P\lambda_{ii+1}}X^\top(i),\right.$$
$$\left. \cdots, \sqrt{\varepsilon_P\lambda_{iN}}X^\top(i)\right].$$

Using the fact that:

$$X(i) + X^\top(i) - \mathbb{I} \le X^\top(i)X(i),$$

and Schur complement again, this implies that the previous inequality is equivalent to the following:

$$
\begin{bmatrix}
J(i) & B_\omega(i) & \begin{matrix} X^\top(i)C_z^\top(i)+ \\ Y^\top(i)D_z^\top(i) \end{matrix} & S_i(X) & S_i(X) \\
B_\omega^\top(i)E^\top & -\gamma^2\mathbb{I} & B_z^\top(i) & 0 & 0 \\
\begin{bmatrix} C_z(i)X(i)+ \\ D_z(i)Y(i) \end{bmatrix} & B_z(i) & -\mathbb{I} & 0 & 0 \\
S_i^\top(X) & 0 & 0 & -\mathbb{I} & 0 \\
S_i^\top(X) & 0 & 0 & 0 & -X_i(X)
\end{bmatrix} < 0,
$$

with

$$
J(i) = X^\top(i)A^\top(i) + A(i)X(i) + Y^\top(i)B^\top(i) + B(i)Y(i) + \lambda_{ii}X^\top(i)E^\top(i),
$$

$$
X_i(X) = \mathrm{diag}\left[X(1) + X^\top(1) - \mathbb{I}, \cdots, X(i-1) + X^\top(i-1) - \mathbb{I}, \right.
$$

$$
\left. X(i+1) + X^\top(i+1) - \mathbb{I}, \cdots, X(N) + X^\top(N) - \mathbb{I} \right].
$$

From this discussion, we get the following theorem.

Theorem 4.2.10 *Let γ and ε_P be given positive scalars. There exists a state feedback controller of the form (4.39) such that the closed-loop state equation of the nominal system (4.1) is piecewise regular, impulse-free and stochastically stable and moreover the closed-loop system satisfies the disturbance rejection of level γ if there exist a set of nonsingular matrices $X = (X(1), \cdots, X(N))$, $X(i) \in \mathbb{P}^{n\times n}$ and a set of matrices $Y = (Y(1), \cdots, Y(N))$, $Y(i) \in \mathbb{P}^{m\times n}$ such that the following set of coupled LMIs holds for each $i \in \mathscr{S}$:*

$$
\begin{bmatrix}
J(i) & B_\omega(i) & \begin{matrix} X^\top(i)C_z^\top(i)+ \\ Y^\top(i)D_z^\top(i) \end{matrix} & S_i(X) & S_i(X) \\
B_\omega^\top(i) & -\gamma^2\mathbb{I} & B_z^\top(i) & 0 & 0 \\
\begin{bmatrix} C_z(i)X(i)+ \\ D_z(i)Y(i) \end{bmatrix} & B_z(i) & -\mathbb{I} & 0 & 0 \\
S_i^\top(X) & 0 & 0 & -\mathbb{I} & 0 \\
S_i^\top(X) & 0 & 0 & 0 & -X_i(X)
\end{bmatrix} < 0, \qquad (4.44)
$$

where

$$
J(i) = X^\top(i)A^\top(i) + A(i)X(i) + Y^\top(i)B^\top(i) + B(i)Y(i) + \lambda_{ii}X^\top(i)E^\top(i),
$$

$$
X_i(X) = \mathrm{diag}\left[X(1) + X^\top(1) - \mathbb{I}, \cdots, X(i-1) + X^\top(i-1) - \mathbb{I}, \right.
$$

$$
\left. X(i+1) + X^\top(i+1) - \mathbb{I}, \cdots, X(N) + X^\top(N) - \mathbb{I} \right],
$$

$$
S_i(X) = \left[\sqrt{\varepsilon_P \lambda_{i1}}X^\top(i), \cdots, \sqrt{\varepsilon_P \lambda_{ii-1}}X^\top(i), \sqrt{\varepsilon_P \lambda_{ii+1}}X^\top(i), \right.
$$

$$
\left. \cdots, \sqrt{\varepsilon_P \lambda_{iN}}X^\top(i) \right],
$$

with the following constraints:

$$
\varepsilon_P\left[X(i) + X^\top(i) \right] \geq X^\top(i)E^\top(i) = E(i)X(i) \geq 0. \qquad (4.45)
$$

The state feedback controller (4.39) gain is given $K(i) = Y(i)X^{-1}(i)$ for each $i \in \mathscr{S}$.

From the practical point of view, the controller that assures that the closed-loop system is piecewise regular, impulse-free and stochastically stable and at the same time guarantees the minimum disturbance rejection is of great interest. This controller can be obtained by solving the following optimization problem:

$$
P : \begin{cases}
\min\limits_{\substack{\nu>0, \\ X=(X(1),\cdots,X(N)), \\ Y=(Y(1),\cdots,Y(N))}} \nu \\[2em]
s.t : \\
\varepsilon_P\left[X(i) + X^\top(i)\right] \geq X^\top(i)E^\top(i) = E(i)X(i), \geq 0 \\[1em]
\begin{bmatrix}
J(i) & B_\omega(i) & \begin{bmatrix} X^\top(i)C_z^\top(i)+ \\ Y^\top(i)D_z^\top(i) \end{bmatrix} & S_i(X) & S_i(X) \\
B_\omega^\top(i) & -\nu\mathbb{I} & B_z^\top(i) & 0 & 0 \\
\begin{bmatrix} C_z(i)X(i)+ \\ D_z(i)Y(i) \end{bmatrix} & B_z(i) & -\mathbb{I} & 0 & 0 \\
S_i^\top(X) & 0 & 0 & -\mathbb{I} & 0 \\
S_i^\top(X) & 0 & 0 & 0 & -\mathcal{X}_i(X)
\end{bmatrix} < 0,
\end{cases}
$$

where the LMI in the constraints is obtained from (4.44) by replacing γ^2 by ν.

The following theorem gives the results on the design of the controller that assures that the closed-loop system is piecewise regular, impulse-free and stochastically stable and simultaneously guarantees the smallest disturbance rejection level.

Theorem 4.2.11 *Let* $\nu > 0$, $X = (X(1), \cdots, X(N))$ *and* $Y = (Y(1), \cdots, Y(N))$ *be the solution of the optimization problem P for a given positive scalar ε_P. Then, the controller (4.39) with $K(i) = Y(i)X^{-1}(i)$ will guarantee that the class of systems we are considering is piecewise regular, impulse-free and stochastically stable and moreover the closed-loop system satisfies the disturbance rejection of level $\sqrt{\nu}$.*

When the term $E^\top(i)P(i)$ is bounded by $\varepsilon(i)\left[P^\top(i)P(i)\right]$, with $\varepsilon(i) > 0$ we get the following results.

Theorem 4.2.12 *Let γ be a given positive scalar. There exists a state feedback controller of the form (4.39) such that the closed-loop state equation of the system (4.1) is piecewise regular, impulse-free and stochastically stable and moreover the closed-loop system satisfies the disturbance rejection of level γ if there exist a set of nonsingular matrices $X = (X(1), \cdots, X(N))$, $X(i) \in \mathbb{R}^{n\times n}$ a set of matrices $Y = (Y(1), \cdots, Y(N))$, $Y(i) \in \mathbb{R}^{m\times n}$, and a set of positive scalars $\varepsilon = (\varepsilon(1), \cdots, \varepsilon(N))$ such that the following set of coupled LMIs holds for every $i \in \mathcal{S}$:*

$$
\begin{bmatrix}
\tilde{J}(i) & B_\omega(i) & \begin{bmatrix} X(i)C_z^\top(i) \\ +Y^\top(i)D_z^\top(i) \end{bmatrix} & S_i(X) \\
B_\omega^\top(i) & -\gamma^2\mathbb{I} & B_z^\top(i) & 0 \\
\begin{bmatrix} D_z(i)Y(i) \\ +C_z(i)X(i) \end{bmatrix} & B_z(i) & -\mathbb{I} & 0 \\
S_i^\top(X) & 0 & 0 & -\mathcal{X}_i(X)
\end{bmatrix} < 0, \tag{4.46}
$$

where

$$\tilde{J}(i) = X^\top(i)A^\top(i) + A(i)X(i) + Y^\top(i)B^\top(i) + B(i)Y(i) + \lambda_{ii}X^\top(i)E^\top(i),$$

$$X_i(X) = \operatorname{diag}\Big[X(1) + X^\top(1) - \varepsilon(1)\mathbb{I}, \cdots, X(i-1) + X^\top(i-1) - \varepsilon(i-1)\mathbb{I},$$

$$X(i+1) + X^\top(i+1) - \varepsilon(i+1)\mathbb{I}, \cdots, X(N) + X^\top(N) - \varepsilon(N)\mathbb{I}\Big],$$

$$S_i(X) = \Big[\sqrt{\lambda_{i1}}X^\top(i), \cdots, \sqrt{\lambda_{ii-1}}X^\top(i), \sqrt{\lambda_{ii+1}}X^\top(i),$$

$$\cdots, \sqrt{\lambda_{iN}}X^\top(i)\Big],$$

with the following constraints:

$$\varepsilon(i)\mathbb{I} \geq X^\top(i)E^\top(i) = E(i)X(i) \geq 0. \tag{4.47}$$

The state feedback controller (4.39) gain is given $K(i) = Y(i)X^{-1}(i)$ *for each* $i \in \mathscr{S}$.

The following theorem summarizes the results when the term $E^\top(i)P(i)$ is bounded by $\left[\frac{1}{4}\varepsilon^{-1}(i)\mathbb{I} + \varepsilon(i)E^\top(i)P(i)P^\top(i)E(i)\right]$, with $\varepsilon(i) > 0$.

Theorem 4.2.13 *Let* γ *be a given positive scalar. There exists a state feedback controller of the form (4.39) such that the closed-loop state equation of the system (4.1) is piecewise regular, impulse-free and stochastically stable and moreover the closed-loop system satisfies the disturbance rejection of level* γ *if there exist a set of nonsingular matrices* $X = (X(1), \cdots, X(N))$, *a set of matrices* $Y = (Y(1), \cdots, Y(N))$, *and a set of positive scalars* $\varepsilon = (\varepsilon(1), \cdots, \varepsilon(N))$ *such that the following set of coupled LMIs holds for every* $i \in \mathscr{S}$ *and for all admissible uncertainties:*

$$\begin{bmatrix} \tilde{J}(i) & B_\omega(i) & \begin{bmatrix} X(i)C_z^\top(i) \\ +Y^\top(i)D_z^\top(i) \end{bmatrix} & Z_i(X) & S_i(X) \\ B_\omega^\top(i) & -\gamma^2\mathbb{I} & B_z^\top(i) & 0 & 0 \\ \begin{bmatrix} D_z(i)Y(i) \\ +C_z(i)X(i) \end{bmatrix} & B_z(i) & -\mathbb{I} & 0 & 0 \\ Z_i^\top(X) & 0 & 0 & -X_i(\varepsilon) & 0 \\ S_i^\top(X) & 0 & 0 & 0 & -X_i(X) \end{bmatrix} < 0, \tag{4.48}$$

where

$$\tilde{J}(i) = X(i)A^\top(i) + A(i)X(i) + Y^\top(i)B^\top(i) + B(i)Y(i) + \lambda_{ii}X^\top(i)E^\top(i),$$

$$Z_i(X) = \left(\sqrt{\lambda_{i1}}X^\top(i), \cdots, \sqrt{\lambda_{ii-1}}X^\top(i), \sqrt{\lambda_{ii+1}}X^\top(i), \cdots, \sqrt{\lambda_{iN}}X^\top(i)\right),$$

$$S_i(X) = \left(\sqrt{\lambda_{i1}}X^\top(i)E^\top(1), \cdots, \sqrt{\lambda_{ii-1}}X^\top(i)E^\top(i-1),\right.$$

$$\left.\sqrt{\lambda_{ii+1}}X^\top(i)E^\top(i+1), \cdots, \sqrt{\lambda_{iN}}X^\top(i)E^\top(N)\right),$$

$$X_i(\varepsilon) = \operatorname{diag}\left[4\varepsilon(1)\mathbb{I}, \cdots, 4\varepsilon(i-1)\mathbb{I}, 4\varepsilon(i+1)\mathbb{I}, \cdots, 4\varepsilon(N)\mathbb{I}\right],$$

$$X_i(X) = \operatorname{diag}\Big[-\varepsilon(1)\mathbb{I} + X^\top(1) + X(1), \cdots, -\varepsilon(i-1)\mathbb{I} + X^\top(i-1) + X(i-1),$$

$$-\varepsilon(i+1)\mathbb{I} + X^\top(i+1) + X(i+1), \cdots, -\varepsilon(N)\mathbb{I} + X^\top(N) + X(N)\Big],$$

with the following constraints:

$$X^\top(i)E^\top(i) = E(i)X(i) \geq 0. \tag{4.49}$$

The state feedback controller (4.39) gain is given $K(i) = Y(i)X^{-1}(i)$ *for each* $i \in \mathscr{S}$.

Let us now return to the state equation (4.1) and consider of the uncertainties are not equal to zero and see how we can synthesize the state feedback controller of the form (4.39) which guarantees that the class of systems we are studying is piecewise regular, impulse-free and robust stochastically stable and at the same time assures the desired disturbance rejection of level γ. Following the same steps as for the nominal case starting from the results of Theorem 4.2.7, we get:

$$
\begin{bmatrix}
\bar{J}_0(i,t) & P^\top(i)B_\omega(i) & \bar{C}_z^\top(i,t) \\
B_\omega^\top(i)P(i) & -\gamma^2\mathbb{I} & B_z^\top(i) \\
\bar{C}_z(i,t) & B_z(i) & -\mathbb{I}
\end{bmatrix} < 0,
\tag{4.50}
$$

with

$$
\bar{J}_0(i,t) = \bar{A}^\top(i,t)P(i) + P^\top(i)\bar{A}(i,t) + \lambda_{ij}E^\top(i)P(i)
$$

$$
+ \sum_{j=1,j\neq i}^N \varepsilon_P \lambda_{ij}\left[P(j) + P^\top(j)\right],
$$

$$
\bar{A}(i,t) = A(i,t) + B(i,t)K(i),
$$

$$
\bar{C}_z(i,t) = C_z(i,t) + D_z(i,t)K(i).
$$

Using now the expressions of $\bar{A}(i,t)$ and $\bar{C}_z(i,t)$ and the ones of their components, we obtain the following inequality:

$$
\begin{bmatrix}
J_1(i) & P^\top(i)B_\omega(i) & \begin{bmatrix} C_z^\top(i) \\ +K^\top(i)D_z^\top(i) \end{bmatrix} \\
B_\omega^\top(i)P(i) & -\gamma^2\mathbb{I} & B_z^\top(i) \\
\begin{bmatrix} D_z(i)K(i) \\ +C_z(i) \end{bmatrix} & B_z(i) & -\mathbb{I}
\end{bmatrix}
$$

$$
+ \begin{bmatrix}
E_A^\top(i)F_A^\top(i)D_A^\top(i)P(i) & 0 & 0 \\
0 & 0 & 0 \\
0 & 0 & 0
\end{bmatrix}
$$

$$
+ \begin{bmatrix}
P^\top(i)D_A(i)F_A(i)E_A(i) & 0 & 0 \\
0 & 0 & 0 \\
0 & 0 & 0
\end{bmatrix}
$$

$$
+ \begin{bmatrix}
P^\top(i)D_B(i)F_B(i)E_B(i)K(i) & 0 & 0 \\
0 & 0 & 0 \\
0 & 0 & 0
\end{bmatrix}
$$

$$
+ \begin{bmatrix}
K^\top(i)E_B^\top(i)F_B^\top(i)D_B^\top(i)P(i) & 0 & 0 \\
0 & 0 & 0 \\
0 & 0 & 0
\end{bmatrix}
$$

$$
+ \begin{bmatrix}
0 & 0 & K^\top(i)E_{D_z}^\top(i)F_{D_z}^\top(i)D_{D_z}^\top(i) \\
0 & 0 & 0 \\
0 & 0 & 0
\end{bmatrix}
+ \begin{bmatrix}
0 & 0 & 0 \\
0 & 0 & 0 \\
D_{D_z}(i)F_{D_z}(i)E_{D_z}(i)K(i) & 0 & 0
\end{bmatrix}
$$

$$
+ \begin{bmatrix}
0 & 0 & E_{C_z}^\top(i)F_{C_z}^\top(i)D_{C_z}^\top(i) \\
0 & 0 & 0 \\
0 & 0 & 0
\end{bmatrix}
+ \begin{bmatrix}
0 & 0 & 0 \\
0 & 0 & 0 \\
D_{C_z}(i)F_{C_z}(i)E_{C_z}(i) & 0 & 0
\end{bmatrix} < 0,
$$

with

$$J_1(i) = A^\top(i)P(i) + P^\top(i)A(i) + K^\top(i)B^\top(i)P(i) + P^\top(i)B(i)K(i)$$
$$+ \lambda_{ij}E^\top(i)P(i) + \sum_{j=1,j\neq i}^{N} \varepsilon_P \lambda_{ij}\left[P(j) + P^\top(j)\right].$$

Notice that:

$$\begin{bmatrix} E_A^\top(i)F_A^\top(i)D_A^\top(i)P(i) \ 0 \ 0 \\ 0 \qquad\qquad 0 \ 0 \\ 0 \qquad\qquad 0 \ 0 \end{bmatrix} = \begin{bmatrix} E_A^\top(i) \\ 0 \\ 0 \end{bmatrix} F_A^\top(i)\left[D_A^\top(i)P(i) \ 0 \ 0\right],$$

$$\begin{bmatrix} P^\top(i)D_A(i)F_A(i)E_A(i) \ 0 \ 0 \\ 0 \qquad\qquad 0 \ 0 \\ 0 \qquad\qquad 0 \ 0 \end{bmatrix} = \begin{bmatrix} P^\top(i)D_A(i) \\ 0 \\ 0 \end{bmatrix} F_A(i)\left[E_A(i) \ 0 \ 0\right],$$

$$\begin{bmatrix} P^\top(i)D_B(i)F_B(i)E_B(i)K(i) \ 0 \ 0 \\ 0 \qquad\qquad 0 \ 0 \\ 0 \qquad\qquad 0 \ 0 \end{bmatrix} = \begin{bmatrix} P(i)D_B(i) \\ 0 \\ 0 \end{bmatrix} F_B(i)\left[E_B(i)K(i) \ 0 \ 0\right],$$

$$\begin{bmatrix} K^\top(i)E_B^\top(i)F_B^\top(i)D_B^\top(i)P(i) \ 0 \ 0 \\ 0 \qquad\qquad 0 \ 0 \\ 0 \qquad\qquad 0 \ 0 \end{bmatrix} = \begin{bmatrix} K^\top(i)E_B^\top(i) \\ 0 \\ 0 \end{bmatrix} F_B^\top(i)\left[D_B^\top(i)P(i) \ 0 \ 0\right],$$

$$\begin{bmatrix} 0 \ 0 \ K^\top(i)E_{D_z}^\top(i)F_{D_z}^\top(i)D_{D_z}^\top(i) \\ 0 \ 0 \qquad\qquad 0 \\ 0 \ 0 \qquad\qquad 0 \end{bmatrix} = \begin{bmatrix} K^\top(i)E_{D_z}^\top(i) \\ 0 \\ 0 \end{bmatrix} F_{D_z}^\top(i)\left[0 \ 0 \ D_{D_z}^\top(i)\right],$$

$$\begin{bmatrix} 0 \qquad\qquad 0 \ 0 \\ 0 \qquad\qquad 0 \ 0 \\ D_{D_z}(i)F_{D_z}(i)E_{D_z}(i)K(i) \ 0 \ 0 \end{bmatrix} = \begin{bmatrix} 0 \\ 0 \\ D_{D_z}(i) \end{bmatrix} F_{D_z}(i)\left[E_{D_z}(i)K(i) \ 0 \ 0\right],$$

$$\begin{bmatrix} 0 \ 0 \ E_{C_z}^\top(i)F_{C_z}^\top(i)D_{C_z}^\top(i) \\ 0 \ 0 \qquad\qquad 0 \\ 0 \ 0 \qquad\qquad 0 \end{bmatrix} = \begin{bmatrix} E_{C_z}^\top(i) \\ 0 \\ 0 \end{bmatrix} F_{C_z}^\top(i)\left[0 \ 0 \ D_{C_z}^\top(i)\right],$$

and

$$\begin{bmatrix} 0 \qquad\qquad 0 \ 0 \\ 0 \qquad\qquad 0 \ 0 \\ D_{C_z}(i)F_{C_z}(i)E_{C_z}(i) \ 0 \ 0 \end{bmatrix} = \begin{bmatrix} 0 \\ 0 \\ D_{C_z}(i) \end{bmatrix} F_{C_z}(i)\left[E_{C_z}(i) \ 0 \ 0\right].$$

Using now Lemma 1.5.1, we get:

$$\begin{bmatrix} P^\top(i)D_A(i)F_A(i)E_A(i) & 0 & 0 \\ 0 & 0 & 0 \\ 0 & 0 & 0 \end{bmatrix}$$

$$+ \begin{bmatrix} E_A^\top(i)F_A^\top(i)D_A^\top(i)P(i) & 0 & 0 \\ 0 & 0 & 0 \\ 0 & 0 & 0 \end{bmatrix}$$

$$\leq \varepsilon_A(i) \begin{bmatrix} P^\top(i)D_A(i) \\ 0 \\ 0 \end{bmatrix} \begin{bmatrix} D_A^\top(i)P(i) & 0 & 0 \end{bmatrix}$$

$$+ \varepsilon_A^{-1}(i) \begin{bmatrix} E_A^\top(i) \\ 0 \\ 0 \end{bmatrix} \begin{bmatrix} E_A(i) & 0 & 0 \end{bmatrix}$$

$$= \begin{bmatrix} \begin{bmatrix} \varepsilon_A(i)P^\top(i)D_A(i)D_A^\top(i)P(i) \\ +\varepsilon_A^{-1}(i)E_A^\top(i)E_A(i) \end{bmatrix} & 0 & 0 \\ 0 & 0 & 0 \\ 0 & 0 & 0 \end{bmatrix},$$

for any $\varepsilon_A(i) > 0$,

$$\begin{bmatrix} P^\top(i)D_B(i)F_B(i)E_B(i)K(i) & 0 & 0 \\ 0 & 0 & 0 \\ 0 & 0 & 0 \end{bmatrix}$$

$$+ \begin{bmatrix} K^\top(i)E_B^\top(i)F_B^\top(i)D_B^\top(i)P(i) & 0 & 0 \\ 0 & 0 & 0 \\ 0 & 0 & 0 \end{bmatrix}$$

$$\leq \varepsilon_B(i) \begin{bmatrix} P^\top(i)D_B(i) \\ 0 \\ 0 \end{bmatrix} \begin{bmatrix} D_B^\top(i)P(i) & 0 & 0 \end{bmatrix}$$

$$+ \varepsilon_B^{-1}(i) \begin{bmatrix} K^\top(i)E_B^\top(i) \\ 0 \\ 0 \end{bmatrix} \begin{bmatrix} E_B(i)K(i) & 0 & 0 \end{bmatrix}$$

$$= \begin{bmatrix} \begin{bmatrix} \varepsilon_B(i)P^\top(i)D_B(i)D_B^\top(i)P(i) \\ +\varepsilon_B^{-1}(i)K^\top(i)E_B^\top(i)E_B(i)K(i) \end{bmatrix} & 0 & 0 \\ 0 & 0 & 0 \\ 0 & 0 & 0 \end{bmatrix},$$

for any $\varepsilon_B(i) > 0$,

$$\begin{bmatrix} 0 & 0 & K^\top(i)E_{D_z}^\top(i)F_{D_z}^\top(i)D_{D_z}^\top(i) \\ 0 & 0 & 0 \\ 0 & 0 & 0 \end{bmatrix}$$

$$+ \begin{bmatrix} & 0 & 0 & 0 \\ & 0 & 0 & 0 \\ D_{D_z}(i)F_{D_z}(i)E_{D_z}(i)K(i) & 0 & 0 \end{bmatrix}$$

$$\leq \varepsilon_{D_z}^{-1}(i) \begin{bmatrix} K^\top(i)E_{D_z}^\top(i) \\ 0 \\ 0 \end{bmatrix} \begin{bmatrix} E_{D_z}(i)K(i) & 0 & 0 \end{bmatrix}$$

$$+ \varepsilon_{D_z}(i) \begin{bmatrix} 0 \\ 0 \\ D_{D_z}(i) \end{bmatrix} \begin{bmatrix} 0 & 0 & D_{D_z}^\top(i) \end{bmatrix}$$

$$= \begin{bmatrix} \varepsilon_{D_z}^{-1}(i)K^\top(i)E_{D_z}^\top(i)E_{D_z}(i)K(i) & 0 & 0 \\ 0 & 0 & 0 \\ 0 & 0 & \varepsilon_{D_z}(i)D_{D_z}(i)D_{D_z}^\top(i) \end{bmatrix},$$

for any $\varepsilon_{D_z}(i) > 0$,

$$\begin{bmatrix} 0 & 0 & E_{C_z}^\top(i)F_{C_z}^\top(i)D_{C_z}^\top(i) \\ 0 & 0 & 0 \\ 0 & 0 & 0 \end{bmatrix}$$

$$+ \begin{bmatrix} & 0 & 0 & 0 \\ & 0 & 0 & 0 \\ D_{C_z}(i)F_{C_z}(i)E_{C_z}(i) & 0 & 0 \end{bmatrix}$$

$$\leq \varepsilon_{C_z}^{-1}(i) \begin{bmatrix} E_{C_z}^\top(i) \\ 0 \\ 0 \end{bmatrix} \begin{bmatrix} E_{C_z}(i) & 0 & 0 \end{bmatrix}$$

$$+ \varepsilon_{C_z}(i) \begin{bmatrix} 0 \\ 0 \\ D_{C_z}(i) \end{bmatrix} \begin{bmatrix} 0 & 0 & D_{C_z}^\top(i) \end{bmatrix}$$

$$= \begin{bmatrix} \varepsilon_{C_z}^{-1}(i)E_{C_z}^\top(i)E_{C_z}(i) & 0 & 0 \\ 0 & 0 & 0 \\ 0 & 0 & \varepsilon_{C_z}(i)D_{C_z}(i)D_{C_z}^\top(i) \end{bmatrix}.$$

for any $\varepsilon_{C_z}(i) > 0$.

Taking into account these inequalities we get:

$$
\begin{bmatrix}
J_1(i) & P^\top(i)B_\omega(i) & C_z^\top(i) + K^\top(i)D_z^\top(i) \\
B_\omega^\top(i)P(i) & -\gamma^2\mathbb{I} & B_z^\top(i) \\
D_z(i)K(i) + C_z(i) & B_z(i) & -\mathbb{I}
\end{bmatrix}
$$

$$
+ \begin{bmatrix}
\begin{bmatrix} \varepsilon_A(i)P^\top(i)D_A(i)D_A^\top(i)P(i) \\ +\varepsilon_A^{-1}(i)E_A^\top(i)E_A(i) \end{bmatrix} & 0\,0 \\
0 & 0\,0 \\
0 & 0\,0
\end{bmatrix}
+ \begin{bmatrix}
\begin{bmatrix} \varepsilon_B(i)P^\top(i)D_B(i)D_B^\top(i)P(i) \\ +\varepsilon_B^{-1}K^\top(i)E_B^\top(i)E_B(i)K(i) \end{bmatrix} & 0\,0 \\
0 & 0\,0 \\
0 & 0\,0
\end{bmatrix}
$$

$$
+ \begin{bmatrix}
\varepsilon_{D_z}^{-1}(i)K^\top(i)E_{D_z}^\top(i)E_{D_z}(i)K(i) & 0 & 0 \\
0 & 0 & 0 \\
0 & 0 & \varepsilon_{D_z}(i)D_{D_z}(i)D_{D_z}^\top(i)
\end{bmatrix}
$$

$$
+ \begin{bmatrix}
\varepsilon_{C_z}^{-1}(i)E_{C_z}^\top(i)E_{C_z}(i) & 0 & 0 \\
0 & 0 & 0 \\
0 & 0 & \varepsilon_{C_z}(i)D_{C_z}(i)D_{C_z}^\top(i)
\end{bmatrix} < 0,
$$

with

$$
J_1(i) = A^\top(i)P(i) + P^\top(i)A(i) + K^\top(i)B^\top(i)P(i) + P^\top(i)B(i)K(i)
$$

$$
+ \lambda_{ij}E^\top(i)P(i) + \sum_{j=1, j\neq i}^{N} \varepsilon_P\lambda_{ij}\left[P(j) + P^\top(j)\right].
$$

Let $J_2(i)$, $\mathcal{W}(i)$ and $\mathcal{T}(i)$ be defined as:

$$
J_2(i) = J_1(i) + \varepsilon_A^{-1}(i)E_A^\top(i)E_A(i) + \varepsilon_B^{-1}(i)K^\top(i)E_B^\top(i)E_B(i)K(i),
$$

$$
\mathcal{W}(i) = \mathrm{diag}[\varepsilon_A^{-1}(i)\mathbb{I}, \varepsilon_B^{-1}(i)\mathbb{I}, \varepsilon_{C_z}(i)\mathbb{I}, \varepsilon_{D_z}(i)\mathbb{I}],
$$

$$
\mathcal{T}(i) = \left[P^\top(i)D_A(i), P^\top(i)D_B(i), K^\top(i)E_{C_z}^\top(i), K^\top(i)E_{D_z}^\top(i)\right],
$$

and using Schur complement, we get the equivalent inequality:

$$
\begin{bmatrix}
J_2(i) & P^\top(i)B_\omega(i) & \begin{bmatrix} C_z^\top(i) \\ +K^\top(i)D_z^\top(i) \end{bmatrix} & \mathcal{T}(i) \\
B_\omega^\top(i)P(i) & -\gamma^2\mathbb{I} & B_z^\top(i) & 0 \\
\begin{bmatrix} D_z(i)K(i) \\ +C_z(i) \end{bmatrix} & B_z(i) & -\mathcal{U}(i) & 0 \\
\mathcal{T}^\top(i) & 0 & 0 & -\mathcal{W}(i)
\end{bmatrix} < 0,
$$

with $\mathcal{U}(i) = \mathbb{I} - \varepsilon_{D_z}(i)D_{D_z}(i)D_{D_z}^\top(i) - \varepsilon_{C_z}(i)D_{C_z}(i)D_{C_z}^\top(i)$.

This matrix inequality is nonlinear in $P(i)$ and $K(i)$. To put it into an LMI form, let $X(i) = P^{-1}(i)$. Pre- and post-multiply this matrix inequality by $\mathrm{diag}[X^\top(i), \mathbb{I}, \mathbb{I}, \mathbb{I}]$ and its transpose, we get:

$$
\begin{bmatrix}
J_3(i) & B_\omega(i) & \begin{bmatrix} X^\top(i)C_z^\top(i) \\ +X^\top(i)K^\top(i)D_z^\top(i) \end{bmatrix} & X(i)\mathcal{T}(i) \\
B_\omega^\top(i) & -\gamma^2\mathbb{I} & B_z^\top(i) & 0 \\
\begin{bmatrix} D_z(i)K(i)X(i) \\ +C_z(i)X(i) \end{bmatrix} & B_z(i) & -\mathcal{U}(i) & 0 \\
\mathcal{T}^\top(i)X(i) & 0 & 0 & -\mathcal{W}(i)
\end{bmatrix} < 0,
$$

with

$$
\begin{aligned}
J_3(i) = {}& X^\top(i)A^\top(i) + A(i)X(i) + X^\top(i)K^\top(i)B^\top(i) \\
& + B(i)K(i)X(i) + \varepsilon_A^{-1}(i)X^\top(i)E_A^\top(i)E_A(i)X(i) \\
& + \varepsilon_B^{-1}(i)X^\top(i)K^\top(i)E_B^\top(i)E_B(i)K(i)X(i) \\
& + \lambda_{ii}X^\top(i)E^\top(i) + \sum_{j=1,j\neq i}^{N} \varepsilon_P \lambda_{ij} X^\top(i)\left[X^{-1}(j) + X^{-\top}(j)\right]X(i)\,.
\end{aligned}
$$

Notice that:

$$
X^\top(i)\mathscr{T}(i) = \left[D_A(i), D_B(i), X^\top(i)K^\top(i)E_{C_z}^\top(i), X^\top(i)K^\top(i)E_{D_z}^\top(i)\right],
$$

and

$$
\sum_{j=1,j\neq i}^{N} \varepsilon_P \lambda_{ij} X^\top(i)\left[X^{-1}(j) + X^{-\top}(j)\right]X(i) \leq \mathcal{S}_i(X)\mathcal{S}_i^\top(X) + \mathcal{S}_i(X)Z_i^{-1}(X)\mathcal{S}_i^\top(X)\,,
$$

with

$$
\begin{aligned}
Z_i(X) = {}& \mathrm{diag}\left[X^\top(1)X(1),\cdots, X^\top(i-1)X(i-1), X^\top(i+1)X(i+1),\right. \\
& \left.\cdots, X^\top(N)X(N)\right], \\
\mathcal{S}_i(X) = {}& \left(\sqrt{\varepsilon_P \lambda_{i1}}X^\top(i),\cdots, \sqrt{\varepsilon_P \lambda_{ii-1}}X^\top(i),\right. \\
& \left.\sqrt{\varepsilon_P \lambda_{ii+1}}X^\top(i),\cdots, \sqrt{\varepsilon_P \lambda_{iN}}X^\top(i)\right).
\end{aligned}
$$

Letting $Y(i) = K(i)X(i)$ and proceeding as before and using the Schur complement we obtain:

$$
\begin{bmatrix}
\tilde{J}(i) & B_\omega(i) & \begin{matrix} X(i)C_z^\top(i) \\ +Y^\top(i)D_z^\top(i) \end{matrix} & \mathcal{R}(i) & \mathcal{S}_i(X) & \mathcal{S}_i(X) \\
B_\omega^\top(i) & -\gamma^2 \mathbb{I} & B_z^\top(i) & 0 & 0 & 0 \\
\begin{matrix} D_z(i)Y(i) \\ +C_z(i)X(i) \end{matrix} & B_z(i) & -\mathcal{U}(i) & 0 & 0 & 0 \\
\mathcal{R}^\top(i) & 0 & 0 & -\mathcal{V}(i) & 0 & 0 \\
\mathcal{S}_i^\top(X) & 0 & 0 & 0 & -\mathbb{I} & 0 \\
\mathcal{S}_i^\top(X) & 0 & 0 & 0 & 0 & -\mathcal{X}_i(X)
\end{bmatrix} < 0,
$$

with

$$
\begin{aligned}
\tilde{J}(i) = {}& X^\top(i)A^\top(i) + A(i)X(i) + Y^\top(i)B^\top(i) + B(i)Y(i) \\
& + \varepsilon_A(i)D_A(i)D_A^\top(i) + \varepsilon_B(i)D_B(i)D_B^\top(i) + \lambda_{ii}X^\top(i)E^\top(i)\,, \\
\mathcal{R}(i) = {}& \left[X^\top(i)E_A^\top(i), Y^\top(i)E_B^\top(i), X^\top(i)E_{C_z}^\top(i), Y^\top(i)E_{D_z}^\top(i)\right], \\
\mathcal{V}(i) = {}& \mathrm{diag}[\varepsilon_A(i)\mathbb{I}, \varepsilon_B(i)\mathbb{I}, \varepsilon_{C_z}(i)\mathbb{I}, \varepsilon_{D_z}(i)\mathbb{I}]\,, \\
\mathcal{S}_i(X) = {}& \left(\sqrt{\varepsilon_P \lambda_{i1}}X^\top(i),\cdots, \sqrt{\varepsilon_P \lambda_{ii-1}}X^\top(i),\right. \\
& \left.\sqrt{\varepsilon_P \lambda_{ii+1}}X^\top(i),\cdots, \sqrt{\varepsilon_P \lambda_{iN}}X^\top(i)\right), \\
\mathcal{X}_i(X) = {}& \mathrm{diag}\left[X^\top(1) + X(1) - \mathbb{I},\cdots, X^\top(i-1) + X(i-1) - \mathbb{I},\right. \\
& \left.X^\top(i+1) + X(i+1) - \mathbb{I},\cdots, X^\top(N) + X(N) - \mathbb{I}\right].
\end{aligned}
$$

The following theorem summarizes the results of this development.

Theorem 4.2.14 *Let γ and ε_P be given positive scalars. There exists a state feedback controller of the form (4.39) such that the closed-loop state equation of the system (4.1) is piecewise regular, impulse-free and stochastically stable and moreover the closed-loop system satisfies the disturbance rejection of level γ if there exist a set of nonsingular matrices $X = (X(1), \cdots, X(N))$, $X(i) \in \mathbb{R}^{n \times n}$ a set of matrices $Y = (Y(1), \cdots, Y(N))$, $Y(i) \in \mathbb{R}^{m \times n}$ and sets of positive scalars $\varepsilon_A = (\varepsilon_A(1), \cdots, \varepsilon_A(N))$, $\varepsilon_B = (\varepsilon_B(1), \cdots, \varepsilon_B(N))$, $\varepsilon_{C_z} = (\varepsilon_{C_z}(1), \cdots, \varepsilon_{C_z}(N))$, and $\varepsilon_{D_z} = (\varepsilon_{D_z}(1), \cdots, \varepsilon_{D_z}(N))$, such that the following set of coupled LMIs holds for every $i \in \mathscr{S}$ and for all admissible uncertainties:*

$$
\begin{bmatrix}
\tilde{J}(i) & B_\omega(i) & \begin{bmatrix} X(i)C_z^\mathsf{T}(i) \\ +Y^\mathsf{T}(i)D_z^\mathsf{T}(i) \end{bmatrix} & \mathcal{R}(i) & S_i(X) & S_i(X) \\
B_\omega^\mathsf{T}(i) & -\gamma^2\mathbb{I} & B_z^\mathsf{T}(i) & 0 & 0 & 0 \\
\begin{bmatrix} D_z(i)Y(i) \\ +C_z(i)X(i) \end{bmatrix} & B_z(i) & -\mathcal{U}(i) & 0 & 0 & 0 \\
\mathcal{R}^\mathsf{T}(i) & 0 & 0 & -\mathcal{V}(i) & 0 & 0 \\
S_i^\mathsf{T}(X) & 0 & 0 & 0 & -\mathbb{I} & 0 \\
S_i^\mathsf{T}(X) & 0 & 0 & 0 & 0 & -\mathcal{X}_i(X)
\end{bmatrix} < 0, \qquad (4.51)
$$

where

$$
\begin{aligned}
\tilde{J}(i) &= X(i)A^\mathsf{T}(i) + A(i)X(i) + Y^\mathsf{T}(i)B^\mathsf{T}(i) + B(i)Y(i) \\
&\quad + \varepsilon_A(i)D_A(i)D_A^\mathsf{T}(i) + \varepsilon_B(i)D_B(i)D_B^\mathsf{T}(i) \\
&\quad + \lambda_{ii}X^\mathsf{T}(i)E^\mathsf{T}(i), \\
\mathcal{U}(i) &= \mathbb{I} - \varepsilon_{D_z}(i)D_{D_z}(i)D_{D_z}^\mathsf{T}(i) - \varepsilon_{C_z}(i)D_{C_z}(i)D_{C_z}^\mathsf{T}(i), \\
\mathcal{R}(i) &= \left[X^\mathsf{T}(i)E_A^\mathsf{T}(i), Y^\mathsf{T}(i)E_B^\mathsf{T}(i), X^\mathsf{T}(i)E_{C_z}^\mathsf{T}(i), Y^\mathsf{T}(i)E_{D_z}^\mathsf{T}(i) \right], \\
\mathcal{V}(i) &= \mathrm{diag}[\varepsilon_A(i)\mathbb{I}, \varepsilon_B(i)\mathbb{I}, \varepsilon_{C_z}(i)\mathbb{I}, \varepsilon_{D_z}(i)\mathbb{I}], \\
S_i(X) &= \left(\sqrt{\varepsilon_P\lambda_{i1}}X^\mathsf{T}(i), \cdots, \sqrt{\varepsilon_P\lambda_{ii-1}}X^\mathsf{T}(i), \right. \\
&\qquad \left. \sqrt{\varepsilon_P\lambda_{ii+1}}X^\mathsf{T}(i), \cdots, \sqrt{\varepsilon_P\lambda_{iN}}X^\mathsf{T}(i) \right), \\
\mathcal{X}_i(X) &= \mathrm{diag}\left[X^\mathsf{T}(1) + X(1) - \mathbb{I}, \cdots, X^\mathsf{T}(i-1) + X(i-1) - \mathbb{I}, \right. \\
&\qquad \left. X^\mathsf{T}(i+1) + X(i+1) - \mathbb{I}, \cdots, X^\mathsf{T}(N) + X(N) - \mathbb{I} \right],
\end{aligned}
$$

with the following constraints:

$$
\varepsilon_P \left[X(i) + X^\mathsf{T}(i) \right] \geq X^\mathsf{T}(i)E^\mathsf{T}(i) = E(i)X(i) \geq 0. \qquad (4.52)
$$

The state feedback controller (4.39) gain is given $K(i) = Y(i)X^{-1}(i)$ for each $i \in \mathscr{S}$.

As it was done for the nominal system, we can determine the controller that assures that the closed-loop system is piecewise regular, impulse-free and stochastically stable and at the same time guarantees the minimum disturbance rejection by solving the following optimization problem:

Pu :
$$
\begin{cases}
\min\limits_{\substack{\nu>0,\\ \varepsilon_A=(\varepsilon_A(1),\cdots,\varepsilon_A(N))>0,\\ \varepsilon_B=(\varepsilon_B(1),\cdots,\varepsilon_B(N))>0,\\ \varepsilon_{C_z}=(\varepsilon_{D_z}(1),\cdots,\varepsilon_{C_z}(N))>0,\\ \varepsilon_{D_z}=(\varepsilon_{D_z}(1),\cdots,\varepsilon_{D_z}(N))>0,\\ X=(X(1),\cdots,X(N)),\\ Y=(Y(1),\cdots,Y(N))}} \nu \\[2mm]
s.t: \\[1mm]
\varepsilon_P\left[X(i)+X^\top(i)\right] \ge X^\top(i)E^\top(i) = E(i)X(i) \ge 0, \\[2mm]
\begin{bmatrix}
\tilde{J}(i) & B_\omega(i) & \begin{matrix} X(i)C_z^\top(i) \\ +Y^\top(i)D_z^\top(i) \end{matrix} & \mathcal{R}(i) & \mathcal{S}_i(X) & \mathcal{S}_i(X) \\
B_\omega^\top(i) & -\nu\mathbb{I} & B_z^\top(i) & 0 & 0 & 0 \\
\begin{bmatrix} D_z(i)Y(i) \\ +C_z(i)X(i) \end{bmatrix} & B_z(i) & -\mathcal{U}(i) & 0 & 0 & 0 \\
\mathcal{R}^\top(i) & 0 & 0 & -\mathcal{V}(i) & 0 & 0 \\
\mathcal{S}_i^\top(X) & 0 & 0 & 0 & -\mathbb{I} & 0 \\
\mathcal{S}_i^\top(X) & 0 & 0 & 0 & 0 & -\mathcal{X}_i(X)
\end{bmatrix} < 0,
\end{cases}
$$

where the LMI that represents the constraint of this optimization problem is obtained from (4.51) by replacing γ^2 by ν.

The following theorem summarizes the results on the design of the controller that stochastically stabilizes the system (4.1) and simultaneously guarantees the smallest disturbance rejection level.

Theorem 4.2.15 *Let* $\nu > 0$, $\varepsilon_A = (\varepsilon_A(1), \cdots, \varepsilon_A(N)) > 0$, $\varepsilon_B = (\varepsilon_B(1), \cdots, \varepsilon_B(N)) > 0$, $\varepsilon_{C_z} = (\varepsilon_{C_z}(1), \cdots, \varepsilon_{C_z}(N)) > 0$, $\varepsilon_{D_z} = (\varepsilon_{D_z}(1), \cdots, \varepsilon_{D_z}(N)) > 0$, $X = (X(1), \cdots, X(N))$, *and* $Y = (Y(1), \cdots, Y(N))$ *be the solution of the optimization problem Pu for a given positive scalar* ε_P. *Then, the controller (4.39) with* $K(i) = Y(i)X^{-1}(i)$ *will guarantee that the closed-loop system will be piecewise regular, impulse-free and stochastically stable and moreover the closed-loop system satisfies the disturbance rejection of level* $\sqrt{\nu}$.

We can similarly develop results for a designing a stabilizing state-feedback controller which assures that the closed-loop system is piecewise regular, impulse-free and stochastically stable and at the same time guarantees the disturbance rejection with a given level γ using the other bounds for the term $E^\top(i)P(i)$.

The following theorem summarizes the results when the term $E^\top(i)P(i)$ is bounded by $\varepsilon(i)\left[P^\top(i)P(i)\right]$, with $\varepsilon(i) > 0$.

Theorem 4.2.16 *Let* γ *be a given positive scalar. There exists a state feedback controller of the form (4.39) such that the closed-loop state equation of the system (4.1) is piecewise regular, impulse-free and stochastically stable and moreover the closed-loop system satisfies the disturbance rejection of level* γ *if there exist a set of nonsingular matrices* $X = (X(1), \cdots, X(N))$, $X(i) \in \mathbb{R}^{n\times n}$ *a set of matrices*

$Y = (Y(1), \cdots, Y(N))$, $Y(i) \in \mathbb{R}^{n \times n}$ and sets of positive scalars $\varepsilon_A = (\varepsilon_A(1), \cdots, \varepsilon_A(N))$, $\varepsilon_B = (\varepsilon_B(1), \cdots, \varepsilon_B(N))$, $\varepsilon_{C_z} = (\varepsilon_{C_z}(1), \cdots, \varepsilon_{C_z}(N))$, $\varepsilon_{D_z} = (\varepsilon_{D_z}(1), \cdots, \varepsilon_{D_z}(N))$, and $\varepsilon = (\varepsilon(1), \cdots, \varepsilon(N))$ such that the following set of coupled LMIs holds for every $i \in \mathscr{S}$ and for all admissible uncertainties:

$$
\begin{bmatrix}
\tilde{J}(i) & B_\omega(i) & \begin{bmatrix} X(i)C_z^\top(i) \\ +Y^\top(i)D_z^\top(i) \end{bmatrix} & \mathcal{R}(i) & S_i(X) \\
B_\omega^\top(i) & -\gamma^2 \mathbb{I} & B_z^\top(i) & 0 & 0 \\
\begin{bmatrix} D_z(i)Y(i) \\ +C_z(i)X(i) \end{bmatrix} & B_z(i) & -\mathcal{U}(i) & 0 & 0 \\
\mathcal{R}^\top(i) & 0 & 0 & -\mathcal{V}(i) & 0 \\
S_i^\top(X) & 0 & 0 & 0 & -\mathcal{X}_i(X)
\end{bmatrix} < 0 , \tag{4.53}
$$

where

$$
\begin{aligned}
\tilde{J}(i) &= X(i)A^\top(i) + A(i)X(i) + Y^\top(i)B^\top(i) + B(i)Y(i) \\
&\quad + \varepsilon_A(i)D_A(i)D_A^\top(i) + \varepsilon_B(i)D_B(i)D_B^\top(i) \\
&\quad + \lambda_{ii}X^\top(i)E^\top(i) , \\
\mathcal{U}(i) &= \mathbb{I} - \varepsilon_{D_z}(i)D_{D_z}(i)D_{D_z}^\top(i) - \varepsilon_{C_z}(i)D_{C_z}(i)D_{C_z}^\top(i) , \\
\mathcal{R}(i) &= \left[X^\top(i)E_A^\top(i), Y^\top(i)E_B^\top(i), X^\top(i)E_{C_z}^\top(i), Y^\top(i)E_{D_z}^\top(i) \right] , \\
\mathcal{V}(i) &= \mathrm{diag}[\varepsilon_A(i)\mathbb{I}, \varepsilon_B(i)\mathbb{I}, \varepsilon_{C_z}(i)\mathbb{I}, \varepsilon_{D_z}(i)\mathbb{I}] , \\
S_i(X) &= \left(\sqrt{\lambda_{i1}}X^\top(i), \cdots, \sqrt{\lambda_{ii-1}}X^\top(i), \right. \\
&\qquad \left. \sqrt{\lambda_{ii+1}}X^\top(i), \cdots, \sqrt{\lambda_{iN}}X^\top(i) \right) , \\
\mathcal{X}_i(X) &= \mathrm{diag}\left[X^\top(1) + X(1) - \varepsilon(1)\mathbb{I}, \cdots, X^\top(i-1) + X(i-1) - \varepsilon(i-1)\mathbb{I}, \right. \\
&\qquad \left. X^\top(i+1) + X(i+1) - \varepsilon(i+1)\mathbb{I}, \cdots, X^\top(N) + X(N) - \varepsilon(N)\mathbb{I} \right] ,
\end{aligned}
$$

with the following constraints:

$$
\varepsilon(i)\mathbb{I} \geq X^\top(i)E^\top(i) = E(i)X(i) \geq 0 . \tag{4.54}
$$

The state feedback controller (4.39) gain is given $K(i) = Y(i)X^{-1}(i)$ for each $i \in \mathscr{S}$.

The following theorem summarizes the results when the term $E^\top(i)P(i)$ is bounded by $\left[\frac{1}{4}\varepsilon^{-1}(i)\mathbb{I} + \varepsilon(i)E^\top(i)P(i)P^\top(i)P(i) \right]$, with $\varepsilon(i) > 0$.

Theorem 4.2.17 Let γ be a given positive scalar. There exists a state feedback controller of the form (4.39) such that the closed-loop state equation of the system (4.1) is piecewise regular, impulse-free and stochastically stable and moreover the closed-loop system satisfies the disturbance rejection of level γ if there exist a set of nonsingular matrices $X = (X(1), \cdots, X(N))$, $X(i) \in \mathbb{R}^{n \times n}$ a set of matrices $Y = (Y(1), \cdots, Y(N))$, $Y(i) \in \mathbb{R}^{n \times n}$ and sets of positive scalars $\varepsilon_A = (\varepsilon_A(1), \cdots, \varepsilon_A(N))$, $\varepsilon_B = (\varepsilon_B(1), \cdots, \varepsilon_B(N))$, $\varepsilon_{C_z} = (\varepsilon_{C_z}(1), \cdots, \varepsilon_{C_z}(N))$, and $\varepsilon_{D_z} = (\varepsilon_{D_z}(1), \cdots,$

$\varepsilon_{D_z}(N))$, *such that the following set of coupled LMIs holds for every* $i \in \mathscr{S}$ *and for all admissible uncertainties:*

$$
\begin{bmatrix}
\tilde{J}(i) & B_\omega(i) & \begin{bmatrix} X(i)C_z^\top(i) \\ +Y^\top(i)D_z^\top(i) \end{bmatrix} & \mathcal{R}(i) & \mathcal{Z}_i(X) & S_i(X) \\
B_\omega^\top(i) & -\gamma^2 \mathbb{I} & B_z^\top(i) & 0 & 0 & 0 \\
\begin{bmatrix} D_z(i)Y(i) \\ +C_z(i)X(i) \end{bmatrix} & B_z(i) & -\mathcal{U}(i) & 0 & 0 & 0 \\
\mathcal{R}^\top(i) & 0 & 0 & -\mathcal{V}(i) & 0 & 0 \\
\mathcal{Z}_i^\top(X) & 0 & 0 & 0 & -\mathcal{X}_i(\varepsilon) & 0 \\
S_i^\top(X) & 0 & 0 & 0 & 0 & -\mathcal{X}_i(X)
\end{bmatrix} < 0, \quad (4.55)
$$

where

$$
\begin{aligned}
\tilde{J}(i) &= X(i)A^\top(i) + A(i)X(i) + Y^\top(i)B^\top(i) + B(i)Y(i) \\
&\quad + \varepsilon_A(i)D_A(i)D_A^\top(i) + \varepsilon_B(i)D_B(i)D_B^\top(i) \\
&\quad + \lambda_{ii}X^\top(i)E^\top(i), \\
\mathcal{U}(i) &= \mathbb{I} - \varepsilon_{D_z}(i)D_{D_z}(i)D_{D_z}^\top(i) - \varepsilon_{C_z}(i)D_{C_z}(i)D_{C_z}^\top(i), \\
\mathcal{R}(i) &= \left[X^\top(i)E_A^\top(i), Y^\top(i)E_B^\top(i), X^\top(i)E_{C_z}^\top(i), Y^\top(i)E_{D_z}^\top(i) \right], \\
\mathcal{V}(i) &= \mathrm{diag}[\varepsilon_A(i)\mathbb{I}, \varepsilon_B(i)\mathbb{I}, \varepsilon_{C_z}(i)\mathbb{I}, \varepsilon_{D_z}(i)\mathbb{I}], \\
\mathcal{Z}_i(X) &= \left(\sqrt{\lambda_{i1}}X^\top(i), \cdots, \sqrt{\lambda_{ii-1}}X^\top(i), \right. \\
&\quad \left. \sqrt{\lambda_{ii+1}}X^\top(i), \cdots, \sqrt{\lambda_{iN}}X^\top(i) \right), \\
S_i(X) &= \left(\sqrt{\lambda_{i1}}X^\top(i)E^\top(1), \cdots, \sqrt{\lambda_{ii-1}}X^\top(i)E^\top(i-1), \right. \\
&\quad \left. \sqrt{\lambda_{ii+1}}X^\top(i)E^\top(i+1), \cdots, \sqrt{\lambda_{iN}}X^\top(i)E^\top(N) \right), \\
\mathcal{X}_i(\varepsilon) &= \mathrm{diag}\left[4\varepsilon(1)\mathbb{I}, \cdots, 4\varepsilon(i-1)\mathbb{I}, 4\varepsilon(i+1)\mathbb{I}, \cdots, 4\varepsilon(N)\mathbb{I} \right], \\
\mathcal{X}_i(X) &= \mathrm{diag}\left[-\varepsilon(1)\mathbb{I} + X^\top(1) + X(1), \cdots, -\varepsilon(i-1)\mathbb{I} + X^\top(i-1) + X(i-1), \right. \\
&\quad \left. -\varepsilon(i+1)\mathbb{I} + X^\top(i+1) + X(i+1), \cdots, -\varepsilon(N)\mathbb{I} + X^\top(N) + X(N) \right],
\end{aligned}
$$

with the following constraints:

$$
X^\top(i)E^\top(i) = E(i)X(i) \geq 0. \quad (4.56)
$$

The state feedback controller (4.39) gain is given $K(i) = Y(i)X^{-1}(i)$ *for each* $i \in \mathscr{S}$.

4.3 $\mathscr{H}_\infty$ Constant Gain Stabilization

Let us now design a constant gain controller that assures that the closed-loop system is piecewise regular, impulse-free and stochastically stable and at the same time guarantees the disturbance rejection of the desired level. For this purpose plugging the constant gain controller in the system state equation, we get the following results in similar way as before.

Theorem 4.3.1 *Let γ and ε_P be given positive scalars and K a given gain. If there exists a nonsingular matrix P such that the following set of LMIs holds for every $i \in \mathscr{S}$*

$$\left[\begin{array}{cc} \bar{J}_0(i) & \left[\begin{array}{c} \bar{C}_z^\top(i)B_z(i) \\ +P^\top B_\omega(i) \end{array} \right] \\ \left[\begin{array}{c} B_z^\top(i)\bar{C}_z(i) \\ +B_\omega^\top(i)P \end{array} \right] & B_z^\top(i)B_z(i) - \gamma^2\mathbb{I} \end{array} \right] < 0, \tag{4.57}$$

where $\bar{J}_0(i) = \bar{A}^\top(i)P + P^\top\bar{A}(i) + \lambda_{ii}E^\top(i)P + \sum_{j=1,j\neq i}^N \varepsilon_P(j)\lambda_{ij}\left[P + P^\top\right] + \bar{C}_z^\top(i)\bar{C}_z(i),$ with the following constraints:

$$\varepsilon_P(i)\left[P + P^\top\right] \geq E^\top(i)P = P^\top E(i) \geq 0, \tag{4.58}$$

then system (4.1) is stochastically stable under the controller (4.39) and satisfies the following

$$\|z\|_2 \leq \left[\gamma^2\|w\|_2^2 + x_0^\top E^\top(r_0)Px_0\right]^{\frac{1}{2}}, \tag{4.59}$$

which means that the system is stochastically stable with γ-disturbance attenuation.

To synthesize the controller with constant gain, let us transform the LMI (4.57) into a form that can be used easily to compute this gain. For this purpose notice that:

$$\left[\begin{array}{cc} \bar{J}_0(i) & \left[\begin{array}{c} \bar{C}_z^\top(i)B_z(i) \\ +P^\top B_\omega(i) \end{array} \right] \\ \left[\begin{array}{c} B_z^\top(i)\bar{C}_z(i) \\ +B_\omega^\top(i)P \end{array} \right] & B_z^\top(i)B_z(i) - \gamma^2\mathbb{I} \end{array} \right] =$$

$$\left[\begin{array}{cc} \bar{J}_1(i) & P^\top B_\omega(i) \\ B_\omega^\top(i)P & -\gamma^2\mathbb{I} \end{array} \right]$$

$$+ \left[\begin{array}{c} \bar{C}_z^\top(i) \\ B_z^\top(i) \end{array} \right] \left[\begin{array}{cc} \bar{C}_z(i) & B_z(i) \end{array} \right],$$

with

$$\bar{J}_0(i) = \bar{A}^\top(i)P + P^\top\bar{A}(i) + \lambda_{ii}E^\top(i)P$$

$$+ \sum_{j=1,j\neq i}^N \varepsilon_P(j)\lambda_{ij}\left[P + P^\top\right] + \bar{C}_z^\top(i)\bar{C}_z(i),$$

$$\bar{J}_1(i) = \bar{A}^\top(i)P + P^\top\bar{A}(i) + \lambda_{ii}E^\top(i)P$$

$$+ \sum_{j=1,j\neq i}^N \varepsilon_P(j)\lambda_{ij}\left[P + P^\top\right].$$

Using now Schur complement, we show that (4.57) is equivalent to the following inequality:

$$\left[\begin{array}{ccc} \bar{J}_1(i) & P^\top B_\omega(i) & \bar{C}_z^\top(i) \\ B_\omega^\top(i)P & -\gamma^2\mathbb{I} & B_z^\top(i) \\ \bar{C}_z(i) & B_z(i) & -\mathbb{I} \end{array} \right] < 0,$$

Since $\bar{A}(i)$ is nonlinear in K and P the previous matrix inequality is then nonlinear and therefore it can not solved using existing linear algorithms. To transform it to an LMI, let $X = P^{-1}$. As we did many times previously let us pre- and post-multiply this inequality by $\mathrm{diag}[X^\top, \mathbb{I}, \mathbb{I}]$ and its transpose, where $\mathbb{I}$ is an appropriate identity matrix, which gives:

$$\begin{bmatrix} \bar{J}_X(i) & B_\omega(i) & X^\top \bar{C}_z^\top(i) \\ B_\omega^\top(i) & -\gamma^2 \mathbb{I} & B_z^\top(i) \\ \bar{C}_z(i)X & B_z(i) & -\mathbb{I} \end{bmatrix} < 0,$$

with $\bar{J}_X(i) = X^\top \bar{A}^\top(i) + \bar{A}(i)X + \lambda_{ii} X^\top E^\top(i) + \sum_{j=1, j\neq i}^N \varepsilon_P(j)\lambda_{ij}[X + X^\top]$.
Notice that

$$X^\top \bar{A}^\top(i) + \bar{A}(i)X = X^\top A^\top(i) + A(i)X + Y^\top B^\top(i) + B(i)Y$$
$$X^\top [C_z(i) + D_z(i)K]^\top = X^\top C_z^\top(i) + Y^\top D_z^\top(i),$$

where $Y = KX$.

Using Schur complement again, this implies that the previous inequality is equivalent to the following:

$$\begin{bmatrix} J(i) & B_\omega(i) & \begin{bmatrix} X^\top C_z^\top(i) \\ +Y^\top D_z^\top(i) \end{bmatrix} \\ B_\omega^\top(i) & -\gamma^2 \mathbb{I} & B_z^\top(i) \\ \begin{bmatrix} C_z(i)X \\ +D_z(i)Y \end{bmatrix} & B_z(i) & -\mathbb{I} \end{bmatrix} < 0,$$

with $J(i) = X^\top A^\top(i) + A(i)X + Y^\top B^\top(i) + B(i)Y + \lambda_{ii} X^\top E^\top(i) + \sum_{j=1, j\neq i}^N \varepsilon_P(j)\lambda_{ij}[X + X^\top]$.
The following theorem gives a procedure to design a $\mathscr{H}_\infty$ constant gain controller.

Theorem 4.3.2 *Let γ be a positive constant and $\varepsilon_P = (\varepsilon_P(1), \cdots, \varepsilon_P(N))$ be a set of positive scalars. There exist a state feedback controller of the form (4.39) that guarantees that the closed-loop state equation of the system is piecewise regular, impulsefree and stochastically stable and moreover the closed-loop satisfies the disturbance rejection, if there exit a nonsingular matrix $X \in \mathbb{R}^{n\times n}$ and a matrix $Y \in \mathbb{R}^{m\times n}$, such that the following set of LMIs holds for every $i \in \mathscr{S}$:*

$$\begin{bmatrix} J(i) & B_\omega(i) & \begin{bmatrix} X^\top C_z^\top(i) \\ +Y^\top D_z^\top(i) \end{bmatrix} \\ B_\omega^\top(i) & -\gamma^2 \mathbb{I} & B_z^\top(i) \\ \begin{bmatrix} C_z(i)X \\ +D_z(i)Y \end{bmatrix} & B_z(i) & -\mathbb{I} \end{bmatrix} < 0, \tag{4.60}$$

where

$$J(i) = X^\top A^\top(i) + A(i)X + Y^\top B^\top(i) + B(i)Y + \lambda_{ii} X^\top E^\top(i)$$
$$+ \sum_{j=1, j\neq i}^N \varepsilon_P(j)\lambda_{ij}[X + X^\top],$$

with the following constraints:

$$\varepsilon_P(i)\left[X + X^\top\right] \geq X^\top E^\top(i) = E(i)X \geq 0.\tag{4.61}$$

The state feedback controller gain is given by $K = YX^{-1}$.

From the practical point of view, the controller that assures that the closed-loop system is piecewise regular, impulse-free and stochastically stable and at the same time guarantees the minimum disturbance rejection is of great interest. This controller can be obtained by solving the following optimization problem for a given set of positive scalars $\varepsilon_P = (\varepsilon(1), \cdots, \varepsilon_P(N))$:

$$\mathrm{P}: \begin{cases} \min_{\substack{v>0, \\ X, \\ Y}} v \\[4pt] s.t: \\[4pt] \varepsilon_P(i)\left[X + X^\top\right] \geq X^\top E^\top(i) = E(i)X \geq 0, \\[4pt] \begin{bmatrix} J(i) & B_\omega(i) & \begin{bmatrix} X^\top C_z^\top(i) \\ +Y^\top D_z^\top(i) \end{bmatrix} \\ B_\omega^\top(i) & -v\mathbb{I} & B_z^\top(i) \\ \begin{bmatrix} C_z(i)X \\ +D_z(i)Y \end{bmatrix} & B_z(i) & -\mathbb{I} \end{bmatrix} < 0, \end{cases}$$

where the LMI in the constraints is obtained from (4.60) by replacing γ^2 by v.

The following theorem gives the results on the design of the controller that stochastically stabilizes the system (4.1) and simultaneously guarantees the smallest disturbance rejection level.

Theorem 4.3.3 *Let $v > 0$, X and Y be the solution of the optimization problem P for a given set of positive scalars $\varepsilon_P = (\varepsilon_P(1), \cdots, \varepsilon_P(N))$. Then, the controller (4.39) with $K = YX^{-1}$ guarantees that closed-loop system is piecewise regular, impulse-free and stochastically stabilizes the class of systems we are considering and moreover the closed-loop system satisfies the disturbance rejection of level $\sqrt{v}$.*

For the other bounds, we can establish easily the following results.

Theorem 4.3.4 *Let γ be a positive constant. There exists a state feedback controller of the form (4.39) that guarantees that the closed-loop state equation of the system is piecewise regular, impulse-free and stochastically stable and moreover the closed-loop satisfies the disturbance rejection, if there exit a nonsingular matrix $X \in \mathbb{R}^{n \times n}$, a matrix $Y \in \mathbb{R}^{m \times n}$ and a set of positive scalars $\varepsilon = (\varepsilon(1), \cdots, \varepsilon(N))$, such that the following set of LMIs holds for every $i \in \mathscr{S}$:*

$$\begin{bmatrix} J(i) & B_\omega(i) & \begin{bmatrix} X^\top C_z^\top(i) \\ +Y^\top D_z^\top(i) \end{bmatrix} \\ B_\omega^\top(i) & -\gamma^2\mathbb{I} & B_z^\top(i) \\ \begin{bmatrix} C_z(i)X \\ +D_z(i)Y \end{bmatrix} & B_z(i) & -\mathbb{I} \end{bmatrix} < 0,\tag{4.62}$$

where

$$J(i) = X^\top A^\top(i) + A(i)X + Y^\top B^\top(i) + B(i)Y + \lambda_{ii}X^\top E^\top(i)$$

$$+ \sum_{j=1,j\neq i}^{N} \varepsilon(j)\lambda_{ij}\mathbb{I},$$

with the following constraints:

$$\varepsilon(i)\mathbb{I} \geq X^\top E^\top(i) = E(i)X \geq 0. \qquad (4.63)$$

The state feedback controller gain is given by $K = YX^{-1}$.

Theorem 4.3.5 *Let γ be a positive constant. There exists a state feedback controller of the form (4.39) that guarantees that the closed-loop state equation of the system is piecewise regular, impulse-free and stochastically stable and moreover the closed-loop satisfies the disturbance rejection, if there exit a nonsingular matrix $X \in \mathbb{R}^{n\times n}$, a matrix $Y \in \mathbb{R}^{m\times n}$ and a set of positive scalars $\varepsilon = (\varepsilon(1), \cdots, \varepsilon(N))$, such that the following set of LMIs holds for every $i \in \mathscr{S}$:*

$$\begin{bmatrix} J(i) & B_\omega(i) & \begin{bmatrix} X^\top C_z^\top(i) \\ +Y^\top D_z^\top(i) \end{bmatrix} & \mathcal{Z}_i(X) & \mathcal{S}_i(X) \\ B_\omega^\top(i) & -\gamma^2\mathbb{I} & B_z^\top(i) & 0 & 0 \\ \begin{bmatrix} C_z(i)X \\ +D_z(i)Y \end{bmatrix} & B_z(i) & -\mathbb{I} & 0 & 0 \\ \mathcal{Z}_i^\top(X) & 0 & 0 & X_i(\varepsilon) & 0 \\ \mathcal{S}_i^\top(X) & 0 & 0 & 0 & X_i(X) \end{bmatrix} < 0, \qquad (4.64)$$

where

$$J(i) = X^\top A^\top(i) + A(i)X + Y^\top B^\top(i) + B(i)Y + \lambda_{ii}X^\top E^\top(i)$$

$$+ \sum_{j=1,j\neq i}^{N} \varepsilon(j)\lambda_{ij}\mathbb{I},$$

$$\mathcal{Z}_i^\top(X) = \left(\sqrt{\lambda_{i1}}X^\top, \cdots, \sqrt{\lambda_{ii-1}}X^\top, \sqrt{\lambda_{ii+1}}X^\top, \cdots, \sqrt{\lambda_{iN}}X^\top \right),$$

$$\mathcal{S}_i^\top(X) = \left(\sqrt{\lambda_{i1}}X^\top E^\top(1), \cdots, \sqrt{\lambda_{ii-1}}X^\top E^\top(i-1), \right.$$

$$\left. \sqrt{\lambda_{ii+1}}X^\top E^\top(i+1), \cdots, \sqrt{\lambda_{iN}}X^\top E^\top(N) \right),$$

$$X_i(\varepsilon) = diag\left[4\varepsilon(1)\mathbb{I}, \cdots, 4\varepsilon(i-1)\mathbb{I}, 4\varepsilon(i+1)\mathbb{I}, \cdots, 4\varepsilon(N)\mathbb{I}\right],$$

$$X_i(X) = diag\left[X^\top + X - \varepsilon(1)\mathbb{I}, \cdots, X^\top + X - \varepsilon(i-1)\mathbb{I},\right.$$

$$\left. X^\top + X - \varepsilon(i+1)\mathbb{I}, \cdots, X^\top + X - \varepsilon(N)\mathbb{I}\right],$$

with the following constraints:

$$\varepsilon(i)\mathbb{I} \geq X^\top E^\top(i) = E(i)X \geq 0. \qquad (4.65)$$

The state feedback controller gain is given by $K = YX^{-1}$.

Let us now return to the state equation (4.1) and consider that the uncertainties are not equal to zero and see how we can synthesize the state feedback controller of the form (4.39) that assures that closed-loop system is piecewise regular, impulse-free and robust stochastically stable the class of systems we are studying and at the same time guarantees the desired disturbance rejection of level γ. Following the same steps for the nominal case starting from the results of Theorem 4.2.7, we get:

$$\begin{bmatrix} \bar{J}_0(i,t) & P^\top B_\omega(i) & \bar{C}_z^\top(i,t) \\ B_\omega^\top(i)P & -\gamma^2 \mathbb{I} & B_z^\top(i) \\ \bar{C}_z(i,t) & B_z(i) & -\mathbb{I} \end{bmatrix} < 0, \tag{4.66}$$

with

$$\bar{J}_0(i,t) = \bar{A}^\top(i,t)P + P^\top \bar{A}(i,t) + \lambda_{ii}E^\top(i)P + \sum_{j=1,j\neq i}^{N} \varepsilon_P(j)\lambda_{ij}\left[P + P^\top\right].$$

$$\bar{A}(i,t) = A(i,t) + B(i,t)K,$$

$$\bar{C}_z(i,t) = C_z(i,t) + D_z(i,t)K.$$

Using now the expression of $\bar{A}(i,t)$ and $\bar{C}_z(i,t)$ and the expressions of their components, we obtain the following inequality:

$$\begin{bmatrix} J_1(i) & P^\top B_\omega(i) & \begin{bmatrix} C_z^\top(i) \\ +K^\top D_z^\top(i) \end{bmatrix} \\ B_\omega^\top(i)P & -\gamma^2 \mathbb{I} & B_z^\top(i) \\ \begin{bmatrix} D_z(i)K \\ +C_z(i) \end{bmatrix} & B_z(i) & -\mathbb{I} \end{bmatrix}$$

$$+ \begin{bmatrix} E_A^\top(i)F_A^\top(i)D_A^\top(i)P & 0 & 0 \\ 0 & 0 & 0 \\ 0 & 0 & 0 \end{bmatrix} + \begin{bmatrix} P^\top D_A(i)F_A(i)E_A(i) & 0 & 0 \\ 0 & 0 & 0 \\ 0 & 0 & 0 \end{bmatrix}$$

$$+ \begin{bmatrix} P^\top D_B(i)F_B(i)E_B(i)K & 0 & 0 \\ 0 & 0 & 0 \\ 0 & 0 & 0 \end{bmatrix} + \begin{bmatrix} K^\top E_B^\top(i)F_B^\top(i)D_B^\top(i)P & 0 & 0 \\ 0 & 0 & 0 \\ 0 & 0 & 0 \end{bmatrix}$$

$$+ \begin{bmatrix} 0 & 0 & K^\top E_{D_z}^\top(i)F_{D_z}^\top(i)D_{D_z}^\top(i) \\ 0 & 0 & 0 \\ 0 & 0 & 0 \end{bmatrix} + \begin{bmatrix} 0 & 0 & 0 \\ 0 & 0 & 0 \\ D_{D_z}(i)F_{D_z}(i)E_{D_z}(i)K & 0 & 0 \end{bmatrix}$$

$$+ \begin{bmatrix} 0 & 0 & E_{C_z}^\top(i)F_{C_z}^\top(i)D_{C_z}^\top(i) \\ 0 & 0 & 0 \\ 0 & 0 & 0 \end{bmatrix} + \begin{bmatrix} 0 & 0 & 0 \\ 0 & 0 & 0 \\ D_{C_z}(i)F_{C_z}(i)E_{C_z}(i) & 0 & 0 \end{bmatrix} < 0,$$

with $J_1(i) = A^\top(i)P + P^\top A(i) + K^\top B^\top(i)P + P^\top B(i)K + \lambda_{ii}E^\top(i)P + \sum_{j=1, j\neq i}^{N} \varepsilon_P(j)\lambda_{ij}$ $[P + P^\top]$.

Notice that:

$$\begin{bmatrix} E_A^\top(i)F_A^\top(i)D_A^\top(i)P\ 0\ 0 \\ 0 \qquad\qquad 0\ 0 \\ 0 \qquad\qquad 0\ 0 \end{bmatrix} = \begin{bmatrix} E_A^\top(i) \\ 0 \\ 0 \end{bmatrix} F_A^\top(i)\left[D_A^\top(i)P\ 0\ 0 \right],$$

$$\begin{bmatrix} P^\top D_A(i)F_A(i)E_A(i)\ 0\ 0 \\ 0 \qquad\qquad 0\ 0 \\ 0 \qquad\qquad 0\ 0 \end{bmatrix} = \begin{bmatrix} P^\top D_A(i) \\ 0 \\ 0 \end{bmatrix} F_A(i)\left[E_A(i)\ 0\ 0 \right],$$

$$\begin{bmatrix} P^\top D_B(i)F_B(i)E_B(i)K\ 0\ 0 \\ 0 \qquad\qquad 0\ 0 \\ 0 \qquad\qquad 0\ 0 \end{bmatrix} = \begin{bmatrix} PD_B(i) \\ 0 \\ 0 \end{bmatrix} F_B(i)\left[E_B(i)K\ 0\ 0 \right],$$

$$\begin{bmatrix} K^\top E_B^\top(i)F_B^\top(i)D_B^\top(i)P\ 0\ 0 \\ 0 \qquad\qquad 0\ 0 \\ 0 \qquad\qquad 0\ 0 \end{bmatrix} = \begin{bmatrix} K^\top E_B^\top(i) \\ 0 \\ 0 \end{bmatrix} F_B^\top(i)\left[D_B^\top(i)P\ 0\ 0 \right],$$

$$\begin{bmatrix} 0\ 0\ K^\top E_{D_z}^\top(i)F_{D_z}^\top(i)D_{D_z}^\top(i) \\ 0\ 0 \qquad\qquad 0 \\ 0\ 0 \qquad\qquad 0 \end{bmatrix} = \begin{bmatrix} K^\top E_{D_z}^\top(i) \\ 0 \\ 0 \end{bmatrix} F_{D_z}^\top(i)\left[0\ 0\ D_{D_z}^\top(i) \right],$$

$$\begin{bmatrix} 0 \qquad\qquad 0\ 0 \\ 0 \qquad\qquad 0\ 0 \\ D_{D_z}(i)F_{D_z}(i)E_{D_z}(i)K\ 0\ 0 \end{bmatrix} = \begin{bmatrix} 0 \\ 0 \\ D_{D_z}(i) \end{bmatrix} F_{D_z}(i)\left[E_{D_z}(i)K\ 0\ 0 \right],$$

$$\begin{bmatrix} 0\ 0\ E_{C_z}^\top(i)F_{C_z}^\top(i)D_{C_z}^\top(i) \\ 0\ 0 \qquad\qquad 0 \\ 0\ 0 \qquad\qquad 0 \end{bmatrix} = \begin{bmatrix} E_{C_z}^\top(i) \\ 0 \\ 0 \end{bmatrix} F_{C_z}^\top(i)\left[0\ 0\ D_{C_z}^\top(i) \right],$$

and

$$\begin{bmatrix} 0 \qquad\qquad 0\ 0 \\ 0 \qquad\qquad 0\ 0 \\ D_{C_z}(i)F_{C_z}(i)E_{C_z}(i)\ 0\ 0 \end{bmatrix} = \begin{bmatrix} 0 \\ 0 \\ D_{C_z}(i) \end{bmatrix} F_{C_z}(i)\left[E_{C_z}(i)\ 0\ 0 \right].$$

Using now Lemma 1.5.1, we get:

$$\begin{bmatrix} P^\top D_A(i)F_A(i)E_A(i) & 0 & 0 \\ 0 & 0 & 0 \\ 0 & 0 & 0 \end{bmatrix}$$

$$+ \begin{bmatrix} E_A^\top(i)F_A^\top(i)D_A^\top(i)P & 0 & 0 \\ 0 & 0 & 0 \\ 0 & 0 & 0 \end{bmatrix}$$

$$\leq \varepsilon_A(i) \begin{bmatrix} P^\top D_A(i) \\ 0 \\ 0 \end{bmatrix} \begin{bmatrix} D_A^\top(i)P & 0 & 0 \end{bmatrix}$$

$$+ \varepsilon_A^{-1}(i) \begin{bmatrix} E_A^\top(i) \\ 0 \\ 0 \end{bmatrix} \begin{bmatrix} E_A(i) & 0 & 0 \end{bmatrix}$$

$$= \begin{bmatrix} \begin{bmatrix} \varepsilon_A(i)P^\top D_A(i)D_A^\top(i)P \\ +\varepsilon_A^{-1}(i)E_A^\top(i)E_A(i) \end{bmatrix} & 0 & 0 \\ 0 & 0 & 0 \\ 0 & 0 & 0 \end{bmatrix},$$

for any $\varepsilon_A(i) > 0$,

$$\begin{bmatrix} P^\top D_B(i)F_B(i)E_B(i)K & 0 & 0 \\ 0 & 0 & 0 \\ 0 & 0 & 0 \end{bmatrix}$$

$$+ \begin{bmatrix} K^\top E_B^\top(i)F_B^\top(i,t)D_B^\top(i)P & 0 & 0 \\ 0 & 0 & 0 \\ 0 & 0 & 0 \end{bmatrix}$$

$$\leq \varepsilon_B(i) \begin{bmatrix} P^\top D_B(i) \\ 0 \\ 0 \end{bmatrix} \begin{bmatrix} D_B^\top(i)P & 0 & 0 \end{bmatrix}$$

$$+ \varepsilon_B^{-1}(i) \begin{bmatrix} K^\top E_B^\top(i) \\ 0 \\ 0 \end{bmatrix} \begin{bmatrix} E_B(i)K & 0 & 0 \end{bmatrix}$$

$$= \begin{bmatrix} \begin{bmatrix} \varepsilon_B(i)P^\top D_B(i)D_B^\top(i)P \\ +\varepsilon_B^{-1}(i)K^\top E_B^\top(i)E_B(i)K \end{bmatrix} & 0 & 0 \\ 0 & 0 & 0 \\ 0 & 0 & 0 \end{bmatrix},$$

for any $\varepsilon_B(i) > 0$,

$$\begin{bmatrix} 0 & 0 & K^\mathsf{T} E_{D_z}^\mathsf{T}(i) F_{D_z}^\mathsf{T}(i) D_{D_z}^\mathsf{T}(i) \\ 0 & 0 & 0 \\ 0 & 0 & 0 \end{bmatrix}$$

$$+ \begin{bmatrix} 0 & 0 & 0 \\ 0 & 0 & 0 \\ D_{D_z}(i) F_{D_z}(i) E_{D_z}(i) K & 0 & 0 \end{bmatrix}$$

$$\leq \varepsilon_{D_z}^{-1}(i) \begin{bmatrix} K^\mathsf{T} E_{D_z}^\mathsf{T}(i) \\ 0 \\ 0 \end{bmatrix} \begin{bmatrix} E_{D_z}(i) K & 0 & 0 \end{bmatrix}$$

$$+ \varepsilon_{D_z}(i) \begin{bmatrix} 0 \\ 0 \\ D_{D_z}(i) \end{bmatrix} \begin{bmatrix} 0 & 0 & D_{D_z}^\mathsf{T}(i) \end{bmatrix}$$

$$= \begin{bmatrix} \varepsilon_{D_z}^{-1}(i) K^\mathsf{T} E_{D_z}^\mathsf{T}(i) E_{D_z}(i) K & 0 & 0 \\ 0 & 0 & 0 \\ 0 & 0 & \varepsilon_{D_z}(i) D_{D_z}(i) D_{D_z}^\mathsf{T}(i) \end{bmatrix},$$

for any $\varepsilon_{D_z}(i) > 0$, and

$$\begin{bmatrix} 0 & 0 & E_{C_z}^\mathsf{T}(i) F_{C_z}^\mathsf{T}(i) D_{C_z}^\mathsf{T}(i) \\ 0 & 0 & 0 \\ 0 & 0 & 0 \end{bmatrix}$$

$$+ \begin{bmatrix} 0 & 0 & 0 \\ 0 & 0 & 0 \\ D_{C_z}(i) F_{C_z}(i) E_{C_z}(i) & 0 & 0 \end{bmatrix}$$

$$\leq \varepsilon_{C_z}^{-1}(i) \begin{bmatrix} E_{C_z}^\mathsf{T}(i) \\ 0 \\ 0 \end{bmatrix} \begin{bmatrix} E_{C_z}(i) & 0 & 0 \end{bmatrix}$$

$$+ \varepsilon_{C_z}(i) \begin{bmatrix} 0 \\ 0 \\ D_{C_z}(i) \end{bmatrix} \begin{bmatrix} 0 & 0 & D_{C_z}^\mathsf{T}(i) \end{bmatrix}$$

$$= \begin{bmatrix} \varepsilon_{C_z}^{-1}(i) E_{C_z}^\mathsf{T}(i) E_{C_z}(i) & 0 & 0 \\ 0 & 0 & 0 \\ 0 & 0 & \varepsilon_{C_z}(i) D_{C_z}(i) D_{C_z}^\mathsf{T}(i) \end{bmatrix},$$

for any $\varepsilon_{C_z}(i) > 0$.

Taking into account these inequalities we get:

$$
\begin{bmatrix}
J_1(i) & P^\top B_\omega(i) & C_z^\top(i) + K^\top D_z^\top(i) \\
B_\omega^\top(i)P & -\gamma^2 \mathbb{I} & B_z^\top(i) \\
D_z(i)K + C_z(i) & B_z(i) & -\mathbb{I}
\end{bmatrix}
$$
$$
+ \begin{bmatrix}
\begin{bmatrix} \varepsilon_A(i)P^\top D_A(i)D_A^\top(i)P \\ +\varepsilon_A^{-1}(i)E_A^\top(i)E_A(i) \end{bmatrix} & 0 & 0 \\
0 & 0 & 0 \\
0 & 0 & 0
\end{bmatrix}
$$
$$
+ \begin{bmatrix}
\begin{bmatrix} \varepsilon_B(i)P^\top D_B(i)D_B^\top(i)P \\ +\varepsilon_B^{-1}K^\top E_B^\top(i)E_B(i)K \end{bmatrix} & 0 & 0 \\
0 & 0 & 0 \\
0 & 0 & 0
\end{bmatrix}
$$
$$
+ \begin{bmatrix}
\varepsilon_{D_z}^{-1}(i)K^\top E_{D_z}^\top(i)E_{D_z}(i)K & 0 & 0 \\
0 & 0 & 0 \\
0 & 0 & \varepsilon_{D_z}(i)D_{D_z}(i)D_{D_z}^\top(i)
\end{bmatrix}
$$
$$
+ \begin{bmatrix}
\varepsilon_{C_z}^{-1}(i)E_{C_z}^\top(i)E_{C_z}(i) & 0 & 0 \\
0 & 0 & 0 \\
0 & 0 & \varepsilon_{C_z}(i)D_{C_z}(i)D_{C_z}^\top(i)
\end{bmatrix} < 0 ,
$$

with

$$
J_1(i) = A^\top(i)P + P^\top A(i) + K^\top B^\top(i) + P^\top B(i)K + \lambda_{ii}E^\top(i)P
$$
$$
+ \sum_{j=1, j\neq i}^{N} \varepsilon_P(j)\lambda_{ij}\left[P + P^\top\right] .
$$

Let $J_2(i)$, $\mathcal{W}(i)$ and $\mathcal{T}(i)$ be defined as:

$$
J_2(i) = J_1(i) + \varepsilon_A^{-1}(i)E_A^\top(i)E_A(i) + \varepsilon_B^{-1}(i)K^\top E_B^\top(i)E_B(i)K ,
$$
$$
\mathcal{W}(i) = \text{diag}[\varepsilon_A^{-1}(i)\mathbb{I}, \varepsilon_B^{-1}(i)\mathbb{I}, \varepsilon_{C_z}(i)\mathbb{I}, \varepsilon_{D_z}(i)\mathbb{I}] ,
$$
$$
\mathcal{T}(i) = \left[P^\top D_A(i), P^\top D_B(i), K^\top E_{C_z}^\top(i), K^\top E_{D_z}^\top(i)\right] .
$$

and using Schur complement we get the equivalent inequality:

$$
\begin{bmatrix}
J_2(i) & P^\top B_\omega(i) & \begin{bmatrix} C_z^\top(i) \\ +K^\top D_z^\top(i) \end{bmatrix} & \mathcal{T}(i) \\
B_\omega^\top(i)P & -\gamma^2 \mathbb{I} & B_z^\top(i) & 0 \\
\begin{bmatrix} D_z(i)K \\ +C_z(i) \end{bmatrix} & B_z(i) & -\mathcal{U}(i) & 0 \\
\mathcal{T}^\top(i) & 0 & 0 & -\mathcal{W}(i)
\end{bmatrix} < 0 ,
$$

with $\mathcal{U}(i) = \mathbb{I} - \varepsilon_{D_z}(i)D_{D_z}(i)D_{D_z}^\top(i) - \varepsilon_{C_z}(i)D_{C_z}(i)D_{C_z}^\top(i)$.

This matrix inequality is nonlinear in P and K. To put it into an LMI form, let $X = P^{-1}$. Pre- and post-multiply this matrix inequality by $\text{diag}[X^\top, \mathbb{I}, \mathbb{I}, \mathbb{I}]$ and its transpose, we get:

$$
\begin{bmatrix}
J_3(i) & B_\omega(i) & \begin{bmatrix} X^\top C_z^\top(i) \\ +X^\top K^\top D_z^\top(i) \end{bmatrix} & X^\top \mathcal{T}(i) \\
B_\omega^\top(i) & -\gamma^2 \mathbb{I} & B_z^\top(i) & 0 \\
\begin{bmatrix} D_z(i)KX \\ +C_z(i)X \end{bmatrix} & B_z(i) & -\mathcal{U}(i) & 0 \\
\mathcal{T}^\top(i)X & 0 & 0 & -\mathcal{W}(i)
\end{bmatrix} < 0,
$$

with

$$
\begin{aligned}
J_3(i) = {} & X^\top A^\top(i) + A(i)X + X^\top K^\top B^\top(i) + B(i)KX \\
& + \varepsilon_A^{-1}(i)X^\top E_A^\top(i)E_A(i)X + \varepsilon_B^{-1}(i)X^\top K^\top E_B^\top(i)E_B(i)KX \\
& + \lambda_{ii}X^\top E^\top(i) + \sum_{j=1, j\neq i}^{N} \varepsilon_P(j)\lambda_{ij}\left[X + X^\top\right].
\end{aligned}
$$

Notice that:

$$
X^\top \mathcal{T}(i) = \left[D_A(i), D_B(i), X^\top K^\top E_{C_z}^\top(i), X^\top K^\top E_{D_z}^\top(i)\right].
$$

Letting $Y = KX$ and using the Schur complement we obtain:

$$
\begin{bmatrix}
\tilde{J}(i) & B_\omega(i) & \begin{bmatrix} X^\top C_z^\top(i) \\ +Y^\top D_z^\top(i) \end{bmatrix} & \mathcal{R}(i) \\
B_\omega^\top(i) & -\gamma^2 \mathbb{I} & B_z^\top(i) & 0 \\
\begin{bmatrix} D_z(i)Y \\ +C_z(i)X \end{bmatrix} & B_z(i) & -\mathcal{U}(i) & 0 \\
\mathcal{R}^\top(i) & 0 & 0 & -\mathcal{V}(i)
\end{bmatrix} < 0,
$$

with

$$
\begin{aligned}
\tilde{J}(i) = {} & X^\top A^\top(i) + A(i)X + Y^\top B^\top(i) + B(i)Y \\
& + \varepsilon_A(i)D_A(i)D_A^\top(i) + \varepsilon_B(i)D_B(i)D_B^\top(i) \\
& + \lambda_{ii}X^\top E^\top(i) + \sum_{j=1, j\neq i}^{N} \varepsilon_P(j)\lambda_{ij}\left[X + X^\top\right],
\end{aligned}
$$

$$
\mathcal{R}(i) = \left(X^\top E_A^\top(i), Y^\top E_B^\top(i), X^\top E_{C_z}^\top(i), Y^\top E_{D_z}^\top(i)\right),
$$

$$
\mathcal{V}(i) = \text{diag}[\varepsilon_A(i)\mathbb{I}, \varepsilon_B(i)\mathbb{I}, \varepsilon_{C_z}(i)\mathbb{I}, \varepsilon_{D_z}(i)\mathbb{I}].
$$

The following theorem summarizes the results of this development.

Theorem 4.3.6 *Let γ be a positive constant and $\varepsilon_P = (\varepsilon_P(1), \cdots, \varepsilon_P(N))$ be a set of positive scalars. There exists a state feedback controller with constant gain that* *assures that the closed-loop system is piecewise regular, impulse-free and stochastically stable and moreover the closed-loop system satisfies the disturbance rejection of level γ, if there exist a nonsingular matrix $X \in \mathbb{R}^{n \times n}$ and a matrix $Y \in \mathbb{R}^{m \times n}$ and sets of positive scalars $\varepsilon_A = (\varepsilon_A(1), \cdots, \varepsilon_A(N))$, $\varepsilon_B = (\varepsilon_B(1), \cdots, \varepsilon_B(N))$, $\varepsilon_{C_z} = (\varepsilon_{C_z}(1), \cdots, \varepsilon_{C_z}(N))$, and $\varepsilon_{D_z} = (\varepsilon_{D_z}(1), \cdots, \varepsilon_{D_z}(N))$, such that the following set of LMIs holds for every $i \in \mathscr{S}$ and for all admissible uncertainties:*

$$
\begin{bmatrix}
\tilde{J}(i) & B_\omega(i) & \begin{bmatrix} X^\top C_z^\top(i) \\ +Y^\top D_z^\top(i) \end{bmatrix} & \mathcal{R}(i) \\
B_\omega^\top(i) & -\gamma^2 \mathbb{I} & B_z^\top(i) & 0 \\
\begin{bmatrix} D_z(i)Y \\ +C_z(i)X \end{bmatrix} & B_z(i) & -\mathcal{U}(i) & 0 \\
\mathcal{R}^\top(i) & 0 & 0 & -\mathcal{V}(i)
\end{bmatrix} < 0, \tag{4.67}
$$

where

$$
\begin{aligned}
\tilde{J}(i) = {}& X^\top A^\top(i) + A(i)X + Y^\top B^\top(i) + B(i)Y \\
& + \varepsilon_A(i) n D_A(i) D_A^\top(i) + \varepsilon_B(i) D_B(i) D_B^\top(i) \\
& + \lambda_{ii} X^\top E^\top(i) + \sum_{j=1, j \neq i}^{N} \varepsilon_P(j) \lambda_{ij} \left[X + X^\top \right],
\end{aligned}
$$

$$
\mathcal{U}(i) = \mathbb{I} - \varepsilon_{D_z}(i) D_{D_z}(i) D_{D_z}^\top(i) - \varepsilon_{C_z}(i) D_{C_z}(i) D_{C_z}^\top(i),
$$

$$
\mathcal{R}(i) = \left[X^\top E_A^\top(i), Y^\top (E_B^\top(i), X^\top E_{C_z}^\top(i), Y^\top E_{D_z}^\top(i) \right],
$$

$$
\mathcal{V}(i) = \mathrm{diag}[\varepsilon_A(i)\mathbb{I}, \varepsilon_B(i)\mathbb{I}, \varepsilon_{C_z}(i)\mathbb{I}, \varepsilon_{D_z}(i)\mathbb{I}],
$$

with the following constraints:

$$
\varepsilon_P(i) \left[X + X^\top \right] \geq X^\top E^\top(i) = E(i)X \geq 0. \tag{4.68}
$$

The state feedback controller gain is given by $K = YX^{-1}$.

As it was done for the nominal system, we can determine the controller, that assures that closed-loop system is piecewise regular, impulse-free and stochastically stable and at the same time guarantees the minimum disturbance rejection, by solving the following optimization problem for a given set of positive scalars $\varepsilon_P = (\varepsilon_P(1), \cdots, \varepsilon_P(N))$:

$$
\text{Pu}: \begin{cases}
\min\limits_{\substack{v > 0, \\ \varepsilon_A = (\varepsilon_A(1), \cdots, \varepsilon_A(N)) > 0, \\ \varepsilon_B = (\varepsilon_B(1), \cdots, \varepsilon_B(N)) > 0, \\ \varepsilon_{C_z} = (\varepsilon_{D_z}(1), \cdots, \varepsilon_{C_z}(N)) > 0, \\ \varepsilon_{D_z} = (\varepsilon_{D_z}(1), \cdots, \varepsilon_{D_z}(N)) > 0, \\ X, Y}} v \\[4pt]
\text{s.t}: \\
\varepsilon_P(i)[X + X^\top] \geq X^\top E^\top(i) = E(i)X \geq 0, \\[4pt]
\begin{bmatrix}
\tilde{J}(i) & B_\omega(i) & \begin{bmatrix} X^\top C_z^\top(i) \\ +Y^\top D_z^\top(i) \end{bmatrix} & \mathcal{R}(i) \\
B_\omega^\top(i) & -v\mathbb{I} & B_z^\top(i) & 0 \\
\begin{bmatrix} D_z(i)Y \\ +C_z(i)X \end{bmatrix} & B_z(i) & -\mathcal{U}(i) & 0 \\
\mathcal{R}^\top(i) & 0 & 0 & -\mathcal{V}(i)
\end{bmatrix} < 0,
\end{cases}
$$

where the LMI that represents the constraint of this optimization problem is obtained from (4.67) by replacing γ^2 by v.

The following theorem summarizes the results on the design of the controller that assures that the closed-loop system is piecewise regular, impulse free and stochastically stable and simultaneously guarantees the smallest disturbance rejection level.

Theorem 4.3.7 *Let* $v > 0$, $\varepsilon_A = (\varepsilon_A(1), \cdots, \varepsilon_A(N)) > 0$, $\varepsilon_B = (\varepsilon_B(1), \cdots, \varepsilon_B(N)) > 0$, $\varepsilon_{C_z} = (\varepsilon_{C_z}(1), \cdots, \varepsilon_{C_z}(N)) > 0$, $\varepsilon_{D_z} = (\varepsilon_{D_z}(1), \cdots, \varepsilon_{D_z}(N)) > 0$, X, *and* Y *be the solution of the optimization problem Pu for a given set of positive scalars* $\varepsilon_P = (\varepsilon_P(1), \cdots, \varepsilon_P(N))$. *Then, the controller (4.39) with* $K = YX^{-1}$ *will guarantee that closed-loop system is piecewise regular, impulse-free and stochastically stable and moreover the closed-loop system satisfies the disturbance rejection of level* $\sqrt{v}$.

For this case, we can also extend the results for the other approaches. The following theorem summarizes the results when $E^\top(r(t))P$ is bounded by $\varepsilon(i)[P^\top P]$, for $\varepsilon(i) > 0$.

Theorem 4.3.8 *Let* γ *be a positive constant. There exists a state feedback controller with constant gain that assures that the closed-loop system is piecewise regular, impulse-free and stochastically stable and moreover the closed-loop system satisfies the disturbance rejection of level* γ, *if there exist a nonsingular matrix* $X \in \mathbb{R}^{n \times n}$ *and a matrix* $Y \in \mathbb{R}^{m \times n}$ *and sets of positive scalars* $\varepsilon_A = (\varepsilon_A(1), \cdots, \varepsilon_A(N))$, $\varepsilon_B = (\varepsilon_B(1), \cdots, \varepsilon_B(N))$, $\varepsilon_{C_z} = (\varepsilon_{C_z}(1), \cdots, \varepsilon_{C_z}(N))$, $\varepsilon_{D_z} = (\varepsilon_{D_z}(1), \cdots, \varepsilon_{D_z}(N))$, *and* $\varepsilon = (\varepsilon(1), \cdots, \varepsilon(N))$ *such that the following set of LMIs holds for every* $i \in \mathscr{S}$ *and for all admissible uncertainties:*

$$
\begin{bmatrix}
\tilde{J}(i) & B_\omega(i) & \begin{matrix} X^\top C_z^\top(i) \\ +Y^\top D_z^\top(i) \end{matrix} & \mathcal{R}(i) \\
B_\omega^\top(i) & -\gamma^2 \mathbb{I} & B_z^\top(i) & 0 \\
\begin{matrix} D_z(i)Y \\ +C_z(i)X \end{matrix} & B_z(i) & -\mathcal{U}(i) & 0 \\
\mathcal{R}^\top(i) & 0 & 0 & -\mathcal{V}(i)
\end{bmatrix} < 0,
\tag{4.69}
$$

where

$$
\begin{aligned}
\tilde{J}(i) ={}& X^\top A^\top(i) + A(i)X + Y^\top B^\top(i) + B(i)Y \\
&+ \varepsilon_A(i)D_A(i)D_A^\top(i) + \varepsilon_B(i)D_B(i)D_B^\top(i) \\
&+ \lambda_{ii}X^\top E^\top(i) + \sum_{j=1, j\neq i}^{N} \varepsilon(j)\lambda_{ij}\mathbb{I},
\end{aligned}
$$

$$
\mathcal{U}(i) = \mathbb{I} - \varepsilon_{D_z}(i)D_{D_z}(i)D_{D_z}^\top(i) - \varepsilon_{C_z}(i)D_{C_z}(i)D_{C_z}^\top(i),
$$

$$
\mathcal{R}(i) = \left[X^\top E_A^\top(i), Y^\top E_B^\top(i), X^\top E_{C_z}^\top(i), Y^\top E_{D_z}^\top(i)\right],
$$

$$
\mathcal{V}(i) = \mathrm{diag}[\varepsilon_A(i)\mathbb{I}, \varepsilon_B(i)\mathbb{I}, \varepsilon_{C_z}(i)\mathbb{I}, \varepsilon_{D_z}(i)\mathbb{I}],
$$

with the following constraints:

$$
\varepsilon(i)\mathbb{I} \geq X^\top E^\top(i) = E(i)X \geq 0.
\tag{4.70}
$$

The state feedback controller gain is given by $K = YX^{-1}$.

The following theorem summarizes the results when $E^\top(r(t))P$ is bounded by $\left[\frac{1}{4}\varepsilon^{-1}(i)\mathbb{I} + \varepsilon(i)E^\top(i)PP^\top E(i)\right]$, for $\varepsilon(i) > 0$.

Theorem 4.3.9 *Let γ be a positive constant. There exists a state feedback controller with constant gain that assures that the closed-loop system is piecewise regular, impulse-free and stochastically stable and moreover the closed-loop system satisfies the disturbance rejection of level γ, if there exist a nonsingular matrix $X \in \mathbb{R}^{n\times n}$ and a matrix $Y \in \mathbb{R}^{m\times n}$ and sets of positive scalars $\varepsilon_A = (\varepsilon_A(1), \cdots, \varepsilon_A(N))$, $\varepsilon_B = (\varepsilon_B(1), \cdots, \varepsilon_B(N))$, $\varepsilon_{C_z} = (\varepsilon_{C_z}(1), \cdots, \varepsilon_{C_z}(N))$, $\varepsilon_{D_z} = (\varepsilon_{D_z}(1), \cdots, \varepsilon_{D_z}(N))$, and $\varepsilon = (\varepsilon(1), \cdots, \varepsilon(N))$ such that the following set of LMIs holds for every $i \in \mathscr{S}$ and for all admissible uncertainties:*

$$\begin{bmatrix} \tilde{J}(i) & B_\omega(i) & \begin{matrix} X^\top C_z^\top(i) \\ +Y^\top D_z^\top(i) \end{matrix} & \mathcal{R}(i) & \mathcal{Z}_i(X) & \mathcal{S}_i(X) \\ B_\omega^\top(i) & -\gamma^2 \mathbb{I} & B_z^\top(i) & 0 & 0 & 0 \\ \begin{matrix} D_z(i)Y \\ +C_z(i)X \end{matrix} & B_z(i) & -\mathcal{U}(i) & 0 & 0 & 0 \\ \mathcal{R}^\top(i) & 0 & 0 & -\mathcal{V}(i) & 0 & 0 \\ \mathcal{Z}_i^\top(X) & 0 & 0 & 0 & -\mathcal{X}_i(\varepsilon) & 0 \\ \mathcal{S}_i^\top(X) & 0 & 0 & 0 & 0 & -\mathcal{X}_i(X) \end{bmatrix} < 0, \qquad (4.71)$$

where

$$\begin{aligned} \tilde{J}(i) = {}& X^\top A^\top(i) + A(i)X + Y^\top B^\top(i) + B(i)Y \\ & + \varepsilon_A(i)D_A(i)D_A^\top(i) + \varepsilon_B(i)D_B(i)D_B^\top(i) \\ & + \lambda_{ii}X^\top E^\top(i), \\ \mathcal{U}(i) = {}& \mathbb{I} - \varepsilon_{D_z}(i)D_{D_z}(i)D_{D_z}^\top(i) - \varepsilon_{C_z}(i)D_{C_z}(i)D_{C_z}^\top(i), \\ \mathcal{R}(i) = {}& \left[X^\top E_A^\top(i), Y^\top E_B^\top(i), Y^\top E_{C_z}^\top(i), Y^\top E_{D_z}^\top(i)\right], \\ \mathcal{V}(i) = {}& \mathrm{diag}[\varepsilon_A(i)\mathbb{I}, \varepsilon_B(i)\mathbb{I}, \varepsilon_{C_z}(i)\mathbb{I}, \varepsilon_{D_z}(i)\mathbb{I}], \\ \mathcal{Z}_i(X) = {}& \left(\sqrt{\lambda_{i1}}X^\top, \cdots, \sqrt{\lambda_{ii-1}}X^\top, \sqrt{\lambda_{ii+1}}X^\top, \cdots, \sqrt{\lambda_{iN}}X^\top\right), \\ \mathcal{S}_i(X) = {}& \left(\sqrt{\lambda_{i1}}X^\top E^\top(1), \cdots, \sqrt{\lambda_{ii-1}}X^\top E^\top(i-1),\right. \\ & \left.\sqrt{\lambda_{ii+1}}X^\top E^\top(i+1), \cdots, \sqrt{\lambda_{iN}}X^\top E^\top(N)\right), \\ \mathcal{X}_i(\varepsilon) = {}& \mathrm{diag}[4\varepsilon(1)\mathbb{I}, \cdots, 4\varepsilon(i-1)\mathbb{I}, 4\varepsilon(i+1)\mathbb{I}, \cdots, 4\varepsilon(N)\mathbb{I}], \\ \mathcal{X}_i(X) = {}& \mathrm{diag}\left[X^\top + X - \varepsilon(1)\mathbb{I}, \cdots, X^\top + X - \varepsilon(i-1)\mathbb{I},\right. \\ & \left. X^\top + X - \varepsilon(i+1)\mathbb{I}, \cdots, X^\top + X - \varepsilon(N)\mathbb{I}\right], \end{aligned}$$

with the following constraints:

$$\varepsilon(i)\mathbb{I} \geq X^\top E^\top(i) = E(i)X \geq 0. \qquad (4.72)$$

The state feedback controller gain is given by $K = YX^{-1}$.

4.4 Numerical Examples

In this section, we will present some numerical examples to show the validness of the developed results in this chapter.

Example 4.4.1 *To show the usefulness of the design theorems for nominal systems, let us consider a two modes system with the following data:*

- *mode # 1:*

$$A(1) = \begin{bmatrix} 1.0 & 0.0 & 1.0 \\ 0.0 & 0.0 & 1.0 \\ 0.0 & -1.0 & -1 \end{bmatrix}, \; B(1) = \begin{bmatrix} 0.0 & 0.2 \\ 1.0 & 0.0 \\ -0.1 & 1.0 \end{bmatrix},$$

$$B_w(1) = \begin{bmatrix} 0.01 \\ 0.0 \\ 0.01 \end{bmatrix}, \; B_z(1) = \begin{bmatrix} 0.0 \end{bmatrix},$$

$$C_z(1) = \begin{bmatrix} 1.0 & 1.0 & 0.0 \end{bmatrix}, \; D_z(1) = \begin{bmatrix} 1.0 & 0.0 \end{bmatrix}.$$

- *mode # 2:*

$$A(2) = \begin{bmatrix} 1.0 & 0.0 & 1.0 \\ 0.0 & 0.0 & 1.0 \\ 0.0 & 1.0 & 1.0 \end{bmatrix}, \; B(2) = \begin{bmatrix} 0.0 & 0.2 \\ 1.2 & 0.0 \\ -0.1 & 1.2 \end{bmatrix},$$

$$B_w(2) = \begin{bmatrix} -0.01 \\ 0.0 \\ 0.01 \end{bmatrix}, \; B_z(2) = \begin{bmatrix} 0.0 \end{bmatrix},$$

$$C_z(2) = \begin{bmatrix} 1.0 & 1.0 & 0.0 \end{bmatrix}, \; D_z(2) = \begin{bmatrix} 1.0 & 0.0 \end{bmatrix}.$$

The singular matrix E is given by:

$$E = \begin{bmatrix} 1 & 0 & 0 \\ 0 & 1 & 0 \\ 0 & 0 & 0 \end{bmatrix},$$

The switching between the two modes is described by:

$$\Lambda = \begin{bmatrix} -1 & 1 \\ 1.1 & -1.1 \end{bmatrix}.$$

Letting $\gamma = 0.1$ and solving the set of coupled LMIs of Theorem 4.2.12 gives:

$$X(1) = \begin{bmatrix} 0.3580 & -0.0159 & 0.0 \\ -0.0159 & 0.3363 & 0.0 \\ -0.2285 & 0.1218 & 0.7955 \end{bmatrix},$$

$$X(2) = \begin{bmatrix} 0.1847 & -0.0074 & 0.0 \\ -0.0074 & 0.2207 & 0.0 \\ -0.1845 & 0.1193 & 0.7902 \end{bmatrix},$$

$$Y(1) = \begin{bmatrix} -0.0190 & -0.5247 & -0.0543 \\ -1.2846 & -0.2454 & 0.1673 \end{bmatrix},$$

$$Y(2) = \begin{bmatrix} 0.1109 & -0.3247 & -0.0571 \\ -0.3063 & -0.9225 & -1.1982 \end{bmatrix},$$

which gives the following gains:

$$K(1) = \begin{bmatrix} -0.1651 & -1.5433 & -0.0683 \\ -3.4967 & -0.9709 & 0.2103 \end{bmatrix},$$

$$K(2) = \begin{bmatrix} 0.4715 & -1.4162 & -0.0723 \\ -3.3119 & -3.4714 & -1.5163 \end{bmatrix}.$$

Letting again $\gamma = 0.1$ *and solving the set of coupled LMIs of Theorem 4.2.13 gives:*

$$X(1) = \begin{bmatrix} 0.3110 & -0.0115 & 0.0 \\ -0.0115 & 0.4254 & 0.0 \\ -0.0485 & 0.0428 & 0.3563 \end{bmatrix},$$

$$X(2) = \begin{bmatrix} 0.1576 & 0.0071 & 0.0 \\ 0.0071 & 0.4156 & 0.0 \\ -0.1853 & 0.1768 & 0.6356 \end{bmatrix},$$

$$Y(1) = \begin{bmatrix} -0.0792 & -0.4386 & -0.0465 \\ -0.4280 & 0.1158 & -0.0755 \end{bmatrix},$$

$$Y(2) = \begin{bmatrix} 0.0860 & -0.4485 & -0.0380 \\ -0.1841 & -1.0932 & -0.9581 \end{bmatrix},$$

which gives the following gains:

$$K(1) = \begin{bmatrix} -0.3127 & -1.0263 & -0.1304 \\ -1.3997 & 0.2558 & -0.2120 \end{bmatrix},$$

$$K(2) = \begin{bmatrix} 0.5233 & -1.0625 & -0.0598 \\ -2.8525 & -1.9405 & -1.5073 \end{bmatrix}.$$

Example 4.4.2 *To show the usefulness of the design theorems for uncertain systems, let us consider the previous example with the following extra data:*

- *mode # 1:*

$$D_A(1) = \begin{bmatrix} 0.1 \\ 0.0 \\ 0.0 \end{bmatrix}, \quad E_A(1) = \begin{bmatrix} 0.1 & 0.0 & 0.0 \end{bmatrix},$$

$$D_B(1) = \begin{bmatrix} 0.1 \\ 0.1 \\ 0.3 \end{bmatrix}, \quad E_B(1) = \begin{bmatrix} 0.3 & 0.1 & 0.1 \end{bmatrix},$$

$$D_{C_z}(1) = \begin{bmatrix} 0.1 \\ 0.2 \\ 0.1 \end{bmatrix}, \quad E_{C_z}(1) = \begin{bmatrix} 0.1 & 0.1 & 0.1 \end{bmatrix},$$

$$D_{D_z}(1) = \begin{bmatrix} 0.1 \\ 0.1 \\ 0.1 \end{bmatrix}, \quad E_{D_z}(1) = \begin{bmatrix} 0.1 & 0.2 & 0.1 \end{bmatrix}.$$

- *mode # 2:*

$$D_A(2) = \begin{bmatrix} 0.3 \\ 0.2 \\ 0 \end{bmatrix}, \ E_A(2) = \begin{bmatrix} 0.1 \ 0.2 \ 0.3 \end{bmatrix},$$

$$D_B(2) = \begin{bmatrix} 0.2 \\ 0.2 \\ 0.1 \end{bmatrix}, \ E_B(2) = \begin{bmatrix} 0.1 \ 0.2 \ 0.1 \end{bmatrix},$$

$$D_{C_z}(2) = \begin{bmatrix} 0.1 \\ 0.2 \\ 0.1 \end{bmatrix}, \ E_{C_z}(2) = \begin{bmatrix} 0.2 \ 0.2 \ 0.1 \end{bmatrix},$$

$$D_{D_z}(2) = \begin{bmatrix} 0.1 \\ 0.1 \\ 0.1 \end{bmatrix}, \ E_{D_z}(2) = \begin{bmatrix} 0.2 \ 0.1 \ 0.1 \end{bmatrix}.$$

Letting $\gamma = 0.1$ and solving the set of coupled LMIs of Theorem 4.2.16 gives:

$$\varepsilon(1) = 0.4317, \ \varepsilon(2) = 0.2447,$$
$$\varepsilon_A(1) = 0.5479, \ \varepsilon_A(2) = 0.5325,$$
$$\varepsilon_B(1) = 0.5470, \ \varepsilon_B(2) = 0.5430,$$
$$\varepsilon_{C_z}(1) = 0.5439, \ \varepsilon_{C_z}(2) = 0.5483,$$
$$\varepsilon_{D_z}(1) = 0.5459, \ \varepsilon_{D_z}(2) = 0.5496,$$

$$X(1) = \begin{bmatrix} 0.3514 & -0.0158 & 0.0 \\ -0.0158 & 0.3301 & 0.0 \\ -0.2272 & 0.1205 & 0.7891 \end{bmatrix},$$

$$X(2) = \begin{bmatrix} 0.1809 & -0.0073 & 0.0 \\ -0.0073 & 0.2164 & 0.0 \\ -0.1828 & 0.1177 & 0.7838 \end{bmatrix},$$

$$Y(1) = \begin{bmatrix} -0.0177 & -0.5231 & -0.0505 \\ -1.2775 & -0.2578 & 0.1654 \end{bmatrix},$$

$$Y(2) = \begin{bmatrix} 0.1079 & -0.3243 & -0.0551 \\ -0.3047 & -0.9194 & -1.1888 \end{bmatrix},$$

which gives the following gains:

$$K(1) = \begin{bmatrix} -0.1621 & -1.5692 & -0.0640 \\ -3.5462 & -1.0269 & 0.2096 \end{bmatrix},$$

$$K(2) = \begin{bmatrix} 0.4675 & -1.4445 & -0.0703 \\ -3.3602 & -3.5365 & -1.5168 \end{bmatrix}.$$

Letting again $\gamma = 0.1$ and solving the set of coupled LMIs of Theorem 4.2.17 gives:

$$\varepsilon(1) = 0.3687, \ \varepsilon(2) = 0.0973,$$
$$\varepsilon_A(1) = 0.6688, \ \varepsilon_A(2) = 0.6499,$$

$$\varepsilon_B(1) = 0.6689, \quad \varepsilon_B(2) = 0.6677,$$
$$\varepsilon_{C_z}(1) = 0.6694, \quad \varepsilon_{C_z}(2) = 0.6686,$$
$$\varepsilon_{D_z}(1) = 0.6701, \quad \varepsilon_{D_z}(2) = 0.6718,$$

$$X(1) = \begin{bmatrix} 0.2147 & 0.0068 & 0.0 \\ 0.0068 & 0.3121 & 0.0 \\ -0.0948 & 0.0373 & 0.0205 \end{bmatrix},$$

$$X(2) = \begin{bmatrix} 0.1459 & 0.0071 & 0.0 \\ 0.0071 & 0.4553 & 0.0 \\ -0.1956 & 0.1930 & 0.6854 \end{bmatrix},$$

$$Y(1) = \begin{bmatrix} -0.0815 & -0.3797 & -0.0278 \\ -0.0865 & 0.3269 & -0.3169 \end{bmatrix},$$

$$Y(2) = \begin{bmatrix} 0.0915 & -0.4531 & -0.0385 \\ -0.2275 & -1.1698 & -0.9903 \end{bmatrix},$$

which gives the following gains:

$$K(1) = \begin{bmatrix} -0.9447 & -1.0346 & -1.3531 \\ -7.3251 & 3.0530 & -15.4523 \end{bmatrix},$$

$$K(2) = \begin{bmatrix} 0.5995 & -0.9808 & -0.0562 \\ -3.4041 & -1.9042 & -1.4447 \end{bmatrix}.$$

4.5 Notes

This chapter dealt with the $\mathscr{H}_\infty$ stabilization problem of the singular class of systems with random abrupt changes. The stochastic $\mathscr{H}_\infty$ stabilizability and the robust stochastic $\mathscr{H}_\infty$ stabilizability problems have been considered and LMI conditions were developed. A state feedback controller that assures that the closed-loop state equation either for the nominal system or the uncertain is piecewise regular, impulse-free and stochastically stable and at the same time guarantees the disturbance rejection with a certain given level $\gamma > 0$ is designed in the LMI setting. The conditions we developed in this chapter are tractable using commercial optimization tools. The content of this chapter is mainly based on the work of the author and his coauthors [2, 21, 22, 18, 23, 20].

5

Output Feedback Stabilization

In the previous, chapters we have always assumed that we have the complete access to the state vector. But this assumption may be restrictive in some applications for reasons like the non availability of the technology to measure some state variables or limitations in the budget of the control design. An alternative consists of using the output feedback stabilization that can be static or dynamic. This problem has attracted a lot of researchers and interesting results have been reported in the literature. For more details on the subject we refer the reader to Boukas and Liu [27], [9, 36, 38, 75, 80, 86, 82, 83, 84, 131, 110, 119, 120, 115, 113, 130, 27] and the references therein.

For the class of singular systems, few results have been reported in the literature. Among these references we quote [15] and the references therein.

The goal of this chapter is to present the output stabilization using a static output feedback controller of the class of linear singular systems with random abrupt changes in the dynamics. The stabilization of the nominal and the uncertain systems are treated and LMI results are developed. Numerical examples are also provided to show the effectiveness of the developed results.

The chapter is organized as follows. In Sect. 5.1, the output feedback control for the class of singular systems with random abrupt changes is stated. Section 5.2 gives results on the design of static output feedback controller for nominal and uncertain systems. Conditions in the LMI setting are developed to synthesize static output feedback controllers. In Sect. 5.3, the $\mathscr{H}_\infty$ static output feedback control is considered and LMI conditions are developed.

5.1 Problem Statement

Let us consider a dynamical singular system defined in a fundamental probability space $(\Omega, \mathcal{F}, \mathbb{P})$ and assume that its state equation is described by the following differential-algebraic equations:

$$\begin{cases} E(r_t)\dot{x}(t) = A(r_t, t)x(t) + B(r_t, t)u(t), x(0) = x_0, \\ y(t) = C(r_t)x(t), \end{cases} \tag{5.1}$$

where $x(t) \in \mathbb{R}^n$ is the state vector, $x_0 \in \mathbb{R}^n$ is the initial state, $u(t) \in \mathbb{R}^m$ is the control input, $y(t) \in \mathbb{R}^p$ is the output of the system at time t, $\{r_t, t \geq 0\}$ is the continuous-time Markov process taking values in a finite space $\mathscr{S} = \{1, 2, \cdots, N\}$ and describes the evolution of the mode at time t, $E(i)$ is a known singular matrix with rank $(E(i)) = n_r \leq n$, for all $i \in \mathscr{S}$, $A(r_t, t) \in \mathbb{R}^{n \times n}$, $B(r_t, t) \in \mathbb{R}^{n \times m}$ and $C(r_t, t) \in \mathbb{R}^{p \times n}$ are matrices with the following forms for every $i \in \mathscr{S}$:

$$A(i, t) = A(i) + D_A(i)F_A(i)E_A(i),$$
$$B(i, t) = B(i) + D_B(i)F_B(i)E_B(i),$$

with $A(i) \in \mathbb{R}^{n \times n}$, $D_A(i) \in \mathbb{R}^{n \times n_D}$, $E_A(i) \in \mathbb{R}^{n_E \times n}$, $B(i) \in \mathbb{R}^{n \times m}$, $D_B(i) \in \mathbb{R}^{n \times m_D}$, $E_B(i) \in \mathbb{R}^{m_E \times m}$, and $C(i) \in \mathbb{R}^{p \times n}$ are real known matrices with appropriate dimensions, and $F_A(i) \in \mathbb{R}^{n_D \times n_E}$, $F_B(i) \in \mathbb{R}^{m_D \times m_E}$ and $F_C(i) \in \mathbb{R}^{p_D \times p_E}$ are unknown real matrices that satisfy the following:

$$\begin{cases} F_A^\top(i)F_A(i) \leq \mathbb{I}, \\ F_B^\top(i)F_B(i) \leq \mathbb{I}. \end{cases} \tag{5.2}$$

The Markov process $\{r_t, t \geq 0\}$ beside taking values in the finite set $\mathscr{S}$, represents the switching between the different modes and its state equation is described by the following probability transitions:

$$\mathbb{P}\left[r_{t+h} = j | r_t = i\right] = \begin{cases} \lambda_{ij}h + o(h), & \text{when } r_t \text{ jumps from } i \text{ to } j, \\ 1 + \lambda_{ii}h + o(h), & \text{otherwise}, \end{cases} \tag{5.3}$$

where λ_{ij} is the transition rate from mode i to mode j with $\lambda_{ij} \geq 0$ when $i \neq j$ and $\lambda_{ii} = -\sum_{j=1, j \neq i}^N \lambda_{ij}$ and $o(h)$ is such that $\lim_{h \to 0} \frac{o(h)}{h} = 0$.

As it was done previously, we will assume here when it is necessary that the transition matrix, Λ, belongs to a polytope, i. e.,

$$\Lambda = \sum_{k=1}^{\kappa} \alpha_k \Lambda_k, \tag{5.4}$$

with κ is a positive given integer, $0 \leq \alpha_k \leq 1$, with $\sum_{k=1}^{\kappa} \alpha_k = 1$ and Λ_k is a known transition matrix and its expression is given by:

$$\Lambda_k = \begin{bmatrix} \lambda_{11}^k & \cdots & \lambda_{1N}^k \\ \vdots & \ddots & \vdots \\ \lambda_{N1}^k & \cdots & \lambda_{NN}^k \end{bmatrix}, \tag{5.5}$$

where λ_{ij}^k keeps the same meaning as previous.

Remark 5.1.1 *The uncertainties satisfying the condition (5.2), (5.4) are referred to as admissible. The uncertainty term, in (5.2), is supposed to depend on the system's mode, r_t.*

Remark 5.1.2 *The matrix $E(i) = n_r \leq n$, for all $i \in \mathcal{S}$, is supposed to be singular which makes the state equation (5.1) different from the one usually used to describe the behavior of the time-invariant dynamical systems as it is the normal practice.*

Remark 5.1.3 *Notice that when $E(i) = n_r \leq n$, for all $i \in \mathcal{S}$, is not singular, (5.1) can be transformed easily to the class of Markov jump linear systems and the results developed in the literature (see Mariton [95], Boukas and Liu [27], Boukas [14] and the references therein), can be used to check the stochastic stability, of this class of systems and to design the appropriate controller.*

The goal of this chapter is to develop LMI conditions to design a static output feedback controller that renders the closed-loop dynamics piecewise regular, impulse-free and stochastically stable and also in case of the presence of external disturbance guarantees the disturbance rejection of a desired level.

5.2 Static Output Feedback

All the results we developed in the previous chapters assumed the complete access to the state vector. Let us now drop this assumption and see how we can design a static output feedback controller that guarantees that the closed-loop system is piecewise regular, impulse-free and stochastically stable. The controller we will use has the following form:

$$u(t) = F(r_t)y(t) = F(r_t)C(r_t)x(t), \tag{5.6}$$

where $F(i) \in \mathbb{R}^{m \times p}$, $\forall i \in \mathcal{S}$ is a design parameter that we have to determine.

Plugging the controller (5.6) in the nominal system state equation (5.1) gives:

$$E(r_t)\dot{x}(t) = [A(r_t) + B(r_t)F(r_t)C(r_t)]\, x(t)$$
$$= A_{cl}(r_t)x(r_t),$$

with $A_{cl}(i) = A(i) + B(i)F(i)C(i)$.

Based on the results of the Chap. 2, the closed-loop system is piecewise regular, impulse-free and stochastically stable if there exists a set of nonsingular matrices, $P = (P(1), \cdots, P(N))$, $P(i) \in \mathbb{R}^{n \times n}$ such that the following hold for a given positive scalar ε_P:

$$\begin{cases} \varepsilon_P \left[P(i) + P^\top(i) \right] \geq E^\top(i)P(i) = P^\top(i)E(i) \geq 0, \\ P^\top(i)A_{cl}(i) + A_{cl}^\top(i)P(i) + \lambda_{ii}E^\top(i)P(i) \\ \quad + \sum_{j=1, j \neq i}^{N} \varepsilon_P \lambda_{ij} \left[P(j) + P^\top(j) \right] < 0, \end{cases}$$

which gives for the second matrix inequality:

$$P^\top(i)A(i) + A^\top(i)P(i) + P^\top(i)B(i)F(i)C(i) + \left[P^\top(i)B(i)F(i)C(i) \right]^\top$$
$$+ \lambda_{ii}E^\top(i)P(i) + \sum_{j=1, j \neq i}^{N} \varepsilon_P \lambda_{ij} \left[P(j) + P^\top(j) \right] < 0.$$

This inequality matrix is nonlinear in the design parameters $P(i)$ and $F(i)$. To put it into the LMI form let $X(i) = P^{-1}(i)$. Pre- and post-multiply this inequality respectively by $X^{\top}(i)$ and $X(i)$ give:

$$A(i)X(i) + X^{\top}(i)A^{\top}(i) + B(i)F(i)C(i)X(i) + X^{\top}(i)C^{\top}(i)F^{\top}(i)B^{\top}(i)$$

$$+ \lambda_{ii}X^{\top}(i)E^{\top}(i) + \sum_{j=1, j\neq i}^{N} \varepsilon_P \lambda_{ij} X^{\top}(i) \left[X^{-1}(j) + X^{-\top}(j) \right] X(i) < 0.$$

Similarly, the first condition can be transformed to:

$$\varepsilon_P \left[X(i) + X^{\top}(i) \right] \geq X^{\top}(i)E^{\top} = EX(i) \geq 0.$$

Using the fact that:

$$X^{-1}(j) + X^{-\top}(j) \leq \mathbb{I} + X^{-1}(j)X^{-\top}(j)$$

$$= \mathbb{I} + \left[X^{\top}(j)X(j) \right]^{-1}.$$

Letting

$$Z_i(X) = \operatorname{diag}\left[X^{\top}(1)X(1), \cdots, X^{\top}(i-1)X(i-1), \right.$$

$$\left. X^{\top}(i+1)X(i+1), \cdots, X^{\top}(N)X(N) \right],$$

we have:

$$\sum_{j=1, j\neq i}^{N} \varepsilon_P \lambda_{ij} X^{\top}(i) \left[X^{-1}(j) + X^{-\top}(j) \right] X(i) \leq S_i(X)S_i^{\top}(X)$$

$$+ S_i(X)Z_i^{-1}(X)S_i^{\top}(X),$$

where $S_i(X)$ is defined as follows:

$$S_i(X) = \left[\sqrt{\varepsilon_P \lambda_{i1}} X^{\top}(i), \cdots, \sqrt{\varepsilon_P \lambda_{ii-1}} X^{\top}(i), \right.$$

$$\left. \sqrt{\varepsilon_P \lambda_{ii+1}} X^{\top}(i), \cdots, \sqrt{\varepsilon_P \lambda_{iN}} X^{\top}(i) \right].$$

Now if we let $F(i) = G(i)Y^{-1}(i)$ and $Y(i)C(i) = C(i)X(i)$ hold for every $i \in \mathscr{S}$ for some appropriate matrices that we have to determine and using the fact that:

$$X^{\top}(i)X(i) \geq X^{\top}(i) + X(i) - \mathbb{I},$$

we get:

$$A(i)X(i) + X^{\top}(i)A^{\top}(i) + B(i)G(i)C(i) + C^{\top}(i)G^{\top}(i)B^{\top}(i)$$

$$+ \lambda_{ii}X^{\top}(i)E^{\top}(i) + S_i(X)S_i^{\top}(X) + S_i(X)\mathcal{Z}_i^{-1}(X)S_i^{\top}(X) < 0,$$

with

$$\mathcal{Z}_i(X) = \operatorname{diag}\left[X^{\top}(1) + X(1) - \mathbb{I}, \cdots, X^{\top}(i-1) + X(i-1) - \mathbb{I}, \right.$$

$$\left. X^{\top}(i+1) + X(i+1) - \mathbb{I}, \cdots, X^{\top}(N) + X(N) - \mathbb{I} \right].$$

Finally using Schur complement gives:

$$\begin{bmatrix} J(i) & S_i(X) & S_i(X) \\ S_i^\top(X) & -\mathbb{I} & 0 \\ S_i^\top(X) & 0 & -\mathcal{Z}_i(X) \end{bmatrix} < 0,$$

with $J(i) = A(i)X(i) + X^\top(i)A^\top(i) + B(i)G(i)C(i) + C^\top(i)G^\top(i)B^\top(i) + \lambda_{ii}X^\top(i)E^\top(i)$.
The following theorem summarizes the results of this development.

Theorem 5.2.1 *Let ε_P be a given positive scalar. There exists a static output feedback controller of the form (5.6) such that the closed-loop system (5.1) is piecewise regular, impulse-free and stochastically stable if there exist sets of nonsingular matrices $X = (X(1), \cdots, X(N))$ with $X(i) \in \mathbb{R}^{n\times n}$, and $Y = (Y(1), \cdots, Y(N))$ with $Y(i) \in \mathbb{R}^{p\times p}$, and a set of matrices $G = (G(1), \cdots, G(N))$, with $G(i) \in \mathbb{R}^{m\times p}$ such that the following holds for each $i \in \mathscr{S}$:*

$$\begin{bmatrix} J(i) & S_i(X) & S_i(X) \\ S_i^\top(X) & -\mathbb{I} & 0 \\ S_i^\top(X) & 0 & -\mathcal{Z}_i(X) \end{bmatrix} < 0, \tag{5.7}$$

where

$$\begin{aligned} J(i) &= A(i)X(i) + X^\top(i)A^\top(i) + B(i)G(i)C(i) + C^\top(i)G^\top(i)B^\top(i) \\ &\quad + \lambda_{ii}X^\top(i)E^\top(i), \\ S_i(X) &= \left[\sqrt{\varepsilon_P\lambda_{i1}}X^\top(i), \cdots, \sqrt{\varepsilon_P\lambda_{ii-1}}X^\top(i), \right. \\ &\quad \left. \sqrt{\varepsilon_P\lambda_{ii+1}}X^\top(i), \cdots, \sqrt{\varepsilon_P\lambda_{iN}}X^\top(i) \right], \\ \mathcal{Z}_i(X) &= diag\left[X^\top(1) + X(1) - \mathbb{I}, \cdots, X^\top(i-1) + X(i-1) - \mathbb{I}, \right. \\ &\quad \left. X^\top(i+1) + X(i+1) - \mathbb{I}, \cdots, X^\top(N) + X(N) - \mathbb{I} \right], \end{aligned}$$

with the following constraints:

$$\begin{cases} \varepsilon_P\left[X(i) + X^\top(i) \right] \geq E(i)X(i) = X^\top(i)E^\top(i) \geq 0, \\ Y(i)C(i) = C(i)X(i). \end{cases} \tag{5.8}$$

The controller gain is given by $F(i) = G(i)Y^{-1}(i)$, $i \in \mathscr{S}$.

Remark 5.2.1 *The conditions (5.8) may be difficult to solve using the LMI toolbox of Matlab. To avoid this, we can replace these conditions by the following ones:*

$$[Y(i)C(i) - C(i)X(i)]^\top [Y(i)C(i) - C(i)X(i)] \leq \beta_1\mathbb{I}$$
$$\left[X^\top(i)E^\top(i) - E(i)X(i)\right]^\top \left[X^\top(i)E^\top(i) - E(i)X(i)\right] \leq \beta_2\mathbb{I},$$

that give the following LMIs:

$$\begin{bmatrix} -\beta_1\mathbb{I} & [Y(i)C(i) - C(i)X(i)]^\top \\ [Y(i)C(i) - C(i)X(i)] & -\mathbb{I} \end{bmatrix} \leq 0, \tag{5.9}$$

$$\begin{bmatrix} -\beta_2\mathbb{I} & [X^\top(i)E^\top(i) - E(i)X(i)]^\top \\ [X^\top(i)E^\top(i) - E(i)X(i)] & -\mathbb{I} \end{bmatrix} \leq 0. \tag{5.10}$$

Therefore the design of the stabilizing output feedback controller is brought to the following problem:

$$\min_{\substack{\beta_1 \geq 0, \\ \beta_2 \geq 0}} \beta_1 + \beta_2$$

$$s.t. :$$

$$(5.7), (5.9) - (5.10).$$

For the second upper bound of the term $E^\top(j)P(j)$, i.e.,

$$\varepsilon(j)P^\top(j)P(j) \geq E^\top(j)P(j) \geq 0$$

for any $\varepsilon(j) > 0$.

The closed-loop dynamics will be piecewise regular, impulse-free and stochastically stable if there exists a nonsingular set of matrices $P = (P(1), \cdots, P(N))$ such that the following hold for each $i \in \mathscr{S}$:

$$\varepsilon(i)P^\top(i)P(i) \geq E^\top(i)P(i) = P^\top(i)E(i) \geq 0$$

$$A^\top(i)P(i) + P^\top(i)A(i) + P^\top(i)B(i)F(i)C(i) + \left[P^\top(i)B(i)F(i)C(i)\right]^\top$$

$$+ \lambda_{ii}E^\top(i)P(i) + \sum_{j=1, j\neq i}^{N} \lambda_{ij}\varepsilon(j)P^\top(j)P(j) < 0.$$

If we let $X(i) = P^{-1}(i)$ and pre- and post-multiplying the second inequality respectively by $X^\top(i)$ and $X(i)$, we get:

$$\varepsilon(i)\mathbb{I} \geq X^\top(i)E^\top(i) = E(i)X(i) \geq 0$$

$$X^\top(i)A^\top(i) + A(i)X(i) + B(i)F(i)C(i)X(i) + [B(i)F(i)C(i)X(i)]^\top$$

$$+ \lambda_{ii}X^\top(i)E^\top(i) + \sum_{j=1, j\neq i}^{N} \lambda_{ij}\varepsilon(j)X^\top(i)\left[X^{-\top}(j)X^{-1}(j)\right]X(i) < 0,$$

Note that:

$$\sum_{j=1, j\neq i}^{N} \lambda_{ij}X^\top(i)\left[\varepsilon^{-1}(j)X(j)X^\top(j)\right]^{-1}X(i) = \mathcal{S}_i(X)Z_i^{-1}(X)\mathcal{S}_i^\top(X)$$

where

$$\mathcal{S}_i(X) = \left[\sqrt{\lambda_{i1}}X^\top(i), \cdots, \sqrt{\lambda_{ii-1}}X^\top(i), \sqrt{\lambda_{ii+1}}X^\top(i), \cdots, \sqrt{\lambda_{iN}}X^\top(i)\right],$$

$$Z_i(X) = \text{diag}\left[\varepsilon^{-1}(1)X(1)X^\top(1), \cdots, \varepsilon^{-1}(i-1)X(i-1)X^\top(i-1),\right.$$

$$\left. \varepsilon^{-1}(i+1)X(i+1)X^\top(i+1), \cdots, \varepsilon^{-1}(N)X(N)X^\top(N)\right].$$

Using the fact that $\varepsilon^{-1}(i)X(i)X^\top(i) \geq X^\top(i) + X(i) - \varepsilon(i)\mathbb{I}$ and defining $\mathcal{X}_i(X)$ as follows:

$$\mathcal{X}_i(X) = \text{diag}\left[-\varepsilon^{-1}(1)\mathbb{I} + X(1) + X^\top(1), \cdots, -\varepsilon^{-1}(i-1)\mathbb{I} + X(i-1) + X^\top(i-1),\right.$$

$$\left. -\varepsilon^{-1}(i+1)\mathbb{I} + X(i+1) + X^\top(i+1), \cdots, -\varepsilon^{-1}(N)\mathbb{I} + X(N) + X^\top(N)\right],$$

and letting $C(i)X(i) = Y(i)X(i)$ and $G(i) = F(i)Y(i)$, we get the following result.

Theorem 5.2.2 *There exists a static output feedback controller of the form (5.6) such that the closed-loop system (5.1) is piecewise regular, impulse-free and stochastically stable if there exist sets of nonsingular matrices $X = (X(1), \cdots, X(N))$, with $X(i) \in \mathbb{R}^{n \times n}$, and $Y = (Y(1), \cdots, Y(N))$ with $Y(i) \in \mathbb{R}^{p \times p}$, a set of matrices $G = (G(1), \cdots, G(N))$ with $G(i) \in \mathbb{R}^{m \times p}$, and a set of positive scalars $\varepsilon = (\varepsilon(1), \cdots, \varepsilon(N))$, such that the following holds for each $i \in \mathscr{S}$:*

$$\begin{bmatrix} J(i) & S_i(X) \\ S_i^\top(X) & -X_i(X) \end{bmatrix} < 0, \tag{5.11}$$

where

$$J(i) = A(i)X(i) + X^\top(i)A^\top(i) + B(i)G(i)C(i) + C^\top(i)G^\top(i)B^\top(i)$$
$$+ \lambda_{ii}X^\top(i)E^\top(i),$$

$$S_i(X) = \left[\sqrt{\lambda_{i1}}X^\top(i), \cdots, \sqrt{\lambda_{ii-1}}X^\top(i), \sqrt{\lambda_{ii+1}}X^\top(i), \cdots, \sqrt{\lambda_{iN}}X^\top(i) \right],$$

$$X_i(X) = diag\left[-\varepsilon(1)\mathbb{I} + X(1) + X^\top(1), \cdots, -\varepsilon(i-1)\mathbb{I} + X(i-1) + X^\top(i-1),\right.$$
$$\left. -\varepsilon(i+1)\mathbb{I} + X(i+1) + X^\top(i+1), \cdots, -\varepsilon(N)\mathbb{I} + X(N) + X^\top(N)\right],$$

with the following constraints:

$$\begin{cases} \varepsilon(i)\mathbb{I} \geq E(i)X(i) = X^\top(i)E^\top(i) \geq 0, \\ Y(i)C(i) = C(i)X(i). \end{cases} \tag{5.12}$$

The controller gain is given by $F(i) = G(i)Y^{-1}(i)$, $i \in \mathscr{S}$.

For the third upper bound of the term $E^\top(j)P(j)$, i.e.,

$$\frac{1}{4}\varepsilon^{-1}(j)\mathbb{I} + \varepsilon(j)E^\top(j)P(j)P^\top(j)E(j) \geq E^\top(j)P(j) \geq 0.$$

The closed-loop dynamics will be piecewise regular, impulse-free and stochastically stable if there exists a nonsingular set of matrices $P = (P(1), \cdots, P(N))$ such that the following hold for each $i \in \mathscr{S}$:

$$E^\top(i)P(i) - P^\top(i)E(i) \geq 0,$$

$$A^\top(i)P(i) + P^\top(i)A(i) + P^\top(i)B(i)F(i)C(i) + \left[P^\top(i)B(i)F(i)C(i)\right]^\top$$

$$+ \lambda_{ii}E^\top(i)P(i) + \sum_{j=1,j\neq i}^{N} \lambda_{ij}\left[\frac{1}{4}\varepsilon^{-1}(j)\mathbb{I} + \varepsilon(j)E^\top(j)P(j)P^\top(j)E(j)\right] < 0.$$

If we let $X(i) = P^{-1}(i)$ and pre- and post-multiplying the second inequality respectively by $X^\top(i)$ and $X(i)$, we get:

$$X^\top(i)E^\top(i) = E(i)X(i) \geq 0,$$

$$X^\top(i)A^\top(i) + A(i)X(i) + B(i)F(i)C(i)X(i) + [B(i)F(i)C(i)X(i)]^\top$$

$$+ \lambda_{ii}X^\top(i)E^\top(i) + \sum_{j=1,j\neq i}^{N} \lambda_{ij}X^\top(i)\left[\frac{1}{4}\varepsilon^{-1}(j)\mathbb{I}\right]X(i)$$

$$+ \sum_{j=1,j\neq i}^{N} \lambda_{ij}X^\top(i)\left[\varepsilon(j)E^\top(j)X^{-1}(j)X^{-\top}(j)E(j)\right]X(i) < 0.$$

Note that:

$$\sum_{j=1, j\neq i}^{N} \lambda_{ij} X^{\top}(i)\left[\frac{1}{4}\varepsilon^{-1}(j)\mathbb{I}\right]X(i) = \mathcal{Z}_i(X)\mathcal{X}_i^{-1}(\varepsilon)\mathcal{Z}_i^{\top}(X),$$

$$\sum_{j=1, j\neq i}^{N} \lambda_{ij} X^{\top}(i)\left[\varepsilon(j)E^{\top}(j)X^{-1}(j)X^{-\top}(j)E(j)\right]X(i) = \mathcal{S}_i(X)\mathcal{X}_i^{-1}(X)\mathcal{S}_i^{\top}(X),$$

where

$$\mathcal{Z}_i(X) = \left[\sqrt{\lambda_{i1}}X^{\top}(i), \cdots, \sqrt{\lambda_{ii-1}}X^{\top}(i), \sqrt{\lambda_{ii+1}}X^{\top}(i), \cdots, \sqrt{\lambda_{iN}}X^{\top}(i)\right],$$

$$\mathcal{S}_i(X) = \left[\sqrt{\lambda_{i1}}X^{\top}(i)E^{\top}(1), \cdots, \sqrt{\lambda_{ii-1}}X^{\top}(i)E^{\top}(i-1),\right.$$
$$\left.\sqrt{\lambda_{ii+1}}X^{\top}(i)E^{\top}(i+1), \cdots, \sqrt{\lambda_{iN}}X^{\top}(i)E^{\top}(N)\right],$$

$$\mathcal{X}_i(\varepsilon) = \mathrm{diag}\left[4\varepsilon(1)\mathbb{I}, \cdots, 4\varepsilon(i-1)\mathbb{I}, 4\varepsilon(i+1)\mathbb{I}, \cdots, 4\varepsilon(N)\mathbb{I}\right],$$

$$\mathcal{X}_i(X) = \mathrm{diag}\left[\varepsilon^{-1}(1)X^{\top}(1)X(1), \cdots, \varepsilon^{-1}(i-1)X^{\top}(i-1)X(i-1),\right.$$
$$\left.\varepsilon^{-1}(i+1)X^{\top}(i+1)X(i+1), \cdots, \varepsilon^{-1}(N)X^{\top}(N)X(N)\right].$$

Using the fact that $\varepsilon^{-1}(j)X(j)X^{\top}(j) \geq X^{\top}(j) + X(j) - \varepsilon(j)\mathbb{I}$ and defining $\mathcal{X}_i(X)$ as follows:

$$\mathcal{X}_i(X) = \mathrm{diag}\left[-\varepsilon(1)\mathbb{I} + X(1) + X^{\top}(1), \cdots, -\varepsilon(i-1)\mathbb{I} + X(i-1) + X^{\top}(i-1),\right.$$
$$\left.-\varepsilon(i+1)\mathbb{I} + X(i+1) + X^{\top}(i+1), \cdots, -\varepsilon(N)\mathbb{I} + X(N) + X^{\top}(N)\right],$$

and letting $C(i)X(i) = Y(i)X(i)$ and $G(i) = F(i)Y(i)$, we get the following result.

Theorem 5.2.3 *There exists a static output feedback controller of the form (5.6) such that the closed-loop system (5.1) is piecewise regular, impulse-free and stochastically stable if there exist sets of nonsingular matrices $X = (X(1), \cdots, X(N))$ with $X(i) \in \mathbb{R}^{n\times n}$, and $Y = (Y(1), \cdots, Y(N))$ with $Y(i) \in \mathbb{R}^{p\times p}$, a set of matrices $G = (G(1), \cdots, G(N))$ with $G(i) \in \mathbb{R}^{m\times p}$, and a set of positive scalars $\varepsilon = (\varepsilon(1), \cdots, \varepsilon(N))$, such that the following holds for each $i \in \mathscr{S}$:*

$$\begin{bmatrix} J(i) & \mathcal{Z}_i(X) & \mathcal{S}_i(X) \\ \mathcal{Z}_i^{\top}(X) & -\mathcal{X}_i(\varepsilon) & 0 \\ \mathcal{S}_i^{\top}(X) & 0 & -\mathcal{X}_i(X) \end{bmatrix} < 0, \tag{5.13}$$

where

$$J(i) = A(i)X(i) + X^{\top}(i)A^{\top}(i) + B(i)G(i)C(i) + C^{\top}(i)G^{\top}(i)B^{\top}(i)$$
$$+ \lambda_{ii}X^{\top}(i)E^{\top}(i),$$

$$\mathcal{Z}_i(X) = \left[\sqrt{\lambda_{i1}}X^{\top}(i), \cdots, \sqrt{\lambda_{ii-1}}X^{\top}(i), \sqrt{\lambda_{ii+1}}X^{\top}(i), \cdots, \sqrt{\lambda_{iN}}X^{\top}(i)\right],$$

$$\mathcal{S}_i(X) = \left[\sqrt{\lambda_{i1}}X^{\top}(i)E^{\top}(1), \cdots, \sqrt{\lambda_{ii-1}}X^{\top}(i)E^{\top}(i-1),\right.$$
$$\left.\sqrt{\lambda_{ii+1}}X^{\top}(i)E^{\top}(i+1), \cdots, \sqrt{\lambda_{iN}}X^{\top}(i)E^{\top}(N)\right],$$

$$\mathcal{X}_i(\varepsilon) = \mathrm{diag}\left[4\varepsilon(1)\mathbb{I}, \cdots, 4\varepsilon(i-1)\mathbb{I}, 4\varepsilon^{-1}(i+1)\mathbb{I}, \cdots, 4\varepsilon^{-1}(N)\mathbb{I}\right],$$

$$\mathcal{X}_i(X) = \mathrm{diag}\left[-\varepsilon(1)\mathbb{I} + X(1) + X^{\top}(1), \cdots, -\varepsilon(i-1)\mathbb{I} + X(i-1) + X^{\top}(i-1),\right.$$
$$\left.-\varepsilon(i+1)\mathbb{I} + X(i+1) + X^{\top}(i+1), \cdots, -\varepsilon(N)\mathbb{I} + X(N) + X^{\top}(N)\right],$$

with the following constraints:

$$\begin{cases} E(i)X(i) = X^\top(i)E^\top(i) \geq 0, \\ Y(i)C(i) = C(i)X(i). \end{cases} \tag{5.14}$$

The controller gain is given by $F(i) = G(i)Y^{-1}(i),\ i \in \mathscr{S}.$

Let us now consider the effect of the uncertainties. Based on the results of Chap. 2, system (5.1) is piecewise regular, impulse-free and robust stochastically stable if there exist a set of nonsingular matrices $P = (P(1), \cdots, P(N))$, and a set of positive scalars $\varepsilon_A = (\varepsilon_A(1), \cdots, \varepsilon_A(N))$, such that the following coupled LMIs hold for every $i \in \mathscr{S}$:

$$\begin{cases} \varepsilon_P \left[P(i) + P^\top(i) \right] \geq E^\top(i)P(i) = P^\top(i)E(i) \geq 0, \\ \begin{bmatrix} J_u(i) & P^\top(i)D_A(i) \\ D_A^\top(i)P(i) & -\varepsilon_A^{-1}(i)\mathbb{I} \end{bmatrix} < 0, \end{cases} \tag{5.15}$$

with

$$J_u(i) = P^\top(i)\bar{A}(i) + \bar{A}^\top(i)P(i) + \lambda_{ii}E^\top(i)P(i) + \varepsilon_A^{-1}(i)E_A^\top(i)E_A(i)$$

$$+ \sum_{j=1, j \neq i}^{N} \varepsilon_P \lambda_{ij} \left[P(j) + P^\top(j) \right],$$

$$\bar{A}(i) = A(i) + B(i)F(i)C(i) + D_B(i)F_B(i)E_B(i)F(i)C(i).$$

As we did for the nominal system, we can establish the following result for uncertain system using Lemma 1.5.1 and Schur complement. This result allows the design of a static output feedback that guarantees that the closed-loop system is piecewise regular, impulse-free and robust stochastically stable.

Corollary 5.2.1 *Let* ε_P *be a given positive scalar. There exists a state feedback controller of the form (5.6) such that the closed-loop system (5.1) is piecewise regular, impulse-free and stochastically stable if there exist sets of nonsingular matrices* $X = (X(1), \cdots, X(N))$, $X(i) \in \mathbb{R}^{n \times n}$ *and* $Y = (Y(1), \cdots, Y(N))$, $Y(i) \in \mathbb{R}^{p \times p}$ *a set of matrices* $G = (G(1), \cdots, G(N))$, $G(i) \in \mathbb{R}^{m \times p}$ *and sets of positive scalars* $\varepsilon_A = (\varepsilon_A(1), \cdots, \varepsilon_A(N))$, *and* $\varepsilon_B = (\varepsilon_B(1), \cdots, \varepsilon_B(N))$, *such that the following holds for each* $i \in \mathscr{S}$:

$$\begin{bmatrix} J_X(i) & X^\top(i)E_A^\top(i) & C^\top(i)G^\top(i)E_B^\top(i) & S_i(X) & S_i(X) \\ E_A(i)X(i) & -\varepsilon_A(i)\mathbb{I} & 0 & 0 & 0 \\ E_B(i)G(i)C(i) & 0 & -\varepsilon_B(i)\mathbb{I} & 0 & 0 \\ S_i^\top(X) & 0 & 0 & -\mathbb{I} & 0 \\ S_i^\top(X) & 0 & 0 & 0 & -\mathcal{X}_i(X) \end{bmatrix} < 0, \tag{5.16}$$

where

$$J_X(i) = A(i)X(i) + X^\top(i)A^\top(i) + B(i)G(i)C(i) + C^\top(i)G^\top(i)B^\top(i)$$

$$+ \lambda_{ii}X^\top(i)E^\top(i) + \varepsilon_A(i)D_A(i)D_A^\top(i) + \varepsilon_B(i)D_B(i)D_B^\top(i),$$

$$\mathcal{X}_i(X) = \mathrm{diag}\left[X^\top(1) + X(1) - \mathbb{I}, \cdots, X^\top(i-1) + X(i-1) - \mathbb{I}, \right.$$

$$\left. X^\top(i+1) + X(i+1) - \mathbb{I}, \cdots, X^\top(N) + X(N) - \mathbb{I} \right],$$

$$\mathcal{S}_i(X) = \left[\sqrt{\varepsilon_P \lambda_{i1}} X^\top(i), \cdots, \sqrt{\varepsilon_P \lambda_{ii-1}} X^\top(i), \right.$$
$$\left. \sqrt{\varepsilon_P \lambda_{ii+1}} X^\top(i), \cdots, \sqrt{\varepsilon_P \lambda_{iN}} X^\top(i) \right],$$

with the following constraints:

$$\begin{cases} \varepsilon_P \left[X(i) + X^\top(i) \right] \geq X^\top(i) E^\top(i) = E(i)X(i) \geq 0, \\ Y(i)C(i) = C(i)X(i). \end{cases} \tag{5.17}$$

The controller gain is given by $F(i) = G(i)Y^{-1}(i)$, $i \in \mathcal{S}$.

If now we consider the effect of the uncertainties on the transition matrix similar results can be established. For this purpose notice that:

$$\mu_{ij} = \sum_{k=1}^{\kappa} \alpha_k \lambda_{ij}^k .$$

Using this following corollary gives such results.

Corollary 5.2.2 *Let ε_P be a given positive scalar. There exists a state feedback controller of the form (5.6) such that the closed-loop system (5.1) is piecewise regular, impulse-free and stochastically stable if there exist sets of nonsingular matrices $X = (X(1), \cdots, X(N))$, $X(i) \in \mathbb{R}^{n \times n}$ and $Y = (Y(1), \cdots, Y(N))$, $Y(i) \in \mathbb{R}^{p \times p}$ and a set of matrices $G = (G(1), \cdots, G(N))$, $G(i) \in \mathbb{R}^{m \times p}$ and sets of positive scalars $\varepsilon_A = (\varepsilon_A(1), \cdots, \varepsilon_A(N))$, and $\varepsilon_B = (\varepsilon_B(1), \cdots, \varepsilon_B(N))$, such that the following holds for each $i \in \mathcal{S}$:*

$$\begin{bmatrix} J_X(i) & X^\top(i)E_A^\top(i) & C^\top(i)G^\top(i)E_B^\top(i) & \mathcal{S}_i(X) & \mathcal{S}_i(X) \\ E_A(i)X(i) & -\varepsilon_A(i)\mathbb{I} & 0 & 0 & 0 \\ E_B^\top(i)G(i)C(i) & 0 & -\varepsilon_B(i)\mathbb{I} & 0 & 0 \\ \mathcal{S}_i^\top(X) & 0 & 0 & -\mathbb{I} & 0 \\ \mathcal{S}_i^\top(X) & 0 & 0 & 0 & -\mathcal{X}_i(X) \end{bmatrix} < 0, \tag{5.18}$$

where

$$J_X(i) = A(i)X(i) + X^\top(i)A^\top(i) + B(i)G(i)C(i) + C^\top(i)G^\top(i)B^\top(i)$$
$$+ \mu_{ii}X^\top(i)E^\top(i) + \varepsilon_A(i)D_A(i)D_A^\top(i) + \varepsilon_B(i)D_B(i)D_B^\top(i),$$

$$\mathcal{X}_i(X) = \text{diag}\left[X^\top(1) + X(1) - \mathbb{I}, \cdots, X^\top(i-1) + X(i-1) - \mathbb{I}, \right.$$
$$\left. X^\top(i+1) + X(i+1) - \mathbb{I}, \cdots, X^\top(N) + X(N) - \mathbb{I} \right],$$

$$\mathcal{S}_i(X) = \left[\sqrt{\varepsilon_P \mu_{i1}} X^{\dagger}(i), \cdots, \sqrt{\varepsilon_P \mu_{ii-1}} X^{\dagger}(i), \right.$$
$$\left. \sqrt{\varepsilon_P \mu_{ii+1}} X^\top(i), \cdots, \sqrt{\varepsilon_P \mu_{iN}} X^\top(i) \right],$$

with the following constraints:

$$\begin{cases} \varepsilon_P \left[X(i) + X^\top(i) \right] \geq X^\top(i) E^\top(i) = E(i)X(i) \geq 0, \\ Y(i)C(i) = C(i)X(i). \end{cases} \tag{5.19}$$

The controller gain is given by $F(i) = G(i)Y^{-1}(i)$, $i \in \mathcal{S}$.

For the second upper bound of the term $E^\top(i)P(i)$, the closed-loop system (5.1) is piecewise regular, impulse-free and stochastically stable if there exist sets of non-singular matrices $X = (X(1), \cdots, X(N))$, and $Y = (Y(1), \cdots, Y(N))$, a set of matrices $G = (G(1), \cdots, G(N))$, and a set of positive scalars $\varepsilon = (\varepsilon(1), \cdots, \varepsilon(N))$, such that the following holds for each $i \in \mathscr{S}$:

$$\begin{bmatrix} J(i) & S_i(X) \\ S_i^\top(X) & -X_i(X) \end{bmatrix} < 0,$$

where

$$\bar{A}(i) = A(i) + D_A(i)F_A(i)E_A(i)$$
$$J(i) = \bar{A}(i)X(i) + X^\top(i)\bar{A}^\top(i) + B(i)G(i)C(i) + C^\top(i)G^\top(i)B^\top(i)$$
$$+ D_B(i)F_B(i)E_B(i)G(i)C(i) + [D_B(i)F_B(i)E_B(i)G(i)C(i)]^\top$$
$$+ \lambda_{ii}X^\top(i)E^\top(i),$$

with the following constraints:

$$\begin{cases} \varepsilon(i)\mathbb{I} \geq E(i)X(i) = X^\top(i)E^\top(i) \geq 0, \\ Y(i)C(i) = C(i)X(i). \end{cases}$$

Using Lemma 1.5.1 and Schur complement we get the following result.

Theorem 5.2.4 *There exists a static output feedback controller of the form (5.6) such that the closed-loop system (5.1) is piecewise regular, impulse-free and stochastically stable if there exist sets of nonsingular matrices $X = (X(1), \cdots, X(N))$ with $X(i) \in \mathbb{R}^{n \times n}$, and $Y = (Y(1), \cdots, Y(N))$ with $Y(i) \in \mathbb{R}^{p \times p}$, a set of matrices $G = (G(1), \cdots, G(N))$ with $G(i) \in \mathbb{R}^{m \times p}$, and sets of positive scalars $\varepsilon = (\varepsilon(1), \cdots, \varepsilon(N))$, $\varepsilon_A = (\varepsilon_A(1), \cdots, \varepsilon_A(N))$, and $\varepsilon_B = (\varepsilon_B(1), \cdots, \varepsilon_B(N))$, such that the following holds for each $\iota \in \mathscr{S}$:*

$$\begin{bmatrix} J(i) & X^\top(i)E_A^\top(i) & C^\top(i)G^\top(i)E_B^\top(i) & S_i(X) \\ E_A(i)X(i) & -\varepsilon_A(i)\mathbb{I} & 0 & 0 \\ E_B(i)G(i)C(i) & 0 & -\varepsilon_B(i)\mathbb{I} & 0 \\ S_i^\top(X) & 0 & 0 & -X_i(X) \end{bmatrix} < 0, \qquad (5.20)$$

where

$$J(i) = A(i)X(i) + X^\top(i)A^\top(i) + B(i)G(i)C(i) + C^\top(i)G^\top(i)B^\top(i)$$
$$+ \lambda_{ii}X^\top(i)E^\top(i) + \varepsilon_A(i)D_A(i)D_A^\top(i) + \varepsilon_B(i)D_B(i)D_B^\top(i),$$
$$X_i(X) = \mathrm{diag}\left[X^\top(1) + X(1) - \varepsilon(1)\mathbb{I}, \cdots, X^\top(i-1) + X(i-1) - \varepsilon(i-1)\mathbb{I}, \right.$$
$$\left. X^\top(i+1) + X(i+1) - \varepsilon(i+1)\mathbb{I}, \cdots, X^\top(N) + X(N) - \varepsilon(N)\mathbb{I}\right],$$
$$S_i(X) = \left[\sqrt{\lambda_{i1}}X^\top(i), \cdots, \sqrt{\lambda_{ii-1}}X^\top(i), \right.$$
$$\left. \sqrt{\lambda_{ii+1}}X^\top(i), \cdots, \sqrt{\lambda_{iN}}X^\top(i)\right],$$

with the following constraints:

$$\begin{cases} \varepsilon(i)\mathbb{I} \geq E(i)X(i) = X^\top(i)E^\top(i) \geq 0, \\ Y(i)C(i) = C(i)X(i). \end{cases} \qquad (5.21)$$

The controller gain is given by $F(i) = G(i)Y^{-1}(i)$, $i \in \mathscr{S}$.

For the third upper bound of the term $E^\top(i)P(i)$, the closed-loop system (5.1) with the static output feedback controller is piecewise regular, impulse-free and stochastically stable if there exist sets of nonsingular matrices $X = (X(1), \cdots, X(N))$, and $Y = (Y(1), \cdots, Y(N))$, a set of matrices $G = (G(1), \cdots, G(N))$, and a set of positive scalars $\varepsilon = (\varepsilon(1), \cdots, \varepsilon(N))$, such that the following holds for each $i \in \mathscr{S}$:

$$\begin{bmatrix} J(i) & \mathcal{Z}_i(X) & \mathcal{S}_i(X) \\ \mathcal{Z}_i^\top(X) & -\mathcal{X}_i(\varepsilon) & 0 \\ \mathcal{S}_i^\top(X) & 0 & -\mathcal{X}_i(X) \end{bmatrix} < 0,$$

where

$$\bar{A}(i) = A(i) + D_A(i)F_A(i)E_A(i)$$

$$\begin{aligned} J(i) &= \bar{A}(i)X(i) + X^\top(i)\bar{A}^\top(i) + B(i)G(i)C(i) + C^\top(i)G^\top(i)B^\top(i) \\ &\quad + D_B(i)F_B(i)E_B(i)G(i)C(i) + [D_B(i)F_B(i)E_B(i)G(i)C(i)]^\top \\ &\quad + \lambda_{ii}X^\top(i)E^\top(i), \end{aligned}$$

with the following constraints:

$$\begin{cases} E(i)X(i) = X^\top(i)E^\top(i) \geq 0, \\ Y(i)C(i) = C(i)X(i). \end{cases}$$

Using Lemma 1.5.1 and Schur complement we get the following result.

Theorem 5.2.5 *There exists a static output feedback controller of the form (5.6) such that the closed-loop system (5.1) is piecewise regular, impulse-free and stochastically stable if there exist sets of nonsingular matrices $X = (X(1), \cdots, X(N))$ with $X(i) \in \mathbb{R}^{n \times n}$, and $Y = (Y(1), \cdots, Y(N))$ with $Y(i) \in \mathbb{R}^{p \times p}$, a set of matrices $G = (G(1), \cdots, G(N))$ with $G(i) \in \mathbb{R}^{m \times p}$, and sets of positive scalars $\varepsilon = (\varepsilon(1), \cdots, \varepsilon(N))$, $\varepsilon_A = (\varepsilon_A(1), \cdots, \varepsilon_A(N))$, and $\varepsilon_B = (\varepsilon_B(1), \cdots, \varepsilon_B(N))$, such that the following holds for each $i \in \mathscr{S}$:*

$$\begin{bmatrix} J(i) & X^\top(i)E_A^\top(i) & C^\top(i)G^\top(i)E_B^\top(i) & \mathcal{Z}_i(X) & \mathcal{S}_i(X) \\ E_A(i)X(i) & -\varepsilon_A(i)\mathbb{I} & 0 & 0 & 0 \\ E_B(i)G(i)C(i) & 0 & -\varepsilon_B(i)\mathbb{I} & 0 & 0 \\ \mathcal{Z}_i^\top(X) & 0 & 0 & -\mathcal{X}_i(\varepsilon) & 0 \\ \mathcal{S}_i^\top(X) & 0 & 0 & 0 & -\mathcal{X}_i(X) \end{bmatrix} < 0, \qquad (5.22)$$

where

$$\begin{aligned} J(i) &= A(i)X(i) + X^\top(i)A^\top(i) + B(i)G(i)C(i) + C^\top(i)G^\top(i)B^\top(i) \\ &\quad + \lambda_{ii}X^\top(i)E^\top(i) + \varepsilon_A(i)D_A(i)D_A^\top(i) + \varepsilon_B(i)D_B(i)D_B^\top(i), \end{aligned}$$

$$\begin{aligned} \mathcal{X}_i(X) &= \mathrm{diag}\Big[X^\top(1) + X(1) - \varepsilon(1)\mathbb{I}, \cdots, X^\top(i-1) + X(i-1) - \varepsilon(i-1)\mathbb{I}, \\ &\quad X^\top(i+1) + X(i+1) - \varepsilon(i+1)\mathbb{I}, \cdots, X^\top(N) + X(N) - \varepsilon(N)\mathbb{I}\Big], \end{aligned}$$

$$S_i(X) = \left[\sqrt{\lambda_{i1}} X^\top(i) E^\top(1), \cdots, \sqrt{\lambda_{ii-1}} X^\top(i) E^\top(i-1), \right.$$
$$\left. \sqrt{\lambda_{ii+1}} X^\top(i) E^\top(i+1), \cdots, \sqrt{\lambda_{iN}} X^\top(i) E^\top(N) \right],$$

$$X_i(\varepsilon) = \text{diag} \left[4\varepsilon(1)\mathbb{I}, \cdots, 4\varepsilon(i-1)\mathbb{I}, \right.$$
$$\left. 4\varepsilon(i+1)\mathbb{I}, \cdots, 4\varepsilon(N)\mathbb{I} \right],$$

$$Z_i(X) = \left[\sqrt{\lambda_{i1}} X^\top(i), \cdots, \sqrt{\lambda_{ii-1}} X^\top(i), \right.$$
$$\left. \sqrt{\lambda_{ii+1}} X^\top(i), \cdots, \sqrt{\lambda_{iN}} X^\top(i) \right],$$

with the following constraints:

$$\begin{cases} E(i)X(i) = X^\top(i)E^\top(i) \geq 0, \\ Y(i)C(i) = C(i)X(i). \end{cases} \qquad (5.23)$$

The controller gain is given by $F(i) = G(i)Y^{-1}(i)$, $i \in \mathscr{S}$.

5.3 $\mathcal{H}_\infty$ Static Output Feedback Control

Let us now assume that our system has external disturbance and let its behavior be described by the following dynamics:

$$\begin{cases} E(r_t)\dot{x}(t) = A(r_t,t)x(t) + B(r_t,t)u(t) + B_w(r_t)w(t), \ x(0) = x_0, \\ y(t) = C(r_t)x(t), \\ z(t) = C_z(r_t,t)x(t) + D_z(r_t)u(t) + B_z(r_t)w(t), \end{cases} \qquad (5.24)$$

where $x(t) \in \mathbb{R}^n$ is the state vector, $u(t) \in \mathbb{R}^m$ is the control vector, $z(t) \in \mathbb{R}^q$ is the controlled output and $w(t) \in \mathbb{R}^l$ is the system external disturbance, $E(i)$ is a known singular matrix with rank $(E(i)) = n_r \leq n$ for all $i \in \mathscr{S}$, the matrices $A(r_t,t)$, $B(r_t,t)$, $C_z(r_t,t)$ and $D_z(r_t,t)$ are given by when $r_t = i \in \mathscr{S}$:

$$\begin{cases} A(i,t) = A(i) + D_A(i)F_A(i)E_A(i), \\ B(i,t) = B(i) + D_B(i)F_B(i)E_B(i), \\ C_z(i,t) = C_z(i) + D_{C_z}(i)F_{C_z}(i)E_{C_z}(i), \end{cases}$$

with $A(i) \in \mathbb{R}^{n \times n}$, $B(i) \in \mathbb{R}^{n \times m}$, $B_w(i) \in \mathbb{R}^{n \times l}$, $C_z(i) \in \mathbb{R}^{q \times n}$, $D_z(i) \in \mathbb{R}^{q \times m}$, $B_z(i) \in \mathbb{R}^{q \times l}$, $D_A(i) \in \mathbb{R}^{n \times n_D}$, $E_A(i) \in \mathbb{R}^{n_E \times n}$, $D_B(i) \in \mathbb{R}^{n \times m_D}$, $E_B(i) \in \mathbb{R}^{m_E \times m}$, $D_{C_z}(i) \in \mathbb{R}^{q \times q_D}$ and $E_{C_z}(i) \in \mathbb{R}^{q_E \times n}$, are known real matrices with appropriate dimensions, and the matrices $F_A(i) \in \mathbb{R}^{n_D \times n_E}$, $F_B(i) \in \mathbb{R}^{m_D \times m_E}$, $F_{C_z}(i) \in \mathbb{R}^{q_D \times q_E}$ are time-varying unknown matrices satisfying the following:

$$\begin{cases} F_A^\top(i)F_A(i) \leq \mathbb{I}, \\ F_B^\top(i)F_B(i) \leq \mathbb{I}, \\ F_{C_z}^\top(i)F_{C_z}(i) \leq \mathbb{I}. \end{cases}$$

The system disturbance, $w(t)$, is assumed to belong to $\mathscr{L}_2[0, \infty)$ which means that the following holds:

$$\mathbb{E}\left[\int_0^\infty w^\top(t)w(t)dt\right] < \infty. \tag{5.25}$$

This implies that the disturbance has finite energy.

Combining the system dynamics (5.24) and controller (5.6), we get:

$$E(r_t)\dot{x}(t) = [A(r_t, t) + B(r_t, t)F(r_t)C(r_t)] x(t) + B_w(r_t)w(t),$$
$$= A_{cl}(r_t, t)x(t) + B_w(r_t)w(t).$$

Based on the results of Chap. 4, the closed-loop dynamics of the nominal system will be piecewise regular, impulse-free and stochastically stable with disturbance rejection of level $\gamma > 0$ if there exists a set of nonsingular matrices $P = (P(1), \cdots, P(N))$ that satisfies the following for each $i \in \mathscr{S}$:

$$\begin{bmatrix} \bar{J}(i) & P^\top(i)B_w(i) & C_z^\top(i) + C^\top(i)F^\top(i)D_z^\top(i) \\ B_w^\top(i)P(i) & -\gamma^2\mathbb{I} & B_z^\top(i) \\ C_z(i) + D_z(i)F(i)C(i) & B_z(i) & -\mathbb{I} \end{bmatrix} < 0,$$

where

$$\bar{J}(i) = A^\top(i)P(i) + P^\top(i)A(i) + P^\top(i)B(i)F(i)C(i)$$
$$+ C^\top(i)F^\top(i)B^\top(i)P(i) + \lambda_{ii}E^\top(i)P(i)$$
$$+ \sum_{j=1, j\neq i}^{N} \varepsilon_P(j)\left[P^\top(j) + P(j)\right],$$

with the following constraints:

$$\varepsilon_P(i)\left[P^\top(i) + P(i)\right] \geq E^\top(i)P(i) = P^\top(i)E(i) \geq 0.$$

Pre- and post-multiply the first condition respectively by $\mathrm{diag}\,[X^\top(i), \mathbb{I}, \mathbb{I}]$ and $\mathrm{diag}\,[X(i), \mathbb{I}, \mathbb{I}]$, where $X(i) = P^{-1}(i)$, and the second one by $X^\top(i)$ and $X(i)$ respectively, we get:

$$\begin{bmatrix} \widetilde{J}(i) & B_w(i) & X^\top(i)C_z^\top(i) + X^\top(i)C^\top(i)F^\top(i)D_z^\top(i) \\ B_w^\top(i) & -\gamma^2\mathbb{I} & B_z^\top(i) \\ C_z(i)X(i) + D_z(i)F(i)C(i)X(i) & B_z(i) & -\mathbb{I} \end{bmatrix} < 0$$

where

$$\widetilde{J}(i) = X^\top(i)A^\top(i) + A(i)X(i) + B(i)F(i)C(i)X(i)$$
$$+ X^\top(i)C^\top(i)F^\top(i)B^\top(i) + \lambda_{ii}X^\top(i)E^\top(i)$$
$$+ \sum_{j=1, j\neq i}^{N} \varepsilon_P(j)X^\top(i)\left[X^{-\top}(j) + X^{-1}(j)\right]X(i),$$

with the following constraints:

$$\varepsilon_P(i)\left[X^\top(i) + X(i)\right] \geq X^\top(i)E^\top(i) = E(i)X(i) \geq 0.$$

Let $C(i)X(i) = Y(i)C(i)$ holds for an appropriate $Y(i)$ and define $G(i) = F(i)Y(i)$. Based on this we get:

$$\begin{aligned}
\widetilde{J}(i) = \ &X^\top(i)A^\top(i) + A(i)X(i) + B(i)G(i)C(i) \\
&+ C^\top(i)G^\top(i)B^\top(i) + \lambda_{ii}X^\top(i)E^\top(i) \\
&+ \sum_{j=1, j\neq i}^{N} \varepsilon_P(j)X^\top(i)\left[X^{-\top}(j) + X^{-1}(j)\right]X(i)
\end{aligned}$$

Using now the fact that $X^\top(i) + X(i) \leq \mathbb{I} + (X^\top(i)X(i))$ and proceeding as we did before, we get the following results.

Theorem 5.3.1 *Let $\varepsilon_P = (\varepsilon_P(1), \cdots, \varepsilon_P(N))$ be a given set of positive scalars and γ a given positive scalar. There exists a static output feedback controller of the form (5.6) such that the closed-loop nominal system (5.24) is piecewise regular, impulse-free and stochastically stable and guarantees the disturbance rejection of level γ, if there exist sets of nonsingular matrices $X = (X(1), \cdots, X(N))$ with $X(i) \in \mathbb{R}^{n\times n}$, and $Y = (Y(1), \cdots, Y(N))$ with $Y(i) \in \mathbb{R}^{p\times p}$, a set of matrices $G = (G(1), \cdots, G(N))$ with $G(i) \in \mathbb{R}^{m\times p}$ such that the following holds for each $i \in \mathscr{S}$:*

$$\begin{bmatrix}
J(i) & B_w(i) & X^\top(i)C_z^\top(i)+C^\top(i)G^\top(i)D_z^\top(i) & S_i(X) & S_i(X) \\
B_w^\top(i) & -\gamma^2\mathbb{I} & B_z^\top(i) & 0 & 0 \\
C_z(i)X(i)+D_z(i)G(i)C(i) & B_z(i) & -\mathbb{I} & 0 & 0 \\
S_i^\top(X) & 0 & 0 & -\mathbb{I} & 0 \\
S_i^\top(X) & 0 & 0 & 0 & -X_i(X)
\end{bmatrix} < 0, \tag{5.26}$$

where

$$\begin{aligned}
J(i) = \ &A(i)X(i) + X^\top(i)A^\top(i) + B(i)G(i)C(i) + C^\top(i)G^\top(i)B^\top(i) \\
&+ \lambda_{ii}X^\top(i)E^\top(i), \\
S_i(X) = \ &\Big[\sqrt{\lambda_{i1}\varepsilon_P(1)}X^\top(i), \cdots, \sqrt{\lambda_{ii-1}\varepsilon_P(i-1)}X^\top(i), \\
&\sqrt{\lambda_{ii+1}\varepsilon_P(i+1)}X^\top(i), \cdots, \sqrt{\lambda_{iN}\varepsilon_P(N)}X^\top(i)\Big], \\
X_i(X) = \ &\mathrm{diag}\Big[X^\top(1) + X(1) - \mathbb{I}, \cdots, X^\top(i-1) + X(i-1) - \mathbb{I}, \\
&X^\top(i+1) + X(i+1) - \mathbb{I}, \cdots, X^\top(N) + X(N) - \mathbb{I}\Big],
\end{aligned}$$

with the following constraints:

$$\begin{cases}
\varepsilon_P(i)\left[X^\top(i) + X(i)\right] \geq E(i)X(i) = X^\top(i)E^\top(i) \geq 0, \\
Y(i)C(i) = C(i)X(i).
\end{cases} \tag{5.27}$$

The controller gain is given by $F(i) = G(i)Y^{-1}(i)$, $i \in \mathscr{S}$.

For the uncertain system, we can establish the results of the next theorem by following the same steps as in Chaps. 4 and 5. In fact, the uncertain system is piecewise regular, impulse-free and stochastically stable and guarantees the disturbance rejection of level γ if there exists a set of nonsingular matrices $P = (P(1), \cdots, P(N))$ that satisfies the following for each $i \in \mathscr{S}$:

$$
\begin{bmatrix}
\bar{J}(i) & P^\top(i)B_w(i) & W^\top(i) \\
B_w^\top(i)P(i) & -\gamma^2\mathbb{I} & B_z^\top(i) \\
W(i) & B_z(i) & -\mathbb{I}
\end{bmatrix} < 0,
$$

where

$$
\begin{aligned}
\bar{J}(i) = {}& A^\top(i)P(i) + P^\top(i)A(i) + P^\top(i)B(i)F(i)C(i) \\
& + P^\top(i)D_A(i)F_A(i)E_A(i) + E_A^\top(i)F_A^\top(i)D^\top(i)P(i) \\
& + C^\top(i)F^\top(i)E_B^\top(i)F_B^\top(i)D_B^\top(i)P(i) + P^\top(i)D_B(i)F_B(i)E_B(i)F(i)C(i) \\
& + C^\top(i)F^\top(i)B^\top(i)P(i) + \lambda_{ii}E^\top(i)P(i) + \sum_{j=1,j\neq i}^{N} \varepsilon_P(j)\left[P^\top(j) + P(j)\right],
\end{aligned}
$$

$$
W(i) = C_z(i) + D_{C_z}(i)F_{C_z}(i)E_{C_z}(i) + D_z(i)F(i)C(i),
$$

with the following constraints:

$$
\varepsilon_P(i)\left[P^\top(i) + P(i)\right] \geq E^\top(i)P(i) = P^\top(i)E(i) \geq 0.
$$

Notice that:

$$
\begin{bmatrix}
0 & 0 & 0 \\
0 & 0 & 0 \\
D_{C_z}(i)F_{C_z}(i)E_{C_z}(i) & 0 & 0
\end{bmatrix} = \begin{bmatrix}
0 \\
0 \\
D_{C_z}(i)
\end{bmatrix} F_{C_z}(i)\begin{bmatrix} E_{C_z}(i) & 0 & 0 \end{bmatrix},
$$

$$
\begin{bmatrix}
P^\top(i)D_A(i)F_A(i)E_A(i) & 0 & 0 \\
0 & 0 & 0 \\
0 & 0 & 0
\end{bmatrix} = \begin{bmatrix}
P^\top(i)D_A(i) \\
0 \\
0
\end{bmatrix} F_A(i)\begin{bmatrix} E_A(i) & 0 & 0 \end{bmatrix},
$$

$$
\begin{bmatrix}
P^\top(i)D_B(i)F_B(i)E_B(i)F(i)C(i) & 0 & 0 \\
0 & 0 & 0 \\
0 & 0 & 0
\end{bmatrix} = \begin{bmatrix}
P^\top(i)D_B(i) \\
0 \\
0
\end{bmatrix} F_B(i)
$$

$$
\times \begin{bmatrix} E_B(i)F(i)C(i) & 0 & 0 \end{bmatrix}.
$$

Using Lemma 1.5.1, for some positive sets of scalars $\varepsilon_A = (\varepsilon_A(1), \cdots, \varepsilon_A(N))$, $\varepsilon_B = (\varepsilon_B(1), \cdots, \varepsilon_B(N))$ and $\varepsilon_{C_z} = \left(\varepsilon_{C_z}(1), \cdots, \varepsilon_{C_z}(N)\right)$ and proceeding similarly as we did for nominal system, we get the following results

Theorem 5.3.2 *Let $\varepsilon_P = (\varepsilon_P(1), \cdots, \varepsilon_P(N))$ be a given set of positive scalars and γ a given positive scalar. There exists a static output feedback controller of the form (5.6) such that the closed-loop nominal system (5.24) is piecewise regular, impulse-free and stochastically stable and guarantees the disturbance rejection of level γ, if* *there exist sets of nonsingular matrices* $X = (X(1), \cdots, X(N))$ *with* $X(i) \in \mathbb{R}^{n \times n}$, *and* $Y = (Y(1), \cdots, Y(N))$ *with* $Y(i) \in \mathbb{R}^{p \times p}$, *a set of matrices* $G = (G(1), \cdots, G(N))$ *with* $G(i) \in \mathbb{R}^{m \times p}$ *and sets of positive scalars* $\varepsilon_A = (\varepsilon_A(1), \cdots, \varepsilon_A(N))$, $\varepsilon_B = (\varepsilon_B(1), \cdots, \varepsilon_B(N))$ *and* $\varepsilon_{C_z} = \left(\varepsilon_{C_z}(1), \cdots, \varepsilon_{C_z}(N) \right)$ *such that the following holds for each* $i \in \mathscr{S}$:

$$
\left[
\begin{array}{ccc}
J(i) & B_w(i) & X^\top(i)C_z^\top(i) + C^\top(i)G^\top(i)D_z^\top(i) \\
B_w^\top(i) & -\gamma^2 \mathbb{I} & B_z^\top(i) \\
C_z(i)X(i) + D_z(i)G(i)C(i) & B_z(i) & -\mathbb{I} + \varepsilon_{C_z}(i)D_{C_z}(i)D_{C_z}^\top(i) \\
E_A(i)X(i) & 0 & 0 \\
E_B(i)G(i)C(i) & 0 & 0 \\
E_{C_z}(i)X(i) & 0 & 0 \\
S_i^\top(X) & 0 & 0 \\
S_i^\top(X) & 0 & 0 \\
\end{array}
\right.
$$

$$
\left.
\begin{array}{ccccc}
X^\top(i)E_A^\top(i) & C^\top(i)G^\top(i)E_B^\top(i) & X^\top(i)E_{C_z}^\top(i) & S_i(X) & S_i(X) \\
0 & 0 & 0 & 0 & 0 \\
0 & 0 & 0 & 0 & 0 \\
0 & 0 & 0 & 0 & 0 \\
-\varepsilon_A(i)\mathbb{I} & 0 & 0 & 0 & 0 \\
0 & -\varepsilon_B(i)\mathbb{I} & 0 & 0 & 0 \\
0 & 0 & -\varepsilon_{C_z}(i)\mathbb{I} & 0 & 0 \\
0 & 0 & 0 & -\mathbb{I} & 0 \\
0 & 0 & 0 & 0 & -\mathcal{X}_i(X) \\
\end{array}
\right] < 0, \qquad (5.28)
$$

where

$$
\begin{aligned}
J(i) &= A(i)X(i) + X^\top(i)A^\top(i) + B(i)G(i)C(i) + C^\top(i)G^\top(i)B^\top(i) \\
&\quad + \varepsilon_A(i)D_A(i)D_A^\top(i) + \varepsilon_B(i)D_B(i)D_B^\top(i) \\
&\quad + \lambda_{ii}X^\top(i)E^\top(i), \\
S_i(X) &= \left[\sqrt{\lambda_{i1}}\varepsilon_P(1)X^\top(i), \cdots, \sqrt{\lambda_{ii-1}}\varepsilon_P(i-1)X^\top(i), \right. \\
&\quad \left. \sqrt{\lambda_{ii+1}}\varepsilon_P(i+1)X^\top(i), \cdots, \sqrt{\lambda_{iN}}\varepsilon_P(N)X^\top(i) \right], \\
\mathcal{X}_i(X) &= \mathrm{diag}\left[X^\top(1) + X(1) - \mathbb{I}, \cdots, X^\top(i-1) + X(i-1) - \mathbb{I}, \right. \\
&\quad \left. X^\top(i+1) + X(i+1) - \mathbb{I}, \cdots, X^\top(N) + X(N) - \mathbb{I} \right],
\end{aligned}
$$

with the following constraints:

$$
\begin{cases}
\varepsilon_P(i)\left[X^\top(i) + X(i)\right] \geq E(i)X(i) = X^\top(i)E^\top(i) \geq 0, \\
Y(i)C(i) = C(i)X(i).
\end{cases}
\qquad (5.29)
$$

The controller gain is given by $F(i) = G(i)Y^{-1}(i)$, $i \in \mathscr{S}$.

For the other approaches of finding an upper bound of the term $E^\top(i)P(i)$, we can establish the following results either for the nominal or the uncertain system.

Therefore when $E^\top(i)P(i) \leq \varepsilon(i)P^\top(i)P(i)$, proceeding as we did in the past we get the following results.

Theorem 5.3.3 *Let γ a given positive scalar. There exists a static output feedback controller of the form (5.6) such that the closed-loop nominal system (5.24) is piecewise regular, impulse-free and stochastically stable and guarantees the disturbance rejection of level γ, if there exist sets of nonsingular matrices $X = (X(1), \cdots, X(N))$ with $X(i) \in \mathbb{R}^{n \times n}$, and $Y = (Y(1), \cdots, Y(N))$ with $Y(i) \in \mathbb{R}^{p \times p}$, a set of matrices $G = (G(1), \cdots, G(N))$ with $G(i) \in \mathbb{R}^{m \times p}$ and a set of positive scalars $\varepsilon = (\varepsilon(1), \cdots, \varepsilon(N))$ such that the following holds for each $i \in \mathscr{S}$:*

$$
\begin{bmatrix}
J(i) & B_w(i) & X^\top(i)C_z^\top(i) + C^\top(i)G^\top(i)D_z^\top(i) & S_i(X) \\
B_w^\top(i) & -\gamma^2\mathbb{I} & B_z^\top(i) & 0 \\
C_z(i)X(i) + D_z(i)G(i)C(i) & B_z(i) & -\mathbb{I} & 0 \\
S_i^\top(X) & 0 & 0 & -X_i(X)
\end{bmatrix} < 0,
$$

where

$$
J(i) = A(i)X(i) + X^\top(i)A^\top(i) + B(i)G(i)C(i) + C^\top(i)G^\top(i)B^\top(i)
$$
$$
+ \lambda_{ii}X^\top(i)E^\top(i),
$$
$$
S_i(X) = \left[\sqrt{\lambda_{i1}}X^\top(i), \cdots, \sqrt{\lambda_{ii-1}}X^\top(i), \right.
$$
$$
\left. \sqrt{\lambda_{ii+1}}X^\top(i), \cdots, \sqrt{\lambda_{iN}}X^\top(i) \right],
$$
$$
X_i(X) = \mathrm{diag}\left[X^\top(1) + X(1) - \varepsilon(1)\mathbb{I}, \cdots, X^\top(i-1) + X(i-1) - \varepsilon(i-1)\mathbb{I}, \right.
$$
$$
\left. X^\top(i+1) + X(i+1) - \varepsilon(i+1)\mathbb{I}, \cdots, X^\top(N) + X(N) - \varepsilon(N)\mathbb{I} \right],
$$

with the following constraints:

$$
\begin{cases}
\varepsilon(i)\mathbb{I} \geq E(i)X(i) = X^\top(i)E^\top(i) \geq 0, \\
Y(i)C(i) = C(i)X(i).
\end{cases}
\tag{5.30}
$$

The controller gain is given by $F(i) = G(i)Y^{-1}(i)$, $i \in \mathscr{S}$.

Theorem 5.3.4 *Let γ a given positive scalar. There exists a static output feedback controller of the form (5.6) such that the closed-loop nominal system (5.24) is piecewise regular, impulse-free and stochastically stable and guarantees the disturbance rejection of level γ, if there exist sets of nonsingular matrices $X = (X(1), \cdots, X(N))$ with $X(i) \in \mathbb{R}^{n \times n}$, and $Y = (Y(1), \cdots, Y(N))$ with $Y(i) \in \mathbb{R}^{p \times p}$, a set of matrices $G = (G(1), \cdots, G(N))$ with $G(i) \in \mathbb{R}^{m \times p}$ and a sets of positive scalars $\varepsilon = (\varepsilon(1), \cdots, \varepsilon(N))$, $\varepsilon_A = (\varepsilon_A(1), \cdots, \varepsilon_A(N))$, $\varepsilon_B = (\varepsilon_B(1), \cdots, \varepsilon_B(N))$ and $\varepsilon_{C_z} = \left(\varepsilon_{C_z}(1), \cdots, \varepsilon_{C_z}(N)\right)$ such that the following holds for each $i \in \mathscr{S}$:*

$$
\begin{bmatrix}
J(i) & B_w(i) & X^\top(i)C_z^\top(i) + C^\top(i)G^\top(i)D_z^\top(i) \\
B_w^\top(i) & -\gamma^2\mathbb{I} & B_z^\top(i) \\
C_z(i)X(i) + D_z(i)G(i)C(i) & B_z(i) & -\mathbb{I} + \varepsilon_{C_z}(i)D_{C_z}D_{C_z}^\top \\
E_A(i)X(i) & 0 & 0 \\
E_B(i)G(i)C(i) & 0 & 0 \\
E_{C_z}(i)X(i) & 0 & 0 \\
S_i^\top(X) & 0 & 0
\end{bmatrix}
$$

$$
\begin{bmatrix}
X^\top(i)E_A^\top(i) & C^\top(i)G^\top(i)E_B^\top(i) & X^\top(i)E_{C_z}^\top(i) & S_i(X) \\
0 & 0 & 0 & 0 \\
0 & 0 & 0 & 0 \\
-\varepsilon_A(i)\mathbb{I} & 0 & 0 & 0 \\
0 & -\varepsilon_B(i)\mathbb{I} & 0 & 0 \\
0 & 0 & -\varepsilon_{C_z}(i)\mathbb{I} & 0 \\
0 & 0 & 0 & -X_i(X)
\end{bmatrix} < 0, \quad (5.31)
$$

where

$$
J(i) = A(i)X(i) + X^\top(i)A^\top(i) + B(i)G(i)C(i) + C^\top(i)G^\top(i)B^\top(i)
$$
$$
+ \lambda_{ii}X^\top(i)E^\top(i) + \varepsilon_A(i)D_A(i)D_A^\top(i) + \varepsilon_B(i)D_B(i)D_B^\top(i),
$$
$$
S_i(X) = \left[\sqrt{\lambda_{i1}}X^\top(i), \cdots, \sqrt{\lambda_{ii-1}}X^\top(i), \right.
$$
$$
\left. \sqrt{\lambda_{ii+1}}X^\top(i), \cdots, \sqrt{\lambda_{iN}}X^\top(i) \right],
$$
$$
X_i(X) = \mathrm{diag}\left[X^\top(1) + X(1) - \varepsilon(1)\mathbb{I}, \cdots, X^\top(i-1) + X(i-1) - \varepsilon(i-1)\mathbb{I}, \right.
$$
$$
\left. X^\top(i+1) + X(i+1) - \varepsilon(i+1)\mathbb{I}, \cdots, X^\top(N) + X(N) - \varepsilon(N)\mathbb{I} \right],
$$

with the following constraints:

$$
\begin{cases}
\varepsilon(i)\mathbb{I} \geq E(i)X(i) = X^\top(i)E^\top(i) \geq 0, \\
Y(i)C(i) = C(i)X(i).
\end{cases} \quad (5.32)
$$

The controller gain is given by $F(i) = G(i)Y^{-1}(i)$, $i \in \mathscr{S}$.

When we use the following relation

$$
E^\top(i)P(i) \leq \varepsilon^{-1}(i)\frac{1}{4}\mathbb{I} + \varepsilon(i)E^\top(i)P(i)P^\top(i)E(i),
$$

we get the following results.

Theorem 5.3.5 *Let γ a given positive scalar. There exists a static output feedback controller of the form (5.6) such that the closed-loop nominal system (5.24) is piecewise regular, impulse-free and stochastically stable and guarantees the disturbance rejection of level γ, if there exist sets of nonsingular matrices $X = (X(1), \cdots, X(N))$*

with $X(i) \in \mathbb{R}^{n \times n}$, and $Y = (Y(1), \cdots, Y(N))$ with $Y(i) \in \mathbb{R}^{p \times p}$, a set of matrices $G = (G(1), \cdots, G(N))$ with $G(i) \in \mathbb{R}^{m \times p}$ and a set of positive scalars $\varepsilon = (\varepsilon(1), \cdots, \varepsilon(N))$ such that the following holds for each $i \in \mathscr{S}$:

$$
\begin{bmatrix}
J(i) & B_w(i) & X^\top(i)C_z^\top(i)+C^\top(i)G^\top(i)D_z^\top(i) & Z_i(X) & S_i(X) \\
B_w^\top(i) & -\gamma^2 \mathbb{I} & B_z^\top(i) & 0 & 0 \\
C_z(i)X(i)+D_z(i)G(i)C(i) & B_z(i) & -\mathbb{I} & 0 & 0 \\
Z_i^\top(X) & 0 & 0 & -X_i(\varepsilon) & 0 \\
S_i^\top(X) & 0 & 0 & 0 & -X_i(X)
\end{bmatrix} < 0,
$$
(5.33)

where

$$
\begin{aligned}
J(i) &= A(i)X(i) + X^\top(i)A^\top(i) + B(i)G(i)C(i) + C^\top(i)G^\top(i)B^\top(i) \\
&\quad + \lambda_{ii}X^\top(i)E^\top(i), \\
Z_i(X) &= \left[\sqrt{\lambda_{i1}}X^\top(i), \cdots, \sqrt{\lambda_{ii-1}}X^\top(i), \right. \\
&\quad \left. \sqrt{\lambda_{ii+1}}X^\top(i), \cdots, \sqrt{\lambda_{iN}}X^\top(i) \right], \\
X_i(\varepsilon) &= \operatorname{diag}\left[4\varepsilon(1)\mathbb{I}, \cdots, 4\varepsilon(i-1)\mathbb{I}, 4\varepsilon(i+1)\mathbb{I}, \cdots, 4\varepsilon(N)\mathbb{I} \right], \\
S_i(X) &= \left[\sqrt{\lambda_{i1}}X^\top(i)E^\top(1), \cdots, \sqrt{\lambda_{ii-1}}X^\top(i)E^\top(i-1), \right. \\
&\quad \left. \sqrt{\lambda_{ii+1}}X^\top(i)E^\top(i+1), \cdots, \sqrt{\lambda_{iN}}X^\top(i)E^\top(N) \right], \\
X_i(X) &= \operatorname{diag}\left[X^\top(1) + X(1) - \varepsilon(1)\mathbb{I}, \cdots, X^\top(i-1) + X(i-1) - \varepsilon(i-1)\mathbb{I}, \right. \\
&\quad \left. X^\top(i+1) + X(i+1) - \varepsilon(i+1)\mathbb{I}, \cdots, X^\top(N) + X(N) - \varepsilon(N)\mathbb{I} \right],
\end{aligned}
$$

with the following constraints:

$$
\begin{cases}
E(i)X(i) = X^\top(i)E^\top(i) \geq 0, \\
Y(i)C(i) = C(i)X(i).
\end{cases}
$$
(5.34)

The controller gain is given by $F(i) = G(i)Y^{-1}(i)$, $i \in \mathscr{S}$.

For the uncertain system we can establish the results of the next theorem by following the same steps as in Chaps. 4 and 5.

Theorem 5.3.6 *Let γ a given positive scalar. There exists a static output feedback controller of the form (5.6) such that the closed-loop nominal system (5.24) is piecewise regular, impulse-free and stochastically stable and guarantees the disturbance rejection of level γ, if there exist sets of nonsingular matrices $X = (X(1), \cdots, X(N))$ with $X(i) \in \mathbb{R}^{n \times n}$, and $Y = (Y(1), \cdots, Y(N))$ with $Y(i) \in \mathbb{R}^{p \times p}$, a set of matrices $G = (G(1), \cdots, G(N))$ with $G(i) \in \mathbb{R}^{m \times p}$ and a sets of positive scalars $\varepsilon = (\varepsilon(1), \cdots, \varepsilon(N))$, $\varepsilon_A = (\varepsilon_A(1), \cdots, \varepsilon_A(N))$, $\varepsilon_B = (\varepsilon_B(1), \cdots, \varepsilon_B(N))$ and $\varepsilon_{C_z} = \left(\varepsilon_{C_z}(1), \cdots, \varepsilon_{C_z}(N) \right)$ such that the following holds for each $i \in \mathscr{S}$:*

$$
\begin{bmatrix}
J(i) & B_w(i) & X^\top(i)C_z^\top(i) + C^\top(i)G^\top(i)D_z^\top(i) \\
B_w^\top(i) & -\gamma^2\mathbb{I} & B_z^\top(i) \\
C_z(i)X(i) + D_z(i)G(i)C(i) & B_z(i) & -\mathbb{I} + \varepsilon_{C_z}(i)D_{C_z}D_{C_z}^\top \\
E_A(i)X(i) & 0 & 0 \\
E_B(i)G(i)C(i) & 0 & 0 \\
E_{C_z}(i)X(i) & 0 & 0 \\
\mathcal{Z}_i^\top(X) & 0 & 0 \\
\mathcal{S}_i^\top(X) & 0 & 0
\end{bmatrix}
$$

$$
\begin{bmatrix}
X^\top(i)E_A^\top(i) & C^\top(i)G^\top(i)E_B^\top(i) & X^\top(i)E_{C_z}^\top(i) & \mathcal{Z}_i(X) & \mathcal{S}_i(X) \\
0 & 0 & 0 & 0 & 0 \\
0 & 0 & 0 & 0 & 0 \\
-\varepsilon_A(i)\mathbb{I} & 0 & 0 & 0 & 0 \\
0 & -\varepsilon_B(i)\mathbb{I} & 0 & 0 & 0 \\
0 & 0 & -\varepsilon_{C_z}(i)\mathbb{I} & 0 & 0 \\
0 & 0 & 0 & -\mathcal{X}_i(\varepsilon) & 0 \\
0 & 0 & 0 & 0 & -\mathcal{X}_i(X)
\end{bmatrix} < 0, \quad (5.35)
$$

where

$$
J(i) = A(i)X(i) + X^\top(i)A^\top(i) + B(i)G(i)C(i) + C^\top(i)G^\top(i)B^\top(i)
$$
$$
+ \lambda_{ii}X^\top(i)E^\top(i) + \varepsilon_A(i)D_A(i)D_A^\top(i) + \varepsilon_B(i)D_B(i)D_B^\top(i),
$$

$$
\mathcal{Z}_i(X) = \left[\sqrt{\lambda_{i1}}X^\top(i), \cdots, \sqrt{\lambda_{ii-1}}X^\top(i), \right.
$$
$$
\left. \sqrt{\lambda_{ii+1}}X^\top(i), \cdots, \sqrt{\lambda_{iN}}X^\top(i) \right],
$$

$$
\mathcal{X}_i(\varepsilon) = \mathrm{diag}\left[4\varepsilon(1)\mathbb{I}, \cdots, 4\varepsilon(i-1)\mathbb{I}, 4\varepsilon(i+1)\mathbb{I}, \cdots, 4\varepsilon(N)\mathbb{I}\right],
$$

$$
\mathcal{S}_i(X) = \left[\sqrt{\lambda_{i1}}X^\top(i)E^\top(1), \cdots, \sqrt{\lambda_{ii-1}}X^\top(i)E^\top(i-1), \right.
$$
$$
\left. \sqrt{\lambda_{ii+1}}X^\top(i)E^\top(i+1), \cdots, \sqrt{\lambda_{iN}}X^\top(i)E^\top(N) \right],
$$

$$
\mathcal{X}_i(X) = \mathrm{diag}\left[X^\top(1) + X(1) - \varepsilon(1)\mathbb{I}, \cdots, X^\top(i-1) + X(i-1) - \varepsilon(i-1)\mathbb{I}, \right.
$$
$$
\left. X^\top(i+1) + X(i+1) - \varepsilon(i+1)\mathbb{I}, \cdots, X^\top(N) + X(N) - \varepsilon(N)\mathbb{I} \right],
$$

with the following constraints:

$$
\begin{cases}
E(i)X(i) = X^\top(i)E^\top(i) \geq 0, \\
Y(i)C(i) = C(i)X(i).
\end{cases}
\tag{5.36}
$$

The controller gain is given by $F(i) = G(i)Y^{-1}(i)$, $i \in \mathscr{S}$.

5.4 Numerical Examples

In this section we will provide some numerical examples to show of the developed results in this chapter. Two examples are given to show the results of the third approach we developed for the design of a static output feedback controller in the two following cases:

- nominal system
- uncertain system.

Example 5.4.1 *To show the validness of our results, let us consider a two modes system with the following data:*

- *mode # 1:*

$$A(1) = \begin{bmatrix} 0.0 & -1.0 & 1.0 \\ -1.0 & 3.0 & 0.0 \\ 0.0 & 0.0 & 0.0 \end{bmatrix}, \ B(1) = \begin{bmatrix} 0.0 & 0.2 \\ 1.0 & 0.0 \\ -0.1 & 1.0 \end{bmatrix},$$

$$C(1) = \begin{bmatrix} 1.0 & 0.0 & 1.0 \\ 0.3 & 1.0 & 0.0 \end{bmatrix}.$$

- *mode # 2:*

$$A(2) = \begin{bmatrix} 0.0 & 1.5 & 1.5 \\ -1.0 & -3.0 & 0.0 \\ 0.0 & 0.0 & 0.0 \end{bmatrix}, \ B(2) = \begin{bmatrix} 0.0 & -0.2 \\ 1.2 & 0.0 \\ 0.1 & 1.2 \end{bmatrix},$$

$$C(2) = \begin{bmatrix} 1.0 & 0.0 & 1.0 \\ 0.1 & 1.0 & 0.0 \end{bmatrix}.$$

The matrix E is given by:

$$E = \begin{bmatrix} 1.0 & 0.0 & 0.0 \\ 0.0 & 1.0 & 0.0 \\ 0.0 & 0.0 & 0.0 \end{bmatrix}.$$

The switching between the two modes is described by:

$$\Lambda = \begin{bmatrix} -1.0 & 1.0 \\ 1.1 & -1.0 \end{bmatrix}.$$

Solving the set of LMIs (5.13)-(5.14) gives:

$$\varepsilon(1) = 0.3551, \ \varepsilon(2) = 0.4103$$

$$X(1) = \begin{bmatrix} 0.5265 & 0.0364 & 0.0 \\ 0.0364 & 0.6370 & 0.0 \\ -0.0197 & -0.0335 & 0.5059 \end{bmatrix}, \ Y(1) = \begin{bmatrix} 0.5059 & 0.0029 \\ -0.0000 & 0.6479 \end{bmatrix},$$

$$G(1) = \begin{bmatrix} 2.0424 & -2.4290 \\ -0.1198 & -2.1132 \end{bmatrix}, \ X(2) = \begin{bmatrix} 0.7619 & -0.0162 & 0.0 \\ -0.0162 & 0.6019 & 0.0 \\ -0.3189 & -0.0891 & 0.4535 \end{bmatrix},$$

$$Y(2) = \begin{bmatrix} 0.4535 & -0.1053 \\ 0.0000 & 0.6003 \end{bmatrix}, \ G(2) = \begin{bmatrix} -0.1033 & 1.0530 \\ -0.4625 & 0.0353 \end{bmatrix}.$$

This gives the following gains:

$$K(1) = \begin{bmatrix} 4.0370 & -3.7671 \\ -0.2368 & -3.2606 \end{bmatrix}, \ K(2) = \begin{bmatrix} -0.2278 & 1.7142 \\ -1.0198 & -0.1201 \end{bmatrix}.$$

Example 5.4.2 *As a second example, let us consider the same system of the previous example with the following data:*

- *mode # 1:*

$$D_A(1) = \begin{bmatrix} 0.1 \\ 0.1 \\ 0.0 \end{bmatrix},$$

$$E_A(1) = \begin{bmatrix} 0.0 & 0.1 & 0.0 \end{bmatrix},$$

$$D_B(1) = \begin{bmatrix} 0.0 \\ 0.1 \\ 0.1 \end{bmatrix},$$

$$E_B(1) = \begin{bmatrix} 0.1 & 0.0 \end{bmatrix}.$$

- *mode # 2:*

$$D_A(2) = \begin{bmatrix} 0.2 \\ 0.1 \\ 0.0 \end{bmatrix},$$

$$E_A(2) = \begin{bmatrix} 0.0 & 0.2 & 0.0 \end{bmatrix},$$

$$D_B(2) = \begin{bmatrix} 0.0 \\ -0.1 \\ 0.1 \end{bmatrix},$$

$$E_B(2) = \begin{bmatrix} -0.1 & 0.0 \end{bmatrix}.$$

Solving the set of LMIs (5.22)-(5.23) gives:

$$\varepsilon(1) = 0.3594, \ \varepsilon(2) - 0.4185,$$
$$\varepsilon_A(1) = 0.9657, \ \varepsilon_A(2) = 0.9168,$$
$$\varepsilon_B(1) = 0.9442, \ \varepsilon_B(2) = 1.0042,$$

$$X(1) = \begin{bmatrix} 0.5277 & 0.0333 & 0.0 \\ 0.0333 & 0.6287 & 0.0 \\ -0.0228 & -0.0269 & 0.5030 \end{bmatrix}, \ Y(1) = \begin{bmatrix} 0.5030 & 0.0064 \\ -0.0000 & 0.6387 \end{bmatrix},$$

$$G(1) = \begin{bmatrix} 2.0414 & -2.4416 \\ -0.1283 & -2.0879 \end{bmatrix}, \ X(2) = \begin{bmatrix} 0.7796 & -0.0188 & 0.0 \\ -0.0188 & 0.5936 & 0.0 \\ -0.3346 & -0.1019 & 0.4570 \end{bmatrix},$$

$$Y(2) = \begin{bmatrix} 0.4570 & -0.1207 \\ 0.0000 & 0.5917 \end{bmatrix}, \ G(2) = \begin{bmatrix} -0.0824 & 1.0104 \\ -0.4629 & 0.0315 \end{bmatrix}.$$

This gives the following gains:

$$K(1) = \begin{bmatrix} 4.0585 & -3.8635 \\ -0.2551 & -3.2666 \end{bmatrix}, \ K(2) = \begin{bmatrix} -0.1804 & 1.6708 \\ -1.0128 & -0.1534 \end{bmatrix}.$$

Example 5.4.3 *To show the results of $\mathcal{H}_\infty$ static output feedback stabilization let us consider the same system at the previous example with the following data:*

- *mode # 1:*

$$B_w(1) = \begin{bmatrix} 0.1 \\ 0 \\ -0.1 \end{bmatrix},$$

$$C_z(1) = \begin{bmatrix} 1.0 & -1.0 & 0.0 \end{bmatrix},$$

$$D_z(1) = \begin{bmatrix} 0.0 & 1.0 \end{bmatrix},$$

$$B_z(1) = \begin{bmatrix} 0.0 \end{bmatrix}.$$

- *mode # 2:*

$$B_w(2) = \begin{bmatrix} 0.01 \\ 0.0 \\ 0.1 \end{bmatrix},$$

$$C_z(2) = \begin{bmatrix} 1.2 & -1.2 & 0.0 \end{bmatrix},$$

$$D_z(2) = \begin{bmatrix} 0.0 & 1.1 \end{bmatrix},$$

$$B_z(2) = \begin{bmatrix} 0.0 \end{bmatrix}.$$

Solving the set of LMIs (5.33)-(5.34) gives:

$$\varepsilon(1) = 0.0803, \ \varepsilon(2) = 0.1149, \ \gamma = 1.8607,$$

$$X(1) = \begin{bmatrix} 0.1228 & 0.0403 & 0.0 \\ 0.0403 & 0.2450 & 0.0 \\ -0.0097 & -0.0448 & 0.1144 \end{bmatrix}, \ Y(1) = \begin{bmatrix} 0.1144 & -0.0045 \\ -0.0000 & 0.2571 \end{bmatrix},$$

$$G(1) = \begin{bmatrix} 0.5300 & -1.6808 \\ -0.0389 & -0.5342 \end{bmatrix}, \ X(2) = \begin{bmatrix} 0.1889 & -0.0069 & 0.0 \\ -0.0069 & 0.1208 & 0.0 \\ -0.0821 & -0.0582 & 0.1132 \end{bmatrix},$$

$$Y(2) = \begin{bmatrix} 0.1132 & -0.0651 \\ 0.0000 & 0.1201 \end{bmatrix}, \ G(2) = \begin{bmatrix} 0.0889 & -0.0175 \\ -0.1177 & -0.0253 \end{bmatrix}.$$

This gives the following gains:

$$K(1) = \begin{bmatrix} 4.6320 & -6.4566 \\ -0.3403 & -2.0835 \end{bmatrix}, \ K(2) = \begin{bmatrix} 0.7853 & 0.2797 \\ -1.0393 & -0.7740 \end{bmatrix}.$$

Example 5.4.4 *To show the results of robust $\mathcal{H}_\infty$ static output feedback stabilization let us consider the same system at the previous examples. Solving the set of LMIs (5.35)-(5.36) gives:*

$$\varepsilon(1) = 0.0728, \ \varepsilon(2) = 0.1045, \ \gamma = 2.0254,$$

$$\varepsilon_A(1) = 0.1761, \ \varepsilon_A(2) = 0.0855,$$

$$\varepsilon_B(1) = 0.0785, \ \varepsilon_B(2) = 0.6548,$$

$$X(1) = \begin{bmatrix} 0.1111 & 0.0333 & 0.0 \\ 0.0333 & 0.2121 & 0.0 \\ -0.0104 & -0.0477 & 0.1051 \end{bmatrix}, \ Y(1) = \begin{bmatrix} 0.1051 & -0.0144 \\ -0.0000 & 0.2221 \end{bmatrix},$$

$$G(1) = \begin{bmatrix} 0.5036 & -1.5655 \\ -0.0417 & -0.4428 \end{bmatrix}, \ X(2) = \begin{bmatrix} 0.1727 & -0.0070 & 0.0 \\ -0.0070 & 0.1032 & 0.0 \\ -0.0755 & -0.0515 & 0.1030 \end{bmatrix},$$

$$Y(2) = \begin{bmatrix} 0.1030 & -0.0585 \\ 0.0000 & 0.1025 \end{bmatrix}, \ G(2) = \begin{bmatrix} 0.0833 & -0.0333 \\ -0.1063 & -0.0216 \end{bmatrix}.$$

This gives the following gains:

$$K(1) = \begin{bmatrix} 4.7916 & -6.7367 \\ -0.3963 & -2.0193 \end{bmatrix}, \ K(2) = \begin{bmatrix} 0.8090 & 0.1370 \\ -1.0322 & -0.7995 \end{bmatrix}.$$

5.5 Notes

This chapter dealt with the output stabilizability problem of the singular class of systems with random abrupt changes. The stochastic output stabilizability and the robust stochastic output stabilizability problems have been considered and LMI conditions were developed. An output feedback controller that assures that closed-loop state equation either for the nominal system or the uncertain is piecewise regular, impulse-free and stochastically stable is designed in the LMI setting. The conditions we developed in this chapter are tractable using commercial optimization tools. The content of this chapter is mainly based on the work of the author and his coauthors [23].

6

Observer-Based Feedback Stabilization

Previously, we covered the state feedback stabilization that assumes the complete access to the state vector. Practically this assumption is not realistic since in general we don't have such complete access to the state vector for many reasons like the non existence of the appropriate sensors to measure some of the states or the limitation in the control budget. To continue using the state feedback controller an alternate consists of replacing the actual state measurements by its estimate in the expression of the control law. This technique is referred to as the observer-based stabilization.

The stabilization problem has been extensively studied and many interesting results already in the literature either for normal or singular systems. For more details of this, we refer the reader to [27] and the references therein. For the class of singular systems, the stabilization problem has been tackled and results exist also in the literature see for instance [52, 58, 81, 85, 87, 92, 91, 108, 127, 122, 123, 141, 142]. For the observer-based stabilization approach only few results have been reported in the literature among them we quote the works of [39, 47]. For the class of singular systems with random abrupt changes in the dynamics no results exist in the literature up to date.

Our goal in this chapter is to cover the observer-based stabilization. Both the stabilization for nominal systems and uncertain systems are covered. LMIs conditions will be developed to design controllers that make the closed-loop state equation with these controllers piecewise regular, impulse-free and stochastically stable. The rest of this chapter is organized as follows. In Sect. 1, the stabilization problem is formulated. In Sect. 2, the stabilization problem for nominal system is tackled and LMI conditions are established to design such controllers. In Sect. 3, the stabilization problem for uncertain systems are considered and similarly a design procedure of controller that renders the closed-loop system piecewise regular, impulse-free and stochastically stable for all admissible uncertainties.

6.1 Problem Statement

Let us consider a dynamical singular system with random abrupt changes defined in a fundamental probability space $(\Omega, \mathcal{F}, \mathbb{P})$ and assume that its state equation is described by the following differential-algebraic equations:

$$\begin{cases} E(r_t)\dot{x}(t) = A(r_t, t)x(t) + B(r_t, t)u(t), \, x(0) = x_0 \,, \\ y(t) = C_y(r_t, t)x(t) \,, \end{cases} \tag{6.1}$$

where $x(t) \in \mathbb{R}^n$ is the state vector, $x_0 \in \mathbb{R}^n$ is the initial state, $u(t) \in \mathbb{R}^m$ is the control input, $y(t) \in \mathbb{R}^p$ is the output system at time t, $\{r_t, t \geq 0\}$ is the continuous-time Markov process taking values in a finite space $\mathcal{S} = \{1, 2, \cdots, N\}$ and describes the evolution of the mode at time t, $E(i)$ is a known singular matrix with rank $(E(i)) = n_r \leq n$, for all $i \in \mathcal{S}$, $A(r_t, t) \in \mathbb{R}^{n \times n}$, $B(r_t, t) \in \mathbb{R}^{n \times m}$ and $C_y(r_t, t) \in \mathbb{R}^{p \times n}$ are matrices with the following forms for every $i \in \mathcal{S}$:

$$A(i, t) = A(i) + D_A(i)F_A(i)E_A(i) \,,$$
$$B(i, t) = B(i) + D_B(i)F_B(i)E_B(i) \,,$$
$$C_y(i, t) = C_y(i) + D_{C_y}(i)F_{C_y}(i)E_{C_y}(i) \,,$$

with $A(i) \in \mathbb{R}^{n \times n}$, $D_A(i) \in \mathbb{R}^{n \times n_D}$, $E_A(i) \in \mathbb{R}^{n_E \times n}$, $B(i) \in \mathbb{R}^{n \times m}$, $D_B(i) \in \mathbb{R}^{n \times m_D}$, $E_B(i) \in \mathbb{R}^{m_E \times m}$, $C_y(i) \in \mathbb{R}^{p \times n}$, $D_{C_y}(i) \in \mathbb{R}^{p \times p_D}$ and $E_{C_y}(i) \in \mathbb{R}^{p_E \times n}$ are real known matrices with appropriate dimensions, and $F_A(i) \in \mathbb{R}^{n_D \times n_E}$, $F_B(i) \in \mathbb{R}^{m_D \times m_E}$ and $F_{C_y}(i) \in \mathbb{R}^{p_D \times p_E}$ are unknown real matrices that satisfy the following:

$$\begin{cases} F_A^\top(i)F_A(i) \leq \mathbb{I} \,, \\ F_B^\top(i)F_B(i) \leq \mathbb{I} \,, \\ F_{C_y}^\top(i)F_{C_y}(i) \leq \mathbb{I} \,. \end{cases} \tag{6.2}$$

The Markov process $\{r_t, t \geq 0\}$ beside taking values in the finite set $\mathcal{S}$, represents the switching between the different modes and its state equation is described by the following probability transitions:

$$\mathbb{P}\left[r_{t+h} = j | r_t = i\right] = \begin{cases} \lambda_{ij}h + o(h) \,, & \text{when } r_t \text{ jumps from } i \text{ to } j \,, \\ 1 + \lambda_{ii}h + o(h) \,, & \text{otherwise} \,, \end{cases} \tag{6.3}$$

where λ_{ij} is the transition rate from mode i to mode j with $\lambda_{ij} \geq 0$ when $i \neq j$ and $\lambda_{ii} = -\sum_{j=1, j \neq i}^N \lambda_{ij}$ and $o(h)$ is such that $\lim_{h \to 0} \frac{o(h)}{h} = 0$.

Previously we assumed the complete access to the state vector to construct the state feedback control but unfortunately this is not always possible for many reasons like the non measurement of some states or the desire to design a control system with an acceptable cost. An alternative consists of using estimate of the state vector instead of the actual state in the control expression. The controller should be designed to guarantee that the closed-loop state equation of the class of systems we are considering in this chapter is piecewise regular, impulse-free and stochastically stable. Two

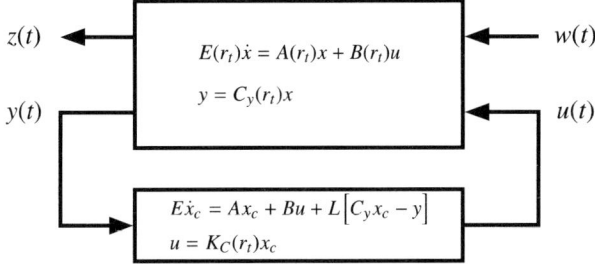

Fig. 6.1. Observer-based output feedback stabilization block diagram

approaches can be considered for this purpose. The first one consists of designing an open loop observer. But unfortunately, this observer will not work since it requires complete knowledge of the dynamics (no uncertainties in the dynamics) and complete knowledge of the initial state vector which are hard to satisfy. The second one consists of using the output measurements which will improve the estimate of the state vector. This technique will be covered in the rest of this chapter.

The structure of the controller is given by the following dynamics:

$$\begin{cases} E(r_t)\dot{x}_c(t) = A(r_t)x_c(t) + B(r_t)u(t) + L(r_t)\left[C_y(r_t)x_c(t) - y(t)\right], \\ u(t) = K(r_t)x_c(t), \end{cases} \tag{6.4}$$

where $x_c(t)$ is the observer state vector, and $K(i)$ and $L(i)$, $i \in \mathscr{S}$ are constant gain matrices for each $i \in \mathscr{S}$ that have to be determined and which constitutes one of our main goal in this chapter.

The block diagram of the closed-loop system under the observer-based state feedback controller is represented by Fig. 6.1.

The problem we will deal with in this chapter consists of determining the gains $K(i)$ and $L(i)$, $i = 1, \cdots, N$ of the controller and the observer that will make the closed-loop augmented system piecewise regular, impulse-free and stochastically stable. LMI conditions to determine such gains are of interest.

6.2 Observer-Based Stabilization of Singular Systems

Let us now concentrate on the design of an observer-based state controller which guarantees that the closed-loop state equation of the class of nominal systems we are considering in this chapter is piecewise regular, impulse-free and stochastically stable.

Before designing the stabilizing controller, we will develop the conditions that assure that the augmented system will be piecewise regular, impulse-free and stochastically stable. For this purpose, let us define the observer error by:

$$e(t) = x(t) - x_c(t). \tag{6.5}$$

Combining now the nominal system state equation and the controller dynamics, we get:

$$E(r_t)\dot{e}(t) = A(r_t)x(t) + B(r_t)K(r_t)x_c(t) - A(r_t)x_c(t) - B(r_t)K(r_t)x_c(t)$$
$$- L(r_t)\left[C_y(r_t)x_c(t) - y(t)\right]$$
$$= \left[A(r_t) + L(r_t)C_y(r_t)\right]e(t).$$

Using now the system and the error dynamics, we get the following augmented one:

$$\widetilde{E}(r_t)\dot{\eta}(t) = \tilde{A}(r_t)\eta(t), \tag{6.6}$$

where

$$\eta(t) = \begin{bmatrix} x(t) \\ e(t) \end{bmatrix},$$

$$\widetilde{E}(r_t) = \begin{bmatrix} E(r_t) & 0 \\ 0 & E(r_t) \end{bmatrix},$$

$$\tilde{A}(i) = \begin{bmatrix} A(i) + B(i)K(i) & -B(i)K(i) \\ 0 & A(i) + L(i)C_y(i) \end{bmatrix}.$$

The following theorem states that when the state feedback controller gains are fixed, the augmented closed-loop of the class of systems we are considering using the estimated state vector is piecewise regular, impulse-free and stochastically stable if some appropriate conditions are satisfied.

Theorem 6.2.1 *Let ε_P be a given positive scalar. Let $K = [K(1), \cdots, K(N)]$, $K(i) \in \mathbb{R}^{n \times n}$, and $L = [L(1), \cdots, L(N)]$, $L(i) \in \mathbb{R}^{n \times p}$ be given sets of constant matrices. If there exist sets of nonsingular matrices $P = [P(1), \cdots, P(N)]$, $P(i) \in \mathbb{R}^{n \times n}$ and $Q = [Q(1), \cdots, Q(N)]$, $Q(i) \in \mathbb{R}^{n \times n}$ satisfying the following for every $i \in S$:*

$$\begin{bmatrix} \mathcal{J}_P(i) & P^\top(i)B(i) \\ B^\top(i)P(i) & -\mathbb{I} \end{bmatrix} < 0$$

$$\begin{bmatrix} \mathcal{J}_Q(i) & K^\top(i) \\ K(i) & -\mathbb{I} \end{bmatrix} < 0,$$

where

$$\mathcal{J}_P(i) = A^\top(i)P(i) + P^\top(i)A(i) + K^\top(i)B^\top(i)P(i)$$
$$+ P^\top(i)B(i)K(i) + \lambda_{ii}E^\top(i)P(i) + \sum_{j=1, j \neq i}^{N} \varepsilon_P \lambda_{ij}\left[P(j) + P^\top(j)\right],$$

$$\mathcal{J}_Q(i) = A^\top(i)Q(i) + Q^\top(i)A(i) + Q^\top(i)L(i)C_y(i)$$
$$+ C_y^\top(i)L^\top(i)Q(i) + \sum_{j=1}^{N} \lambda_{ij}E^\top(j)Q(j),$$

with the following constraints:

$$\varepsilon_P\left[P(i) + P^\top(i)\right] \geq E^\top(i)P(i) = P^\top(i)E(i) \geq 0,$$
$$E^\top(i)Q(i) = Q^\top(i)E(i) \geq 0,$$

then, system (6.1) is piecewise regular, impulse-free and stochastically stable.

Proof: Let (x, e, i) denote respectively the values of the state vector, $x(t)$, the error, $e(t)$, and the mode r_t at time t and consider the Lyapunov function candidate with the following form:

$$V(x(t), e(t), r_t) = x^\top(t)E^\top(r_t)P(r_t)x(t) + e^\top(t)E^\top(r_t)Q(r_t)e(t)$$

$$= \left[x^\top(t) \; e^\top(t) \right] \begin{bmatrix} E^\top(r_t)P(r_t) & 0 \\ 0 & E^\top(r_t)Q(r_t) \end{bmatrix} \begin{bmatrix} x(t) \\ e(t) \end{bmatrix},$$

where $P(i)$ and $Q(i)$, for all $i \in \mathscr{S}$ are nonsingular matrices.

Let $\mathscr{L}$ be the infinitesimal generator of the process $\{(x(t), e(t), r_t), t \geq 0\}$ emanating from the point $(x(t), e(t), i)$ at time t. Then, we get:

$$\mathscr{L}V(x(t), e(t), i) \leq \dot{x}^\top(t)E^\top(i)P(i)x(t) + x^\top(t)E^\top(i)P(i)\dot{x}(t)$$
$$+ \dot{e}^\top(t)E^\top(i)Q(i)e(t) + e^\top(t)E^\top(i)Q(i)\dot{e}(t)$$
$$+ x^\top(t)\lambda_{ii}E^\top(i)P(i)x(t) + \sum_{j=1, j\neq i}^{N} \varepsilon_P \lambda_{ij} x^\top(t)\left[P(j) + P^\top(j)\right]x(t)$$
$$+ \sum_{j=1}^{N} \lambda_{ij} e^\top(t)E^\top(j)Q(j)e(t)$$

$$= \left[[A(i) + B(i)K(i)]\,x(t) - B(i)K(i)e(t)\right]^\top P(i)x(t)$$
$$+ x^\top(t)P^\top(i)\left[[A(i) + B(i)K(i)]\,x(t) - B(i)K(i)e(t)\right]$$
$$+ \left[\left[A(i) + L(i)C_y(i)\right]e(t)\right]^\top Q(i)e(t)$$
$$+ e^\top(t)Q^\top(i)\left[A(i) + L(i)C_y(i)\right]e(t)$$
$$+ \lambda_{ii}E^\top(i)P(i) + \sum_{j=1, j\neq i}^{N} \varepsilon_P \lambda_{ij} x^\top(t)\left[P(j) + P^\top(j)\right]x(t)$$
$$+ \sum_{j=1}^{N} \lambda_{ij} e^\top(t)E^\top(j)Q(j)e(t)$$

$$= x^\top(t)\Big[A^\top(i)P(i) + P^\top(i)A(i) + K^\top(i)B^\top(i)P(i)$$
$$+ P^\top(i)B(i)K(i) + \lambda_{ii}E^\top(i)P(i) + \sum_{j=1, j\neq i}^{N} \varepsilon_P \lambda_{ij}\left[P(j) + P^\top(j)\right]\Big]x(t)$$
$$- 2x^\top(t)P^\top(i)B(i)K(i)e(t) + e^\top(t)\Big[A^\top(i)Q(i)$$
$$+ Q^\top(i)A(i) + Q^\top(i)L(i)C_y(i) + C_y^\top(i)L^\top(i)Q(i)$$
$$+ \sum_{j=1}^{N} \lambda_{ij} E^\top(j)Q(j)\Big]e(t).$$

Noticing that

$$-2x^\top(t)P^\top(i)B(i)K(i)e(t) \leq x^\top(t)P^\top(i)B(i)B^\top(i)P(i)x(t)$$
$$+ e^\top(t)K^\top(i)K(i)e(t),$$

we get:

$$
\begin{aligned}
\mathscr{L} V(x(t), e(t), i) \leq{} & x^\top(t)\Big[A^\top(i)P(i) + P^\top(i)A(i) + K^\top(i)B^\top(i)P(i) \\
& + P^\top(i)B(i)K(i) + P^\top(i)B(i)B^\top(i)P(i) \\
& + \lambda_{ii}E^\top(i)P(i) + \sum_{j=1, j\neq i}^{N} \varepsilon_P \lambda_{ij}\big[P(j) + P^\top(j)\big]\Big]x(t) + e^\top(t)\Big[A^\top(i)Q(i) \\
& + Q^\top(i)A(i) + Q^\top(i)L(i)C_y(i) + C_y^\top(i)L^\top(i)Q(i) \\
& + K^\top(i)K(i) + \sum_{j=1}^{N} \lambda_{ij}E^\top(j)Q(j)\Big]e(t) \\
\leq{} & \big[\, x^\top(t)\ e^\top(t)\,\big]\begin{bmatrix} \bar{\mathscr{J}}_P(i) & 0 \\ 0 & \bar{\mathscr{J}}_Q(i) \end{bmatrix}\begin{bmatrix} x(t) \\ e(t) \end{bmatrix} \\
={} & \big[\, x^\top(t)\ e^\top(t)\,\big]\mathscr{J}(i)\begin{bmatrix} x(t) \\ e(t) \end{bmatrix},
\end{aligned}
$$

with

$$
\begin{aligned}
\bar{\mathscr{J}}_P(i) ={} & A^\top(i)P(i) + P^\top(i)A(i) + K^\top(i)B^\top(i)P(i) \\
& + P^\top(i)B(i)K(i) + P^\top(i)B(i)B^\top(i)P(i) \\
& + \lambda_{ii}E^\top(i)P(i) + \sum_{j=1, j\neq i}^{N} \varepsilon_P \lambda_{ij}\big[P(j) + P^\top(j)\big], \\
\bar{\mathscr{J}}_Q(i) ={} & A^\top(i)Q(i) + Q^\top(i)A(i) + Q^\top(i)L(i)C_y(i) \\
& + C_y^\top(i)L^\top(i)Q(i) + K^\top(i)K(i) \\
& + \sum_{j=1}^{N} \lambda_{ij}E^\top(j)Q(j).
\end{aligned}
$$

Therefore, since $\mathscr{J}(i) < 0$ for all $i \in \mathscr{S}$, we obtain:

$$
\mathscr{L} V(x(t), e(t), i) \leq -\min_{j\in\mathscr{S}}\{\lambda_{min}\left[-\mathscr{J}(j)\right]\}\,\eta^\top(s)\eta(s),
$$

where

$$
\eta(s) = \begin{bmatrix} x(s) \\ e(s) \end{bmatrix}.
$$

Combining this with Dynkin formula, we get:

$$
\begin{aligned}
\mathbb{E}\,[V(x(t), e(t), i)] &- V(x(0), e(0), r_0) \\
&= \mathbb{E}\left[\int_0^t \mathscr{L} V(x(s), e(s), r_s)ds|(r_0, x(0), e(0))\right] \\
&\leq -\min_{j\in\mathscr{S}}\{\lambda_{min}\left[-\mathscr{J}(i)\right]\}\mathbb{E}\left[\int_0^t \eta^\top(s)\eta(s)ds|(r_0, x(0), e(0))\right],
\end{aligned}
$$

which gives in turn:

$$\min_{j \in \mathscr{S}} \{\lambda_{min} \left[-\mathscr{J}(i) \right] \} \mathbb{E} \left[\int_0^t \eta^\top(s) \eta(s) ds | (r_0, x(0), e(0)) \right]$$
$$\leq V(x(0), e(0), r_0) .$$

This implies in turn that the following relation holds for all $t \geq 0$:

$$\mathbb{E} \left[\int_0^t \left[x^\top(s) \; e^\top(s) \right] \begin{bmatrix} x(s) \\ e(s) \end{bmatrix} ds | (r_0, x(0), e(0)) \right]$$
$$\leq \frac{V(x(0), e(0), r_0)}{\min_{j \in \mathscr{S}} \{\lambda_{min} \left[-\mathscr{J}(i) \right] \}} .$$

This proves the theorem.

With this stochastic stability condition, let us focus on the design of the controller gain matrices $K(i)$ and $L(i)$, $i = 1, \cdots, N$. We are now ready to synthesize the observer-based state feedback controller of the form (6.4) that guarantees that the augmented closed-loop state equation of the nominal system (6.1) is piecewise regular, impulse-free and stochastically stable.

Before giving the design algorithm, let us transform our previous conditions in the LMI formalism. For this purpose, notice first of all that $\mathscr{J}_P(i)$ is nonlinear in the design parameters $P(i)$ and $K(i)$. To put it into the LMI form, let $X(i) = P^{-1}(i)$ and pre- and post-multiply $\mathscr{J}_P(i) + P^\top(i)B(i)B^\top(i)P(i)$ by $X^\top(i)$ and $X(i)$ respectively, we get:

$$X^\top(i)A^\top(i) + A(i)X(i) + B(i)K(i)X(i) + X^\top(i)K^\top(i)B^\top(i)$$
$$+ B(i)B^\top(i) + \lambda_{ii}X^\top(i)E^\top(i)$$
$$+ \sum_{j=1, j \neq i}^N \varepsilon_P \lambda_{ij} X^\top(i) \left[X^{-1}(j) + X^{-\top}(j) \right] X(i) < 0 ,$$

Letting $Y_c(i)$, $\mathcal{S}_i(X)$ and $\mathcal{X}_i(X)$ be defined as follows:

$$Y_c(i) = K(i)X(i) ,$$
$$\mathcal{X}_i(X) = \text{diag} \left[X^\top(1) + X(1) - \mathbb{I}, \cdots, X^\top(i-1) + X(i-1) - \mathbb{I}, \right.$$
$$\left. X^\top(i+1) + X(i+1) - \mathbb{I}, \cdots, X^\top(N) + X(N) - \mathbb{I} \right] ,$$
$$\mathcal{S}_i(X) = \left[\sqrt{\varepsilon_P \lambda_{i1}} X^\top(i), \cdots, \sqrt{\varepsilon_P \lambda_{ii-1}} X^\top(i), \sqrt{\varepsilon_P \lambda_{ii+1}} X^\top(i), \right.$$
$$\left. \cdots, \sqrt{\varepsilon_P \lambda_{iN}} X^\top(i) \right] ,$$

and proceeding as before, the previous inequality becomes:

$$\left[\begin{bmatrix} X^\top(i)A^\top(i) + A(i)X(i) \\ +Y_c^\top(i)B^\top(i) + \lambda_{ii}X^\top(i)E^\top(i) \\ +B(i)Y_c(i) + B(i)B^\top(i) \end{bmatrix} \; \mathcal{S}_i(X) \; \mathcal{S}_i(X) \\ \mathcal{S}_i^\top(X) \qquad\qquad -\mathbb{I} \qquad 0 \\ \mathcal{S}_i^\top(X) \qquad\qquad 0 \qquad -\mathcal{X}_i(X) \end{bmatrix} \prec 0 .$$

Let us now transform the condition $\mathscr{J}_Q(i) + K^\top(i)K(i) < 0$ in the LMI form. For this purpose by letting $Y_o(i) = Q^\top(i)L(i)$, we have:

$$
\begin{aligned}
\mathscr{J}_Q(i) + K^\top(i)K(i) &= A^\top(i)Q(i) + Q^\top(i)A(i) + Y_o(i)C_y(i) \\
&\quad + C_y^\top(i)Y_o^\top(i) + K^\top(i)K(i) \\
&\quad + \sum_{j=1}^{N} \lambda_{ij} E^\top(j)Q(j),
\end{aligned}
$$

which gives after using Schur complement:

$$
\begin{bmatrix} J_2(i) & K^\top(i) \\ K(i) & -\mathbb{I} \end{bmatrix} < 0,
$$

where $J_2(i) = A^\top(i)Q(i) + Q^\top(i)A(i) + Y_o(i)C_y(i) + C_y^\top(i)Y_o^\top(i) + \sum_{j=1}^{N} \lambda_{ij} E^\top(j)Q(j)$.

Notice that the condition $\varepsilon_P\left[P(i) + P^\top(i)\right] \geq E^\top(i)P(i) = P^\top(i)E(i) \geq 0$ can be transformed by pre- and post-multiplying this by $X^\top(i)$ and $X(i)$ respectively, to the following: $\varepsilon_P\left[X(i) + X^\top(i)\right] \geq X^\top(i)E^\top(i) = E(i)X(i) \geq 0$.

The following theorem summarizes the results that design the observer-based state feedback controller.

Theorem 6.2.2 *Let ε_P be a given positive scalar. There exists a state feedback controller of the form (6.4) such that the closed-loop system (6.1) is piecewise regular, impulse-free and stochastically stable if there exist a set of nonsingular matrices $X = (X(1), \cdots, X(N))$, $X(i) \in \mathbb{R}^{n \times n}$, and $Q = (Q(1), \cdots, Q(N))$, $Q(i) \in \mathbb{R}^{n \times n}$ and sets of matrices $Y_c = (Y_c(1), \cdots, Y_c(N))$, $Y_c(i) \in \mathbb{R}^{m \times n}$, and $Y_o = (Y_o(1), \cdots, Y_o(N))$, $Y_o(i) \in \mathbb{R}^{n \times p}$ satisfying the following set of coupled matrix inequalities for each $i \in \mathcal{S}$:*

$$
\begin{bmatrix} J_1(i) & S_i(X) & S_i(X) \\ S_i^\top(X) & -\mathbb{I} & 0 \\ S_i^\top(X) & 0 & -X_i(X) \end{bmatrix} < 0, \tag{6.7}
$$

$$
\begin{bmatrix} J_2(i) & K^\top(i) \\ K(i) & -\mathbb{I} \end{bmatrix} < 0, \tag{6.8}
$$

where

$$
\begin{aligned}
J_1(i) &= X^\top(i)A^\top(i) + A(i)X(i) + Y_c^\top(i)B^\top(i) + B(i)Y_c(i) \\
&\quad + B(i)B^\top(i) + \lambda_{ii} X^\top(i)E^\top(i), \\
J_2(i) &= A^\top(i)Q(i) + Q^\top(i)A(i) + Y_o(i)C_y(i) + C_y^\top(i)Y_o^\top(i) \\
&\quad + \sum_{j=1}^{N} \lambda_{ij} E^\top(j)Q(j),
\end{aligned}
$$

with the following constraints:

$$
\varepsilon_P\left[X(i) + X^\top(i)\right] \geq X^\top(i)E^\top(i) = E(i)X(i) \geq 0, \tag{6.9}
$$

$$
E^\top(i)Q(i) = Q^\top(i)E(i) \geq 0. \tag{6.10}
$$

The controller gains are given by:

$$
\begin{cases} K(i) = Y_c(i)X^{-1}(i), \\ L(i) = Q^{-\top}(i)Y_o(i). \end{cases} \tag{6.11}
$$

Remark 6.2.1 *Notice that the conditions of the previous theorem can in some sense be solved in the following one:*

1. *solve the first and the third conditions in $X(i), i \in \mathcal{S}$ to get the gain $K(i), i \in \mathcal{S}$,*
2. *use this gain and solve the second and the fourth conditions in $Q(i), i \in \mathcal{S}$ to get the gain $L(i), i \in \mathcal{S}$.*

The following theorem gives another method to design an observer based state feedback controller of the form (6.4) that assures that the closed-loop system is piecewise regular, impulse-free and stabilizes system (6.1) in the stochastic sense.

Theorem 6.2.3 *Let us suppose that the matrix $B(i)$ is full column rank for each mode. Let ε_P be a given positive scalar. If there exist sets of nonsingular matrices $P = (P(1), \cdots, P(N))$, $P(i) \in \mathbb{R}^{n \times n}$ and $Q = (Q(1), \cdots, Q(N))$, $Q(i) \in \mathbb{R}^{n \times n}$ and sets of matrices $F = (F(1), \cdots, F(N))$, $F(i) \in \mathbb{R}^{m \times n}$, $H = (H(1), \cdots, H(N))$, $H(i) \in \mathbb{R}^{m \times m}$ and $Y_o = (Y_o(1), \cdots, Y_o(N))$, $Y_0(i) \in \mathbb{R}^{n \times p}$, satisfying the following set of coupled LMIs for each $i \in \mathcal{S}$:*

$$\begin{bmatrix} \Phi_1(i) & B(i)H(i) \\ H^\top(i)B^\top(i) & -\mathbb{I} \end{bmatrix} < 0, \tag{6.12}$$

$$\begin{bmatrix} \Phi_2(i) & K^\top(i) \\ K(i) & -\mathbb{I} \end{bmatrix} < 0, \tag{6.13}$$

where

$$\Phi_1(i) = P^\top(i)A(i) + A^\top(i)P(i) + B(i)F(i) + F^\top(i)B^\top(i) + \lambda_{ii}E^\top(i)P(i)$$

$$+ \sum_{j=1, j \neq i}^{N} \varepsilon_P \lambda_{ij} \left[P(j) + P^\top(j) \right],$$

$$\Phi_2(i) = A^\top(i)Q(i) + Q^\top(i)A(i) + Y_o(i)C_y(i) + C_y^\top(i)Y_o^\top(i)$$

$$+ \sum_{j=1}^{N} \lambda_{ij}E^\top(j)Q(j),$$

with the following constraints:

$$\varepsilon_P \left[P(i) + P^\top(i) \right] \geq E^\top(i)P(i) = P^\top(i)E(i) \geq 0, \tag{6.14}$$

$$E^\top(i)Q(i) = Q^\top(i)E(i) \geq 0, \tag{6.15}$$

$$P^\top(i)B(i) = B(i)H(i), \tag{6.16}$$

then the controller gains that guarantee that the closed-loop state equation of the nominal system (6.1) is piecewise regular, impulse-free and stochastically stable are given by:

$$\begin{cases} K(i) = H^{-1}(i)F(i), \\ L(i) = Q^{-T}(i)Y_o(i). \end{cases} \tag{6.17}$$

Proof: Let us consider that the previous stability inequality conditions hold, and suppose that the matrix $B(i)$ is full column rank for each mode, and there exist nonsingular matrices $H(i)$ such that the following hold for each $i \in \mathcal{S}$:

$$P^\top(i)B(i) = B(i)H(i). \tag{6.18}$$

Letting now $F(i) = H(i)K(i)$, we obtain the following:

$$P^\top(i)A(i) + A^\top(i)P(i) + B(i)F(i) + F^\top(i)B^\top(i)$$

$$+ B(i)H(i)H^\top(i)B^\top(i) + \lambda_{ii}E^\top(i)P(i) + \sum_{j=1, j\neq i}^{N} \varepsilon_P \lambda_{ij} \left[P(j) + P^\top(j) \right].$$

Using the Schur complement, we obtain the LMI (6.14) which ends the proof of the theorem.

Similarly we can show the same results for the bounds given by:

$$E^\top(i)P(i) \leq \varepsilon(i)P^\top(i)P(i)$$

for $\varepsilon(i) > 0$ for $i \in \mathscr{S}$.

The following theorem summarizes the results that design the observer-based state feedback controller in this case.

Theorem 6.2.4 *There exists a state feedback controller of the form (6.4) such that the closed-loop system (6.1) is piecewise regular, impulse-free and stochastically stable if there exist a set of nonsingular matrices $X = (X(1), \cdots, X(N))$, $X(i) \in \mathbb{R}^{n\times n}$ and $Q = (Q(1), \cdots, Q(N))$, $Q(i) \in \mathbb{R}^{n\times n}$ and sets of matrices $Y_c = (Y_c(1), \cdots, Y_c(N))$, $Y_c(i) \in \mathbb{R}^{m\times n}$ and $Y_o = (Y_o(1), \cdots, Y_o(N))$, $Y_o(i) \in \mathbb{R}^{n\times p}$ and set of positive scalars $\varepsilon = (\varepsilon(1), \cdots, \varepsilon(N))$ satisfying the following set of coupled matrix inequalities for each $i \in \mathcal{S}$:*

$$\begin{bmatrix} J_1(i) & B(i) & S_i(X) \\ B^\top(i) & -\mathbb{I} & 0 \\ S_i^\top(X) & 0 & -\mathcal{X}_i(X) \end{bmatrix} < 0, \tag{6.19}$$

$$\begin{bmatrix} J_2(i) & K^\top(i) \\ K(i) & -\mathbb{I} \end{bmatrix} < 0, \tag{6.20}$$

where

$$J_1(i) = X^\top(i)A^\top(i) + A(i)X(i) + Y_c^\top(i)B^\top(i) + B(i)Y_c(i)$$
$$+ B(i)B^\top(i) + \lambda_{ii}X^\top(i)E^\top(i),$$

$$J_2(i) = A^\top(i)Q(i) + Q^\top(i)A(i) + Y_o(i)C_y(i) + C_y^\top(i)Y_o^\top(i)$$

$$+ \sum_{j=1}^{N} \lambda_{ij}E^\top(j)Q(j),$$

$$\mathcal{X}_i(X) = diag \left[X^\top(1) + X(1) - \varepsilon(1)\mathbb{I}, \cdots, X^\top(i-1) + X(i-1) - \varepsilon(i-1)\mathbb{I}, \right.$$
$$\left. X^\top(i+1) + X(i+1) - \varepsilon(i+1)\mathbb{I}, \cdots, X^\top(N) + X(N) - \varepsilon(N)\mathbb{I} \right],$$

$$S_i(X) = \left[\sqrt{\lambda_{i1}}X^\top(i), \cdots, \sqrt{\lambda_{ii-1}}X^\top(i), \sqrt{\lambda_{ii+1}}X^\top(i), \right.$$
$$\left. \cdots, \sqrt{\lambda_{iN}}X^\top(i) \right],$$

with the following constraints:

$$\varepsilon(i)\mathbb{I} \geq X^\top(i)E^\top(i) = E(i)X(i) \geq 0, \tag{6.21}$$

$$E^\top(i)Q(i) = Q^\top(i)E(i) \geq 0. \tag{6.22}$$

The controller gains are given by:

$$\begin{cases} K(i) = Y_c(i)X^{-1}(i), \\ L(i) = Q^{-T}(i)Y_o(i). \end{cases} \tag{6.23}$$

Now, if we use the fact that:

$$E^\top(i)P(i) \le \frac{1}{4}\varepsilon^{-1}(i)\mathbb{I} + \varepsilon E^\top(i)P(i)P^\top(i)E(i),$$

for $\varepsilon(i) > 0$ for $i \in \mathscr{S}$.

Following the same steps we did earlier, we get the following theorem that summarizes the results that design the observer-based state feedback controller in this case.

Theorem 6.2.5 *There exists a state feedback controller of the form (6.4) such that the closed-loop system (6.1) is piecewise regular, impulse-free and stochastically stable if there exist a set of nonsingular matrices $X = (X(1), \cdots, X(N))$, $X(i) \in \mathbb{R}^{n \times n}$ and $Q = (Q(1), \cdots, Q(N))$, $Q(i) \in \mathbb{R}^{n \times n}$ and sets of matrices $Y_c = (Y_c(1), \cdots, Y_c(N))$, $Y_c(i) \in \mathbb{R}^{m \times n}$ and $Y_o = (Y_o(1), \cdots, Y_o(N))$, $Y_o(i) \in \mathbb{R}^{n \times p}$ and set of positive scalars $\varepsilon = (\varepsilon(1), \cdots, \varepsilon(N))$ satisfying the following set of coupled matrix inequalities for each $i \in S$:*

$$\begin{bmatrix} J_1(i) & \mathcal{Z}_i(X) & \mathcal{S}_i(X) \\ \mathcal{Z}_i^\top(X) & -\mathcal{X}_i(\varepsilon) & 0 \\ \mathcal{S}_i^\top(X) & 0 & -\mathcal{X}_i(X) \end{bmatrix} < 0, \tag{6.24}$$

$$\begin{bmatrix} J_2(i) & K^\top(i) \\ K(i) & -\mathbb{I} \end{bmatrix} < 0, \tag{6.25}$$

where

$$J_1(i) - X^\top(i)A^\top(i) + A(i)X(i) + Y_c^\top(i)B^\top(i) + B(i)Y_c(i)$$
$$+ B(i)B^\top(i) + \lambda_{ii}X^\top(i)E^\top(i),$$

$$J_2(i) = A^\top(i)Q(i) + Q^\top(i)A(i) + Y_o(i)C_y(i) + C_y^\top(i)Y_o^\top(i)$$
$$+ \sum_{j=1}^{N} \lambda_{ij}E^\top(j)Q(j),$$

$$\mathcal{X}_i(X) = diag\left[X^\top(1) + X(1) - \varepsilon(1)\mathbb{I}, \cdots, X^\top(i-1) + X(i-1) - \varepsilon(i-1)\mathbb{I}, \right.$$
$$\left. X^\top(i+1) + X(i+1) - \varepsilon(i+1)\mathbb{I}, \cdots, X^\top(N) + X(N) - \varepsilon(N)\mathbb{I} \right],$$

$$\mathcal{Z}_i(X) = \left[\sqrt{\lambda_{i1}}X^\top(i), \cdots, \sqrt{\lambda_{ii-1}}X^\top(i), \sqrt{\lambda_{ii+1}}X^\top(i), \right.$$
$$\left. \cdots, \sqrt{\lambda_{iN}}X^\top(i) \right],$$

$$\mathcal{S}_i(X) = \left[\sqrt{\lambda_{i1}}X^\top(i)E^\top(1), \cdots, \sqrt{\lambda_{ii-1}}X^\top(i)E^\top(i-1), \sqrt{\lambda_{ii+1}}X^\top(i)E^\top(i+1), \right.$$
$$\left. \cdots, \sqrt{\lambda_{iN}}X^\top(i)E^\top(N) \right],$$

$$\mathcal{X}_i(\varepsilon) = diag\left[4\varepsilon(1)\mathbb{I}, \cdots, 4\varepsilon(i-1)\mathbb{I}, 4\varepsilon(i+1)\mathbb{I}, \cdots, 4\varepsilon(N)\mathbb{I} \right],$$

with the following constraints:

$$X^{\top}(i)E^{\top}(i) = E(i)X(i) \geq 0, \tag{6.26}$$

$$E^{\top}(i)Q(i) = Q^{\top}(i)E(i) \geq 0. \tag{6.27}$$

The controller gains are given by:

$$\begin{cases} K(i) = Y_c(i)X^{-1}(i), \\ L(i) = Q^{-T}(i)Y_o(i). \end{cases} \tag{6.28}$$

In this section assuming that the state equation has no uncertainties, we established LMI conditions to design the observer-based controller. Practically, this is not realistic and we always have uncertainties in the dynamics. The next section will tackle this problem.

6.3 Robust Observer-Based Stabilization of Singular Systems

Let us now consider the effect of the uncertainties and develop the new conditions that we can use to synthesize an observer-based state feedback controller that guarantees that the closed-loop state equation of the uncertain systems is piecewise regular, impulse-free and stochastically stable. For this purpose, notice that the augmented state equation is given by:

$$E(r_t)\dot{\eta}(t) = \left[\tilde{A}(r_t) + \Delta\tilde{A}(r_t, t)\right]\eta(t), \tag{6.29}$$

where

$$\eta(t) = \begin{bmatrix} x(t) \\ e(t) \end{bmatrix},$$

$$\tilde{A}(i) = \begin{bmatrix} A(i) + B(i)K(i) & -B(i)K(i) \\ 0 & A(i) + L(i)C_y(i) \end{bmatrix},$$

$$\Delta\tilde{A}(i, t) = \begin{bmatrix} \Delta A(i, t) + \Delta B(i, t)K(i) & -\Delta B(i, t)K(i) \\ 0 & \Delta A(i, t) + L(i)\Delta C_y(i, t) \end{bmatrix}.$$

Based on the previous results on stochastic stability, the system will be piecewise regular, impulse-free and stochastically stable if there exist sets of nonsingular matrices $P = (P(1), \cdots, P(N))$, $P(i) \in \mathbb{R}^{n \times n}$ and $Q = (Q(1), \cdots, Q(N))$, $Q(i) \in \mathbb{R}^{n \times n}$ such that the following hold for every $i \in \mathscr{S}$:

$$\begin{cases} \varepsilon_P \left[P(i) + P^{\top}(i)\right] \geq E^{\top}(i)P(i) = P^{\top}(i)E(i) \geq 0, \\ E^{\top}(i)Q(i) = Q^{\top}(i)E(i) \geq 0, \\ \begin{bmatrix} \mathscr{J}_P(i) & P^{\top}(i)B(i, t) \\ B^{\top}(i, t)P(i) & -\mathbb{I} \end{bmatrix} < 0, \\ \begin{bmatrix} \mathscr{J}_Q(i) & K^{\top}(i) \\ K(i) & -\mathbb{I} \end{bmatrix} < 0, \end{cases}$$

where

$$
\mathscr{J}_P(i) = A^\top(i,t)P(i) + P^\top(i)A(i,t) + K^\top(i)B^\top(i,t)P(i)
$$

$$
+ P^\top(i)B(i,t)K(i) + \lambda_{ii}E^\top(i)P(i) + \sum_{j=1,j\neq i}^{N} \varepsilon_P\lambda_{ij}\left[P(j) + P^\top(j)\right],
$$

$$
\mathscr{J}_Q(i) = A^\top(i,t)Q(i) + Q^\top(i)A(i,t) + Q^\top(i)L(i)C_y(i,t)
$$

$$
+ C_y^\top(i,t)L^\top(i)Q(i) + \sum_{j=1}^{N} \lambda_{ij}E^\top(j)Q(j).
$$

First of all notice that:

$$
\begin{bmatrix} 0 & P^\top(i)\Delta B(i) \\ 0 & 0 \end{bmatrix} = \begin{bmatrix} 0 & P^\top(i)D_B(i)F_B(i)E_B(i) \\ 0 & 0 \end{bmatrix}
$$

$$
= \begin{bmatrix} P^\top(i)D_B(i) \\ 0 \end{bmatrix} F_B(i) \begin{bmatrix} 0 & E_B(i) \end{bmatrix}.
$$

Using now Lemma 1.5.1, we get:

$$
\begin{bmatrix} 0 & P^\top(i)D_B(i)F_B(i)E_B(i) \\ 0 & 0 \end{bmatrix} + \begin{bmatrix} 0 & P^\top(i)D_B(i)F_B(i)E_B(i) \\ 0 & 0 \end{bmatrix}^\top
$$

$$
\leq \varepsilon_B^{-1}(i)\begin{bmatrix} P^\top(i)D_B(i)D_B^\top(i)P(i) & 0 \\ 0 & 0 \end{bmatrix} + \varepsilon_B(i)\begin{bmatrix} 0 & 0 \\ 0 & E_B^\top(i)E_B(i) \end{bmatrix}
$$

$$
\leq \begin{bmatrix} \varepsilon_B^{-1}(i)P^\top(i)D_B(i)D_B^\top(i)P(i) & 0 \\ 0 & \varepsilon_B(i)E_B^\top(i)E_B(i) \end{bmatrix},
$$

for any $\varepsilon_B(i) > 0$.

Also we have:

$$
P^\top(i)D_A(i)F_A(i)E_A(i) + E_A^\top(i)F_A^\top(i)D_A^\top(i)P(i)
$$

$$
\leq \varepsilon_A(i)P^\top(i)D_A(i)D_A^\top(i)P(i) + \varepsilon_A^{-1}(i)E_A^\top(i)E_A(i),
$$

for any $\varepsilon_A(i) > 0$,

$$
P^\top(i)D_B(i)F_B(i)E_B(i)K(i) + K^\top(i)E_B^\top(i)F_B^\top(i)D_B^\top(i)P(i)
$$

$$
\leq \varepsilon_{BK}(i)P^\top(i)D_B(i)D_B^\top(i)P(i) + \varepsilon_{BK}^{-1}(i)K^\top(i)E_B^\top(i)E_B(i)K(i),
$$

for any $\varepsilon_{BK}(i) > 0$,

$$
Q^\top(i)D_A(i)F_A(i)E_A(i) + E_A^\top(i)F_A^\top(i)D_A^\top(i)Q(i)
$$

$$
\leq \varepsilon_Q^{-1}(i)Q^\top(i)D_A(i)D_A^\top(i)Q(i) + \varepsilon_Q(i)E_A^\top(i)E_A(i),
$$

for any $\varepsilon_Q(i) > 0$, and

$$
Q^\top(i)L(i)D_{C_y}(i)F_{C_y}(i)E_{C_y}(i) + E_{C_y}^\top(i)F_{C_y}^\top(i)D_{C_y}^\top(i)L^\top(i)Q(i)
$$

$$
\leq \varepsilon_{C_y}^{-1}(i)Q^\top(i)L(i)D_{C_y}(i)D_{C_y}^\top(i)L^\top(i)Q(i) + \varepsilon_{C_y}(i)E_{C_y}^\top(i)E_{C_y}(i),
$$

for any $\varepsilon_{C_y}(i) > 0$.

Based on this and the stochastic stability conditions, we get:

$$\begin{cases} \varepsilon_P\left[P(i) + P^\top(i)\right] \geq E^\top(i)P(i) = P^\top(i)E(i) \geq 0\,, \\ E^\top(i)Q(i) = Q^\top(i)E(i) \geq 0\,, \\ \begin{bmatrix} \mathscr{J}_P(i) & P^\top(i)B(i) \\ B^\top(i)P(i) & \mathbb{I} + c_B(i)E_B^\top(i)E_B(i) \end{bmatrix} < 0\,, \\ \begin{bmatrix} \mathscr{J}_Q(i) & K^\top(i) \\ K(i) & -\mathbb{I} \end{bmatrix} < 0\,, \end{cases}$$

where

$$\begin{aligned} \mathscr{J}_P(i) &= A^\top(i)P(i) + P^\top(i)A(i) + K^\top(i)B^\top(i)P(i) \\ &\quad + P^\top(i)B(i)K(i) + \varepsilon_A(i)P^\top(i)D_A(i)D_A^\top(i)P(i) \\ &\quad + \varepsilon_A^{-1}(i)E_A^\top(i)E_A(i) + \varepsilon_B^{-1}(i)P^\top(i)D_B(i)D_B^\top(i)P(i) \\ &\quad + \varepsilon_{BK}(i)P^\top(i)D_B(i)D_B^\top(i)P(i) \\ &\quad + \varepsilon_{BK}^{-1}(i)K^\top(i)E_B^\top(i)E_B(i)K(i) \\ &\quad + \lambda_{ii}E^\top(i)P(i) + \sum_{j=1,j\neq i}^{N} \lambda_{ij}\varepsilon_P\left[P(j) + P^\top(j)\right]\,, \\ \mathscr{J}_Q(i) &= A^\top(i)Q(i) + Q^\top(i)A(i) + Q^\top(i)L(i)C_y(i) \\ &\quad + C_y^\top(i)L^\top(i)Q(i) + \varepsilon_Q^{-1}(i)Q^\top(i)D_A(i)D_A^\top(i)Q(i) \\ &\quad + \varepsilon_Q(i)E_A^\top(i)E_A(i) \\ &\quad + \varepsilon_{C_y}^{-1}(i)Q^\top(i)L(i)D_{C_y}(i)D_{C_y}^\top(i)L^\top(i)Q(i) \\ &\quad + \varepsilon_{C_y}(i)E_{C_y}^\top(i)E_{C_y}(i) + \sum_{j=1}^{N} \lambda_{ij}E^\top(j)Q(j)\,. \end{aligned}$$

Let us now put the previous inequality matrices in the LMI formalism. For this purpose, let $X(i) = P^{-1}(i)$. Pre- and post-multiply the first inequality by $\mathrm{diag}(X^\top(i), \mathbb{I})$ and its transpose, we get:

$$\begin{bmatrix} X^\top(i)\,\mathscr{J}_P(i)X(i) & B(i) \\ B^\top(i) & -\mathbb{I} + \varepsilon_B(i)E_B^\top(i)E_B(i) \end{bmatrix} < 0\,,$$

where

$$\begin{aligned} X(i)\,\mathscr{J}_P(i)X(i) &= X^\top(i)A^\top(i) + A(i)X(i) + X^\top(i)K^\top(i)B^\top(i) \\ &\quad + B(i)K(i)X(i) + \varepsilon_A(i)D_A(i)D_A^\top(i) \\ &\quad + \varepsilon_A^{-1}(i)X^\top(i)E_A^\top(i)E_A(i)X(i) + \varepsilon_B^{-1}(i)D_B(i)D_B^\top(i) \\ &\quad + \varepsilon_{BK}(i)D_B(i)D_B^\top(i) \\ &\quad + \varepsilon_{BK}^{-1}(i)X^\top(i)K^\top(i)E_B^\top(i)E_B(i)K(i)X(i) \\ &\quad + \lambda_{ii}X^\top(i)E^\top(i) + \sum_{j=1,j\neq i}^{N} \varepsilon_P\lambda_{ij}X^\top(i)\left[X^{-1}(j) + X^{-\top}(j)\right]X(i)\,. \end{aligned}$$

Using the fact that $[X^\top(i)X(i)]^{-1} \geq X^{-\top}(i) + X^{-1}(i) - \mathbb{I}$ and letting $Z_i(X)$ and $Y_c(i)$ be defined as follows:

$$Z_i(X) = \text{diag}\left[X^\top(1)X(1), \cdots, X^\top(i-1)X(i-1),\right.$$
$$\left. X^\top(i+1)X(i+1), \cdots, X^\top(N)X(N)\right],$$
$$Y_c(i) = K(i)X(i),$$

we get:

$$\begin{bmatrix} \mathscr{I}_z(i) & B(i) \\ B^\top(i) & -\mathbb{I} + \varepsilon_B(i)E_B^\top(i)E_B(i) \end{bmatrix} < 0,$$

where

$$\mathscr{I}_z(i) = X^\top(i)A^\top(i) + A(i)X(i) + Y_c^\top(i)B^\top(i)$$
$$+ B(i)Y_c(i) + \varepsilon_A(i)D_A(i)D_A^\top(i)$$
$$+ \varepsilon_A^{-1}(i)X^\top(i)E_A^\top(i)E_A(i)X(i)$$
$$+ \varepsilon_B^{-1}(i)D_B(i)D_B^\top(i) + \varepsilon_{BK}(i)D_B(i)D_B^\top(i)$$
$$+ \varepsilon_{BK}^{-1}(i)Y_c^\top(i)E_B^\top(i)E_B(i)Y_c(i) + \lambda_{ii}X^\top(i)E^\top(i)$$
$$+ \sum_{j=1,j\neq i}^{N} \varepsilon_P\lambda_{ij}X^\top(i)\left[\mathbb{I} + Z_i^{-1}(X)\right]X(i).$$

Using the fact that $X^\top(i)X(i) \geq X^\top(i) + X(i) - \mathbb{I}$, this inequality becomes:

$$\begin{vmatrix} \mathscr{I}_X(i) & B(i) & X^\top(i)E_A^\top(i) \\ B^\top(i) & -\mathbb{I} + \varepsilon_B(i)E_B^\top(i)E_B(i) M & 0 \\ E_A(i)X(i) & 0 & -\varepsilon_A(i)\mathbb{I} \\ E_B(i)Y_c(i) & 0 & 0 \\ D_B^\top(i) & 0 & 0 \\ S_i^\top(X) & 0 & 0 \\ S_i^\top(X) & 0 & 0 \end{vmatrix}$$

$$\left.\begin{matrix} Y_c^\top(i)E_B^\top(i) & D_B(i) & S_i(X) & S_i(X) \\ 0 & 0 & 0 & 0 \\ 0 & 0 & 0 & 0 \\ -\varepsilon_{BK}(i)\mathbb{I} & 0 & 0 & 0 \\ 0 & -\varepsilon_B(i)\mathbb{I} & 0 & 0 \\ 0 & 0 & -\mathbb{I} & 0 \\ 0 & 0 & 0 & -\mathcal{X}_i(X) \end{matrix}\right| < 0,$$

where

$$\mathscr{I}_X(i) = X^\top(i)A^\top(i) + A(i)X(i) + Y_c^\top(i)B^\top(i) + B(i)Y_c(i)$$
$$+ \varepsilon_A(i)D_A(i)D_A^\top(i) + \varepsilon_{BK}(i)D_B(i)D_B^\top(i)$$
$$+ \lambda_{ii}X^\top(i)E^\top(i),$$
$$S_i(X) = \left(\sqrt{\varepsilon_P\lambda_{i1}}X^\top(i), \cdot, \sqrt{\varepsilon_P\lambda_{ii-1}}X^\top(i),\right.$$
$$\left.\sqrt{\varepsilon_P\lambda_{ii+1}}X^\top(i), \cdots, \sqrt{\varepsilon_P\lambda_{iN}}X^\top(i)\right),$$

$$X_i(X) = \mathrm{diag}\left[X^\top(1) + X(1) - \mathbb{I}, \cdots, X^\top(i-1) + X(i-1) - \mathbb{I},\right.$$
$$\left. X^\top(i+1) + X(i+1) - \mathbb{I}, \cdots, X^\top(N) + X(N) - \mathbb{I}\right].$$

For the second inequality, letting $Y_o(i) = Q^\top(i)L(i)$, we get:

$$\begin{bmatrix} \mathscr{J}_w(i) & K^\top(i) \\ K(i) & -\mathbb{I} \end{bmatrix} < 0,$$

where

$$\mathscr{J}_w(i) = A^\top(i)Q(i) + Q^\top(i)A(i) + Y_o(i)C_y(i)$$
$$+ C_y^\top(i)Y_o^\top(i) + \varepsilon_Q^{-1}(i)Q^\top(i)D_A(i)D_A^\top(i)Q(i)$$
$$+ \varepsilon_Q(i)E_A^\top(i)E_A(i)$$
$$+ \varepsilon_{C_y}^{-1}(i)Y_o(i)D_{C_y}(i)D_{C_y}^\top(i)Y_o^\top(i)$$
$$+ \varepsilon_{C_y}(i)E_{C_y}^\top(i)E_{C_y}(i) + \sum_{j=1}^{N}\lambda_{ij}E^\top(j)Q(j).$$

This inequality is equivalent to the following LMI:

$$\begin{bmatrix} \mathscr{J}_Q(i) & K^\top(i) & Q^\top(i)D_A(i) & Y_o(i)D_{C_y}(i) \\ K(i) & -\mathbb{I} & 0 & 0 \\ D_A^\top(i)Q(i) & 0 & -\varepsilon_Q(i)\mathbb{I} & 0 \\ D_{C_y}^\top(i)Y_o^\top(i) & 0 & 0 & -\varepsilon_{C_y}(i)\mathbb{I} \end{bmatrix} < 0,$$

where

$$\mathscr{J}_Q(i) = A^\top(i)Q(i) + Q^\top(i)A(i) + Y_o(i)C_y(i) + C_y^\top(i)Y_o^\top(i)$$
$$+ \varepsilon_Q(i)E_A^\top(i)E_A(i) + \varepsilon_{C_y}(i)E_{C_y}^\top(i)E_{C_y}(i) + \sum_{j=1}^{N}\lambda_{ij}E^\top(j)Q(j).$$

The following theorem gives the design procedure for a robust observer-based state feedback controller that guarantees that the closed-loop state equation of the class of uncertain systems is piecewise regular, impulse-free and stochastically stabilizes.

Theorem 6.3.1 *Let ε_P be a given positive scalar. There exists a state feedback controller of the form (6.29) such that the closed-loop system (6.1) is piecewise regular, impulse-free and stochastically stable if there exist a set of nonsingular matrices $X = (X(1), \cdots, X(N))$, $X(i) \in \mathbb{R}^{n \times n}$, $Q = (Q(1), \cdots, Q(N))$, $Q(i) \in \mathbb{R}^{n \times n}$, and sets of matrices $Y_c = (Y_c(1), \cdots, Y_c(N))$, $Y_c(i) \in \mathbb{R}^{m \times n}$, and $Y_o = (Y_o(1), \cdots, Y_o(N))$, $Y_o(i) \in \mathbb{R}^{n \times p}$, and sets positive scalars $\varepsilon_A = (\varepsilon_A(1), \cdots, \varepsilon_A(N))$, $\varepsilon_B = (\varepsilon_B(1), \cdots, \varepsilon_B(N))$, $\varepsilon_{BK} = (\varepsilon_{BK}(1), \cdots, \varepsilon_{BK}(N))$, $\varepsilon_{C_y} = (\varepsilon_{C_y}(1), \cdots, \varepsilon_{C_y}(N))$,* and $\varepsilon_Q = (\varepsilon_Q(1), \cdots, \varepsilon_Q(N))$, satisfying the following set of coupled LMIs for each $i \in S$:

$$\left\{ \begin{bmatrix} \mathscr{J}_X(i) & B(i) & X^\top(i)E_A^\top(i) & Y_c^\top(i)E_B^\top(i) & D_B(i) & S_i(X) & S_i(X) \\ B^\top(i) & -\mathbb{I} + \varepsilon_B(i)E_B^\top(i)E_B(i) & 0 & 0 & 0 & 0 & 0 \\ E_A(i)X(i) & 0 & -\varepsilon_A(i)\mathbb{I} & 0 & 0 & 0 & 0 \\ E_B(i)Y_c(i) & 0 & 0 & -\varepsilon_{BK}(i)\mathbb{I} & 0 & 0 & 0 \\ D_B^\top(i) & 0 & 0 & 0 & -\varepsilon_B(i)\mathbb{I} & 0 & 0 \\ S_i^\top(X) & 0 & 0 & 0 & 0 & -\mathbb{I} & 0 \\ S_i^\top(X) & 0 & 0 & 0 & 0 & 0 & -\mathcal{X}_i(X) \end{bmatrix} < 0 , \right.$$

$$\left. \begin{bmatrix} \mathscr{J}_Q(i) & K^\top(i) & Q^\top(i)D_A(i) & Y_o(i)D_{C_y}(i) \\ K(i) & -\mathbb{I} & 0 & 0 \\ D_A^\top(i)Q(i) & 0 & -\varepsilon_Q(i)\mathbb{I} & 0 \\ D_{C_y}^\top(i)Y_o^\top(i) & 0 & 0 & -\varepsilon_{C_y}(i)\mathbb{I} \end{bmatrix} < 0 , \right.$$

(6.30)

where

$$\begin{aligned} \mathscr{J}_X(i) &= X(i)A^\top(i) + A(i)X(i) + Y_c^\top(i)B^\top(i) + B(i)Y_c(i) \\ &\quad + \varepsilon_A(i)D_A(i)D_A^\top(i) + \varepsilon_{BK}(i)D_B(i)D_B^\top(i) \\ &\quad + \lambda_{ii}X^\top(i)E^\top(i) , \\ \mathscr{J}_Q(i) &= A^\top(i)Q(i) + Q^\top(i)A(i) + Y_o(i)C_y(i) + C_y^\top(i)Y_o^\top(i) \\ &\quad + \varepsilon_Q(i)E_A^\top(i)E_A(i) + \varepsilon_{C_y}(i)E_{C_y}^\top(i)E_{C_y}(i) + \sum_{j=1}^{N}\lambda_{ij}E^\top(j)Q(j) , \\ S_i(X) &= \left(\sqrt{\varepsilon_P\lambda_{i1}}X^\top(i), \cdot, \sqrt{\varepsilon_P\lambda_{ii-1}}X^\top(i), \right. \\ &\quad \left. \sqrt{\varepsilon_P\lambda_{ii+1}}X^\top(i), \cdots, \sqrt{\varepsilon_P\lambda_{iN}}X^\top(i) \right) , \\ \mathcal{X}_i(X) &= diag\left[X^\top(1) + X(1) - \mathbb{I}, \cdots, X^\top(i-1) + X(i-1) - \mathbb{I}, \right. \\ &\quad \left. X^\top(i+1) + X(i+1) - \mathbb{I}, \cdots, X^\top(N) + X(N) - \mathbb{I} \right] , \end{aligned}$$

with the following constraints:

$$\varepsilon_P\left[X(i) + X^\top(i) \right] \geq X^\top(i)E^\top(i) = E(i)X(i) \geq 0 ,$$
$$E^\top(i)Q(i) = Q^\top(i)E(i) \geq 0 .$$

The controller gains are given by:

$$\begin{cases} K(i) = Y_c(i)X^{-1}(i) , \\ L(i) = Q^{-\top}(i)Y_o(i) . \end{cases} \qquad (6.31)$$

When the following bounds are used:

$$E^\top(i)P(i) \leq \varepsilon(i)P^\top(i)P(i)$$

for any $\varepsilon(i) > 0$, the following theorem, that summarizes the results that design the observer-based state feedback controller, can be obtained.

Theorem 6.3.2 *There exists a state feedback controller of the form (6.29) such that the closed-loop system (6.1) is piecewise regular, impulse-free and stochastically stable if there exist a set of nonsingular matrices $X = (X(1), \cdots, X(N))$, $X(i) \in \mathbb{R}^{n \times n}$, $Q = (Q(1), \cdots, Q(N))$, $Q(i) \in \mathbb{R}^{n \times n}$, and sets of matrices $Y_c = (Y_c(1), \cdots, Y_c(N))$, $Y_c(i) \in \mathbb{R}^{m \times n}$, and $Y_o = (Y_o(1), \cdots, Y_o(N))$, $Y_o(i) \in \mathbb{R}^{n \times p}$, and sets positive scalars $\varepsilon_A = (\varepsilon_A(1), \cdots, \varepsilon_A(N))$, $\varepsilon_B = (\varepsilon_B(1), \cdots, \varepsilon_B(N))$, $\varepsilon_{BK} = (\varepsilon_{BK}(1), \cdots, \varepsilon_{BK}(N))$, $\varepsilon_{C_y} = (\varepsilon_{C_y}(1), \cdots, \varepsilon_{C_y}(N))$, and $\varepsilon_Q = (\varepsilon_Q(1), \cdots, \varepsilon_Q(N))$ and set of positive scalars $\varepsilon = (\varepsilon(1), \cdots, \varepsilon(N))$, satisfying the following set of coupled LMIs for each $i \in \mathcal{S}$:*

$$
\left\{
\begin{array}{l}
\begin{bmatrix}
\mathscr{J}_X(i) & B(i) & X^\top(i)E_A^\top(i) & Y_c^\top(i)E_B^\top(i) & D_B(i) & S_i(X) \\
B^\top(i) & -\mathbb{I} + \varepsilon_B(i)E_B^\top(i)E_B(i) & 0 & 0 & 0 & 0 \\
E_A(i)X(i) & 0 & -\varepsilon_A(i)\mathbb{I} & 0 & 0 & 0 \\
E_B(i)Y_c(i) & 0 & 0 & -\varepsilon_{BK}(i)\mathbb{I} & 0 \\
D_B^\top(i) & 0 & 0 & 0 & -\varepsilon_B(i)\mathbb{I} & 0 \\
S_i^\top(X) & 0 & 0 & 0 & 0 & -\mathcal{X}_i(X)
\end{bmatrix} < 0, \\[2em]
\begin{bmatrix}
\mathscr{J}_Q(i) & K^\top(i) & Q^\top(i)D_A(i) & Y_o(i)D_{C_y}(i) \\
K(i) & -\mathbb{I} & 0 & 0 \\
D_A^\top(i)Q(i) & 0 & -\varepsilon_Q(i)\mathbb{I} & 0 \\
D_{C_y}^\top(i)Y_o^\top(i) & 0 & 0 & -\varepsilon_{C_y}(i)\mathbb{I}
\end{bmatrix} < 0,
\end{array}
\right.
$$

(6.32)

where

$$
\begin{aligned}
\mathscr{J}_X(i) &= X(i)A^\top(i) + A(i)X(i) + Y_c^\top(i)B^\top(i) + B(i)Y_c(i) \\
&\quad + \varepsilon_A(i)D_A(i)D_A^\top(i) + \varepsilon_{BK}(i)D_B(i)D_B^\top(i) \\
&\quad + \lambda_{ii}X^\top(i)E^\top(i), \\
\mathscr{J}_Q(i) &= A^\top(i)Q(i) + Q^\top(i)A(i) + Y_o(i)C_y(i) + C_y^\top(i)Y_o^\top(i) \\
&\quad + \varepsilon_Q(i)A^\top(i)E_A(i) + \varepsilon_{C_y}(i)E_{C_y}^\top(i)E_{C_y}(i) + \sum_{j=1}^N \lambda_{ij}E^\top(j)Q(j), \\
S_i(X) &= \left(\sqrt{\lambda_{i1}}X^\top(i), \cdot, \sqrt{\lambda_{ii-1}}X^\top(i), \right. \\
&\qquad \left. \sqrt{\lambda_{ii+1}}X^\top(i), \cdots, \sqrt{\lambda_{iN}}X^\top(i) \right), \\
\mathcal{X}_i(X) &= diag\left[X^\top(1) + X(1) - \varepsilon(1)\mathbb{I}, \cdots, X^\top(i-1) + X(i-1) - \varepsilon(i-1)\mathbb{I}, \right. \\
&\qquad \left. X^\top(i+1) + X(i+1) - \varepsilon(i+1)\mathbb{I}, \cdots, X^\top(N) + X(N) - \varepsilon(N)\mathbb{I} \right],
\end{aligned}
$$

with the following constraints:

$$
\begin{aligned}
\varepsilon(i)\mathbb{I} &\geq X^\top(i)E^\top(i) = E(i)X(i) \geq 0, \\
E^\top(i)Q(i) &= Q^\top(i)E(i) \geq 0.
\end{aligned}
$$

The controller gains are given by:

$$
\begin{cases}
K(i) = Y_c(i)X^{-1}(i), \\
L(i) = Q^{-\top}(i)Y_o(i).
\end{cases}
$$

(6.33)

To end this section, let us consider the case when

$$E^\top(i)P(i) \leq \frac{1}{4}\varepsilon^{-1}(i)\mathbb{I} + \varepsilon(i)E^\top(i)P(i)P^\top(i)E(i)$$

for any $\varepsilon(i) > 0$.

Following the same steps we did earlier, we get the following theorem that summarizes the results that design the observer-based state feedback controller in this case.

Theorem 6.3.3 *There exists a state feedback controller of the form (6.29) such that the closed-loop system (6.1) is piecewise regular, impulse-free and stochastically stable if there exist a set of nonsingular matrices $X = (X(1), \cdots, X(N))$, $X(i) \in \mathbb{R}^{n \times n}$, $Q = (Q(1), \cdots, Q(N))$, $Q(i) \in \mathbb{R}^{n \times n}$, and sets of matrices $Y_c = (Y_c(1), \cdots, Y_c(N))$, $Y_c(i) \in \mathbb{R}^{m \times n}$, and $Y_o = (Y_o(1), \cdots, Y_o(N))$, $Y_o(i) \in \mathbb{R}^{n \times p}$, and sets positive scalars $\varepsilon_A = (\varepsilon_A(1), \cdots, \varepsilon_A(N))$, $\varepsilon_B = (\varepsilon_B(1), \cdots, \varepsilon_B(N))$, $\varepsilon_{BK} = (\varepsilon_{BK}(1), \cdots, \varepsilon_{BK}(N))$, $\varepsilon_{C_y} = (\varepsilon_{C_y}(1), \cdots, \varepsilon_{C_y}(N))$, and $\varepsilon_Q = (\varepsilon_Q(1), \cdots, \varepsilon_Q(N))$ and set of positive scalars $\varepsilon = (\varepsilon(1), \cdots, \varepsilon(N))$, satisfying the following set of coupled LMIs for each $i \in S$:*

$$\begin{bmatrix} \mathscr{J}_X(i) & B(i) & X^\top(i)E_A^\top(i) & Y_c^\top(i)E_B^\top(i) & D_B(i) & \mathcal{Z}_i(X) & \mathcal{S}_i(X) \\ B^\top(i) & -\mathbb{I} + \varepsilon_B(i)E_B^\top(i)E_B(i) & 0 & 0 & 0 & 0 & 0 \\ E_A(i)X(i) & 0 & -\varepsilon_A(i)\mathbb{I} & 0 & 0 & 0 & 0 \\ E_B(i)Y_c(i) & 0 & 0 & -\varepsilon_{BK}(i)\mathbb{I} & 0 & 0 & 0 \\ D_B^\top(i) & 0 & 0 & 0 & -\varepsilon_B(i)\mathbb{I} & 0 & 0 \\ \mathcal{Z}_i^\top(X) & 0 & 0 & 0 & 0 & -\mathcal{X}_i(\varepsilon) & 0 \\ \mathcal{S}_i^\top(X) & 0 & 0 & 0 & 0 & 0 & -\mathcal{X}_i(X) \end{bmatrix} < 0,$$

$$\begin{bmatrix} \mathscr{J}_Q(i) & K^\top(i) & Q^\top(i)D_A(i) & Y_o(i)D_{C_y}(i) \\ K(i) & -\mathbb{I} & 0 & 0 \\ D_A^\top(i)Q(i) & 0 & -\varepsilon_Q(i)\mathbb{I} & 0 \\ D_{C_y}^\top(i)Y_o^\top(i) & 0 & 0 & -\varepsilon_{C_y}(i)\mathbb{I} \end{bmatrix} < 0,$$

(6.34)

where

$$\begin{aligned} \mathscr{J}_X(i) &= X(i)A^\top(i) + A(i)X(i) + Y_c^\top(i)B^\top(i) + B(i)Y_c(i) \\ &\quad + \varepsilon_A(i)D_A(i)D_A^\top(i) + \varepsilon_{BK}(i)D_B(i)D_B^\top(i) \\ &\quad + \lambda_{ii}X^\top(i)E^\top(i), \\ \mathscr{J}_Q(i) &= A^\top(i)Q(i) + Q^\top(i)A(i) + Y_o(i)C_y(i) + C_y^\top(i)Y_o^\top(i) \\ &\quad + \varepsilon_Q(i)_A^\top(i)E_A(i) + \varepsilon_{C_y}(i)E_{C_y}^\top(i)E_{C_y}(i) + \sum_{j=1,j}^N \lambda_{ij}E^\top(j)Q(j), \\ \mathcal{Z}_i(X) &= \left(\sqrt{\lambda_{i1}}X^\top(i), \cdot, \sqrt{\lambda_{ii-1}}X^\top(i), \right. \\ &\quad \left. \sqrt{\lambda_{ii+1}}X^\top(i), \cdots, \sqrt{\lambda_{iN}}X^\top(i) \right), \\ \mathcal{S}_i(X) &- \left(\sqrt{\lambda_{i1}}X^\top(i)E^\top(1), \cdot, \sqrt{\lambda_{ii-1}}X^\top(i)E^\top(i-1), \right. \\ &\quad \left. \sqrt{\lambda_{ii+1}}X^\top(i)E^\top(i+1), \cdots, \sqrt{\lambda_{iN}}X^\top(i)E^\top(N) \right), \end{aligned}$$

$$X_i(X) = diag\left[X^\mathsf{T}(1) + X(1) - \varepsilon(1)\mathbb{I}, \cdots, X^\mathsf{T}(i-1) + X(i-1) - \varepsilon(i-1)\mathbb{I},\right.$$
$$\left. X^\mathsf{T}(i+1) + X(i+1) - \varepsilon(i+1)\mathbb{I}, \cdots, X^\mathsf{T}(N) + X(N) - \varepsilon(N)\mathbb{I}\right],$$
$$X_i(\varepsilon) = diag\left[4\varepsilon(1)\mathbb{I}, \cdots, 4\varepsilon(i-1)\mathbb{I}, 4\varepsilon(i+1)\mathbb{I}, \cdots, 4\varepsilon(N)\mathbb{I}\right],$$

with the following constraints:

$$X^\mathsf{T}(i)E^\mathsf{T}(i) = E(i)X(i) \geq 0,$$
$$E^\mathsf{T}(i)Q(i) = Q^\mathsf{T}(i)E(i) \geq 0.$$

The controller gains are given by:

$$\begin{cases} K(i) = Y_c(i)X^{-1}(i), \\ L(i) = Q^{-\mathsf{T}}(i)Y_o(i). \end{cases} \tag{6.35}$$

6.4 Numerical Example

Example 6.4.1 *To show the usefulness of the results of this chapter, let us consider the singular system with abrupt changes with the following data:*

- *mode # 1:*

$$E1 = \begin{bmatrix} 1 & 0 & 0 \\ 0 & 1 & 0 \\ 0 & 0 & 0 \end{bmatrix},$$

$$A(1) = \begin{bmatrix} 0.0 & 1.0 & 1.0 \\ -1.0 & 3.0 & 0.0 \\ 0.0 & -1.0 & -1.0 \end{bmatrix}, \ B(1) = \begin{bmatrix} 0.0 & 0.2 \\ 1.0 & 0.0 \\ -0.1 & 1.0 \end{bmatrix}, \ C(1) = \begin{bmatrix} 1.0 & 1.0 & 0.0 \end{bmatrix}.$$

- *mode # 2:*

$$E2 = \begin{bmatrix} 1 & 0 & 0 \\ 0 & 1 & 0 \\ 0 & 0 & 0 \end{bmatrix},$$

$$A(2) = \begin{bmatrix} -1.0 & 0.0 & 1.0 \\ 0.0 & 0.0 & 1.0 \\ 0.0 & 1.0 & -1.0 \end{bmatrix}, \ B(2) = \begin{bmatrix} 0.0 & 0.2 \\ 1.2 & 0.0 \\ -0.1 & 1.2 \end{bmatrix}, \ C(2) = \begin{bmatrix} 1.0 & -1.0 & 0.0 \end{bmatrix}.$$

The switching between the two modes is described by the following transition matrix:

$$\Lambda = \begin{bmatrix} -1 & 1 \\ 1.1 & -1.1 \end{bmatrix}.$$

Solving the conditions of Theorem 6.2.5, we get the following solution:

$$X(1) = \begin{bmatrix} 1.4591 & -0.0889 & 0.0 \\ -0.0889 & 1.5349 & 0.0 \\ -0.4505 & 0.0412 & 1.4030 \end{bmatrix},$$

$$X(2) = \begin{bmatrix} 1.5335 & -0.0365 & 0.0 \\ -0.0365 & 1.5018 & 0.0 \\ 0.1094 & -0.0028 & 1.2567 \end{bmatrix},$$

$$Y_c(1) = \begin{bmatrix} 1.2254 & -6.5597 & 6.0982 \\ -2.0153 & -5.0900 & 0.0000 \end{bmatrix},$$

$$Y_c(2) = \begin{bmatrix} 0.2417 & -1.7021 & 0.0000 \\ -0.9593 & -2.3427 & -0.8067 \end{bmatrix},$$

$$\varepsilon(1) = 0.8098, \varepsilon(2) = 0.8298.$$

which gives:

$$K(1) = \begin{bmatrix} 1.9210 & -4.2790 & 4.3465 \\ -1.5889 & -3.4082 & 0.0000 \end{bmatrix},$$

$$K(2) = \begin{bmatrix} 0.1307 & -1.1302 & 0.0000 \\ -0.6173 & -1.5761 & -0.6419 \end{bmatrix}.$$

For this set of gains, we get based on Theorem 6.2.5:

$$Q(1) = \begin{bmatrix} 104.4319 & 34.1465 & 0.0 \\ 34.1465 & 17.2277 & 0.0 \\ 214.0873 & 122.4839 & 31.1483 \end{bmatrix},$$

$$Q(2) = \begin{bmatrix} 258.6363 & 11.0050 & 0.0 \\ 11.0050 & 38.5357 & 0.0 \\ 253.1275 & 56.4266 & 22.7303 \end{bmatrix},$$

$$Y_o(1) = \begin{bmatrix} 8.9728 \\ 0.0000 \\ 119.3772 \end{bmatrix},$$

$$Y_o(2) = \begin{bmatrix} 320.7276 \\ 69.1864 \\ -0.0000 \end{bmatrix},$$

$$\varepsilon(1) = 0.7595, \varepsilon(2) = 0.7695.$$

which gives in turn the following gains:

$$L(1) = \begin{bmatrix} 0.2442 \\ -0.4839 \\ 4.0574 \end{bmatrix},$$

$$L(2) = \begin{bmatrix} 1.1780 \\ 1.4590 \\ -16.7401 \end{bmatrix}.$$

Example 6.4.2 *To show the usefulness of the results of this chapter in case of uncertain systems, let us consider the same singular system with abrupt changes we use in the previous example with the following data:*

- *mode # 1:*

$$D_A(1) = \begin{bmatrix} 0.1 \\ 0.1 \\ 0.0 \end{bmatrix}, \ E_A(1) = \begin{bmatrix} 0.0 & 0.1 & 0.0 \end{bmatrix},$$

$$D_B(1) = \begin{bmatrix} 0.0 \\ 0.1 \\ 0.1 \end{bmatrix}, \; E_B(1) = \begin{bmatrix} 0.1 & 0.0 \end{bmatrix},$$

$$D_{C_y}(1) = \begin{bmatrix} 0.1 \end{bmatrix}, \; E_{C_y}(1) = \begin{bmatrix} 0.0 & 0.0 & 0.0 \end{bmatrix}.$$

- *mode # 2:*

$$D_A(2) = \begin{bmatrix} 0.1 \\ 0.1 \\ 0.0 \end{bmatrix}, \; E_A(2) = \begin{bmatrix} 0.0 & 0.1 & 0.0 \end{bmatrix},$$

$$D_B(2) = \begin{bmatrix} 0.0 \\ 0.1 \\ 0.1 \end{bmatrix}, \; E_B(2) = \begin{bmatrix} 0.1 & 0.0 \end{bmatrix},$$

$$D_{C_y}(2) = \begin{bmatrix} 0.1 \end{bmatrix}, \; E_{C_y}(2) = \begin{bmatrix} 0.0 & 0.0 & 0.0 \end{bmatrix}.$$

Solving the conditions of Theorem 6.3.3, we get the following solution:

$$X(1) = \begin{bmatrix} 1.4725 & -0.0860 & 0.0 \\ -0.0860 & 1.4998 & 0.0 \\ -0.4640 & 0.0438 & 1.4000 \end{bmatrix},$$

$$X(2) = \begin{bmatrix} 1.5460 & -0.0348 & 0.0 \\ -0.0348 & 1.4858 & 0.0 \\ 0.0879 & -0.0047 & 1.2399 \end{bmatrix},$$

$$Y_c(1) = \begin{bmatrix} 0.0205 & -6.5314 & -0.0253 \\ -2.0190 & 0.9761 & -0.6389 \end{bmatrix},$$

$$Y_c(2) = \begin{bmatrix} 0.1985 & -2.3540 & -0.0394 \\ -1.0611 & -2.2049 & -1.4498 \end{bmatrix},$$

$$\varepsilon_A(1) = 2.4623, \varepsilon_A(2) = 2.4838,$$

$$\varepsilon_B(1) = 2.4369, \varepsilon_B(2) = 2.4320,$$

$$\varepsilon_{BK}(1) = 2.5541, \varepsilon_{BK}(2) = 2.4755,$$

$$\varepsilon(1) = 0.7595, \varepsilon(2) = 0.7695,$$

which gives:

$$K(1) = \begin{bmatrix} -0.2467 & -4.3684 & -0.0180 \\ -1.4811 & 0.5793 & -0.4563 \end{bmatrix},$$

$$K(2) = \begin{bmatrix} 0.0946 & -1.5822 & -0.0318 \\ -0.6537 & -1.5029 & -1.1693 \end{bmatrix}.$$

For this set of gains, we get based on Theorem 6.3.3:

$$\varepsilon_Q(1) = 50.9758, \varepsilon_Q(2) = 113.6511,$$

$$\varepsilon_{C_y}(1) = 49.3512, \varepsilon_{C_y}(2) = 79.8026,$$

$$Q(1) = \begin{bmatrix} 63.2121 & 12.1502 & 0.0 \\ 12.1502 & 4.5289 & 0.0 \\ 76.2083 & 38.9532 & 17.2656 \end{bmatrix},$$

$$Q(2) = \begin{bmatrix} 155.4025 & 54.5968 & 0.0 \\ 54.5968 & 27.0482 & 0.0 \\ 241.0472 & 42.2803 & 19.9238 \end{bmatrix},$$

$$Y_o(1) = \begin{bmatrix} -52.9776 \\ -24.8729 \\ 28.2704 \end{bmatrix},$$

$$Y_o(2) = \begin{bmatrix} 180.1536 \\ 50.1005 \\ 45.0648 \end{bmatrix},$$

which gives in turn the following gains:

$$L(1) = \begin{bmatrix} 0.4492 \\ -6.6972 \\ 14.7643 \end{bmatrix},$$

$$L(2) = \begin{bmatrix} 1.7484 \\ -1.6769 \\ -15.3326 \end{bmatrix}.$$

6.5 Notes

This chapter dealt with the observer-based state stabilization problem of the singular class of systems with random abrupt changes. The stochastic observer-based state stabilization and the robust stochastic observer-based state stabilization problems have been considered and LMI conditions were developed. A state feedback controller that assures that closed-loop state equation either for the nominal system or the uncertain is piecewise regular, impulse-free and stochastically stable is designed in the LMI setting. The conditions we developed in this chapter are tractable using commercial optimization tools. The content of this chapter is mainly based on the work of the author and his coauthors [20].

Part IV

Filtering

The filtering problem is one the most important control problems. It consists of estimating the state vector of dynamical systems using the measurement of the output that is always corrupted by noise. This problem has been widely studied for the class of normal dynamical systems and the literature overflows with results. Among the approaches that have been proposed we quote the popular Kalman filtering and the $\mathscr{H}_\infty$ techniques. The Kalman filtering approach consists of estimating the state vector by minimizing the covariance of the estimation error. This technique requires the complete knowledge of the system dynamics and no uncertainty is tolerated and on the top of this the noise must satisfy some statistical properties that are in general hard to satisfy in practice. To overcome these limitations regarding the system uncertainties and the statistical properties, the $\mathscr{H}_\infty$ filtering has been proposed and it consists of determining a filter that makes the system error asymptotically stable and guarantees that a certain given level of the external disturbance that has finite energy and no need for the statistical properties to be satisfied. For more details of this approach we refer the reader to [27] and the references therein.

The filtering problem of the class of deterministic singular systems has attracted a lot of researchers from the control community and many results have been reported in the literature. Among these contributions, we quote the works of [7, 11, 42, 55, 56, 57, 90, 98, 104, 126, 133, 135, 72, 30, 39, 41, 71, 99, 100, 101, 111, 118, 129, 132, 138] and the references therein. The filtering problem has been tackled for both the continuous-time case and the discrete-time one using different approaches like the Kalman filtering technique and the $\mathscr{H}_\infty$ filtering technique. The robust filtering has also been tackled.

For the class of system we are considering in this volume, to the best of our knowledge no results on this topics have been reported in the literature. Our goal in this part is to design a filter such that the class of linear singular systems with random abrupt changes is not only piecewise regular, impulse-free and stochastically stable but also guarantees a prescribed rejection level of an external disturbance with finite energy.

This part will deal with the filtering problem for the class of linear singular systems with random abrupt changes. Firstly, the $\mathscr{H}_\infty$ filtering problem for nominal singular systems is tackled. LMI conditions are developed to determine the filter gains. Secondly, the robust case is treated and similarly robust sufficient conditions in the LMI setting are established to design the $\mathscr{H}_\infty$ filter.

7

Filtering

Practically it is almost impossible to have the complete access to the state vector either for control or for any other purpose. The reason for that is that sometimes the technology is not available to measure some state variables or even if the technology is there, the resulting control system may be over the budget constraint which may result in costly system. An alternative to overcome these limitations (non measurable state variables, non availability of the technology, etc.) is to use the filtering technique that can provide an estimate of the complete state vector or some non measurable state variables.

The filtering has been applied for many years and continue to be used in many industrial applications ranging from aerospace to economics including engineering, biology, geoscience, management, etc. This problem has attracted a lot of researcher from different communities and interesting results have been reported in the literature. For more details on this topics we refer the reader to Boukas and Liu [27] and the references therein For the class of linear singular systems we quote the works of [7, 11, 42, 55, 56, 57, 90, 98, 104, 126, 133, 135, 72, 30, 39, 41, 71, 99, 100, 101, 111, 118, 129, 132, 138] and the references therein.

The goal of this chapter consists of developing results of $\mathcal{H}_\infty$ filtering for the class of linear singular systems with abrupt changes. The chapter will be organized as follows. In Sect. 1, the problem is stated and some definitions are given. In Sect. 2, the $\mathcal{H}_\infty$ filter is designed for the nominal systems. In Sect. 3, the robust $\mathcal{H}_\infty$ filter is tackled. Most of the results are in the LMI setting.

7.1 Problem Statement

Let us see how we can design a filter that estimates the state of the singular systems we are treating in this chapter. For this purpose, let us assume that the system state equation is described by the following differential-algebraic equations:

$$\begin{cases} E(r_t)\dot{x}(t) = A(r_t, t)x(t) + B(r_t, t)w(t), \; x(0) = x_0 \,, \\ y(t) = C_y(r_t, t)x(t) + D_y(r_t)v(t) \,, \\ z(t) = C_z(r_t, t)x(t) + D_z(r_t)v(t) \,, \end{cases} \tag{7.1}$$

where $x(t) \in \mathbb{R}^n$ is the state vector, $w(t) \in \mathbb{R}^m$ and $v(t) \in \mathbb{R}^l$ are the noise signals, $y(t) \in \mathbb{R}^k$ is the measurement, and $z(t) \in \mathbb{R}^p$ is the signal to be estimated, $E(r_t)$ is a known singular matrix with $0 \leq rank(E(i)) = n_r \leq n$, for all $i \in \mathscr{S}$, the matrices $A(r_t, t) \in \mathbb{R}^{n \times n}$, $B(r_t, t) \in \mathbb{R}^{n \times m}$, $C_y(r_t, t) \in \mathbb{R}^{k \times n}$, and $C_z(r_t, t) \in \mathbb{R}^{p \times n}$ are given by the following expressions when $r_t = i \in \mathscr{S}$:

$$A(i, t) = A(i) + D_A(i)F_A(i)E_A(i),$$
$$B(i, t) = B(i) + D_B(i)F_B(i)E_B(i),$$
$$C_y(i, t) = C_y(i) + D_{C_y}(i)F_{C_y}(i)E_{C_y}(i),$$
$$C_z(i, t) = C_z(i) + D_{C_z}(i)F_{C_z}(i)E_{C_z}(i),$$

where $A(i) \in \mathbb{R}^{n \times n}$, $B(i) \in \mathbb{R}^{n \times m}$, $C_y(i) \in \mathbb{R}^{k \times n}$, $D_y(i) \in \mathbb{R}^{k \times l}$, $C_z(i) \in \mathbb{R}^{p \times n}$, $D_z(i) \in \mathbb{R}^{p \times l}$, $D_A(i) \in \mathbb{R}^{n \times n_D}$, $E_A(i) \in \mathbb{R}^{n_E \times n}$, $D_B(i) \in \mathbb{R}^{n \times m_D}$, $E_B(i) \in \mathbb{R}^{m_E \times n}$, $D_{C_y}(i) \in \mathbb{R}^{k c_y \times m_{C_y}}$, $E_{C_y}(i) \in \mathbb{R}^{m_{C_y} \times l}$, $D_{C_z}(i) \in \mathbb{R}^{p \times q_D}$, and $E_{C_z}(i) \in \mathbb{R}^{q_E \times n}$ are known real matrices with appropriate dimensions and $F_A(i) \in \mathbb{R}^{n_D \times n_E}$, $F_B(i) \in \mathbb{R}^{m_D \times m_E}$, $F_{C_y}(i) \in \mathbb{R}^{m_{C_y} \times m_{C_y}}$, and $F_{C_z}(i) \in \mathbb{R}^{q_D \times q_E}$ are unknown matrices representing parameters uncertainties that will be assumed to satisfy the following assumptions.

Assumption 7.1.1 *Let the uncertainties $F_A(i)$, $F_B(i)$, $F_{C_y}(i)$, and $F_{C_z}(i)$ satisfy the following for every $i \in \mathscr{S}$ and $t \geq 0$:*

$$\begin{cases} F_A^\top(i)F_A(i) \leq \mathbb{I}, \\ F_B^\top(i)F_B(i) \leq \mathbb{I}, \\ F_{C_y}^\top(i)F_{C_y}(i) \leq \mathbb{I}, \\ F_{C_z}^\top(i)F_{C_z}(i) \leq \mathbb{I}. \end{cases} \tag{7.2}$$

Remark 7.1.1 *The uncertainties that satisfy this assumption will be referred to as admissible. In our case we are considering uncertainties that depend on mode, r_t.*

The filtering problem consists of computing an estimate, $\hat{z}(t)$, of the signal, $z(t)$, via a causal Markovian jump linear filter which provides a uniformly small estimation error, $z(t) - \hat{z}(t)$, for all ω satisfying some properties (finite energy or finite power) irrespective to the admissible uncertainties.

There exist in the literature different approaches for designing a filter that estimates the system states. In this chapter we will restrict ourselves to the $\mathscr{H}_\infty$ filtering.

In order to put the $\mathscr{H}_\infty$ filtering problem of the class of systems (7.1) we are considering here in the stochastic setting, let us introduce the space $\mathscr{L}_2[\Omega, \mathcal{F}, \mathbb{P}]$ of $\mathcal{F}$-measurable processes, $z(t) - \hat{z}(t)$, for which the following holds:

$$\|z - \hat{z}\|_2 \triangleq \left\{ \mathbb{E} \left[\int_0^\infty [z(t) - \hat{z}(t)]^\top [z(t) - \hat{z}(t)] dt \right] \right\}^{\frac{1}{2}} < \infty. \tag{7.3}$$

The goal of this chapter is to design a linear n-order filter of the following form:

$$\begin{cases} E(r_t)\dot{\hat{x}}(t) = K_A(r_t)\hat{x}(t) + K_B(r_t)y(t), \hat{x}(0) = 0, \\ \hat{z}(t) = K_C(r_t)\hat{x}(t), \end{cases} \tag{7.4}$$

which gives an estimate of the state vector, $\hat{x}(t)$, at time, t, and can ensure that the extended system $(x(t), x(t)-\hat{x}(t))$ is piecewise regular, impulse-free and stochastically stable and the estimation error, $z(t) - \hat{z}(t)$, is bounded for all noises $\omega(t) \in \mathcal{L}_2[0, \infty)$. The matrices $K_A(i)$, $K_B(i)$ and $K_C(i)$, $i \in \mathcal{S}$ are design parameters that should be determined in order to estimate the state vector properly.

7.2 $\mathcal{H}_\infty$ Filtering for Nominal Systems

Let us now drop the uncertainties from the dynamics and see how we can design the $\mathcal{H}_\infty$ filter with the structure defined by (7.4) for the nominal system. We will consider that $v(t) = w(t)$ for the rest of this chapter.

If we combine the dynamical system's state equation (7.1) with the filter's state equation (7.4), we get the following extended one:

$$\tilde{E}(r_t)\dot{\tilde{x}}(t) = \tilde{A}(r_t)\tilde{x}(t) + \tilde{B}(r_t)\omega(t), \quad \tilde{x}(0) = (x_0^\top, x_0^\top)^\top, \tag{7.5}$$

where

$$\tilde{x}(t) = \begin{bmatrix} x(t) \\ x(t) - \hat{x}(t) \end{bmatrix},$$

$$\tilde{E}(r_t) = \begin{bmatrix} E(r_t) & 0 \\ 0 & E(r_t) \end{bmatrix},$$

$$\tilde{A}(r_t) = \begin{bmatrix} A(r_t) & 0 \\ A(r_t) - K_B(r_t)C_y(r_t) - K_A(r_t) & K_A(r_t) \end{bmatrix},$$

$$\tilde{B}(r_t) = \begin{bmatrix} B(r_t) \\ B(r_t) - K_B(r_t)D_y(r_t) \end{bmatrix}.$$

The estimation error, $e(t) = z(t) - \hat{z}(t)$, satisfies the following:

$$e(t) = \tilde{C}(r_t)\tilde{x}(t) + \tilde{D}(r_t)\omega(t), \tag{7.6}$$

with

$$\tilde{C}(r_t) = \begin{bmatrix} C_z(r_t) - K_C(r_t) & K_C(r_t) \end{bmatrix},$$

$$\tilde{D}(t) = D_z(r_t).$$

Remark 7.2.1 *To get the extended dynamics we computed:*

$$E(r_t)\dot{x}(t) - E(r_t)\dot{\hat{x}}(t) = A(r_t)x(t) + B(r_t)\omega(t) - K_A(r_t)\hat{x}(t) - K_B(r_t)y(t)$$

$$= \left[A(r_t) - K_B(r_t)C_y(r_t) - K_A(r_t) \right] x(t)$$

$$+ K_A(r_t)(x(t) - \hat{x}(t)) + \left[B(r_t) - K_B(r_t)D_y(r_t) \right] \omega(t),$$

for the second component of the state vector, $\tilde{x}(t)$, and

$$e(t) = z(t) - \hat{z}(t)$$

$$= C_z(r_t)x(t) + D_z(r_t)\omega(t) - K_C(r_t)\hat{x}(t) - K_C(r_t)x(t) + K_C(r_t)x(t)$$

$$= \begin{bmatrix} C_z(r_t) - K_C(r_t) & K_C(r_t) \end{bmatrix} \begin{bmatrix} x(t) \\ x(t) - \hat{x}(t) \end{bmatrix} + D_z(r_t)\omega(t)$$

$$= \tilde{C}(r_t)\tilde{x}(t) + \tilde{D}(r_t)\omega(t),$$

for the estimation error equation.

The next theorem states that when the filter (7.4) exists (i.e., we can get a set of gains, $K_A = (K_A(1), \cdots, K_A(N))$, $K_B = (K_B(1), \cdots, K_B(N))$ and $K_C = (K_C(1), \cdots, K_C(N)))$, the extended system is piecewise regular, impulse-free and stochastically stable and the estimation error, $z(t) - \hat{z}(t)$, is bounded for all signals $\omega(t) \in \mathcal{L}_2[0, \infty)$ if some given conditions are satisfied.

Theorem 7.2.1 *Let* $K_A = (K_A(1), \cdots, K_A(N))$, $K_A(i) \in \mathbb{R}^{n \times n}$, $K_B = (K_B(1), \cdots, K_B(N))$, $K_B(i) \in \mathbb{R}^{n \times k}$, *and* $K_C = (K_C(1), \cdots, K_C(N)))$, $K_C(i) \in \mathbb{R}^{p \times n}$, *be given sets of gains. Let* γ *be a given positive constant and R is a given symmetric nonnegative-definite matrix representing the weight of the initial conditions. If there exists a set of nonsingular matrices* $P = (P(1), \cdots, P(N))$, $P(i) \in \mathbb{R}^{n \times n}$, *such that the following set of the coupled LMIs holds:*

$$\begin{bmatrix} \tilde{J}_1(i) & P^\top(i)\tilde{B}(i) & \tilde{C}^\top(i) \\ \tilde{B}^\top(i)P(i) & -\gamma^2 \mathbb{I} & \tilde{D}^\top(i) \\ \tilde{C}(i) & \tilde{D}(i) & -\mathbb{I} \end{bmatrix} < 0, \tag{7.7}$$

$$\begin{bmatrix} \mathbb{I} & \mathbb{I} \end{bmatrix} \tilde{E}^\top(r_0)P(r_0)\begin{bmatrix} \mathbb{I} \\ \mathbb{I} \end{bmatrix} \leq \gamma^2 R, \tag{7.8}$$

where

$$\tilde{J}_1(i) = \tilde{A}^\top(i)P(i) + P^\top(i)\tilde{A}(i) + \sum_{j=1}^{N} \lambda_{ij}E^\top(j)P(j),$$

with the following constraints:

$$E^\top(i)P(i) = P^\top(i)E(i) \geq 0, \tag{7.9}$$

then the extended system is piecewise regular, impulse-free and stochastically stable and, moreover the estimation error satisfies the following:

$$\|z(t) - \hat{z}(t)\|_2 \leq \gamma \left[\|\omega\|_2^2 + x_0^\top R x_0\right]^{\frac{1}{2}}. \tag{7.10}$$

Proof: Let us first of all prove that the extended system is piecewise regular, impulse-free and stochastically stable. For this purpose, from (7.7), we get:

$$\tilde{J}_1(i) = \tilde{A}^\top(i)P(i) + P^\top(i)\tilde{A}(i) + \sum_{j=1}^{N} \lambda_{ij}E^\top(j)P(j) < 0,$$

that combined with the results of Chap. 2, implies that the extended system is piecewise regular, impulse-free and stochastically stable.

Let us now prove the second part of the theorem which indicates that the estimation error is bounded for all signals $\omega(t) \in \mathcal{L}_2[0, \infty)$. To this end, let us define the following $\mathcal{H}_\infty$ performance:

$$J_T = \mathbb{E}\left[\int_0^T \left[e^\top(t)e(t) - \gamma^2 w^\top(t)w(t)\right]dt\right], \forall T > 0, \tag{7.11}$$

and let $\mathcal{L}$ be the infinitesimal generator of the Markov process $\{(\tilde{x}(t), r_t), t \geq 0\}$.

Let us denote by $(\tilde{x}, i)$ the values of the state vector, $\tilde{x}(t)$ and the mode, r_t, at time t and consider the following Lyapunov function candidate as follows:

$$V(\tilde{x}(t), r_t) = \tilde{x}^\top(t)\tilde{E}^\top(r_t)P(r_t)\tilde{x}(t), \qquad (7.12)$$

where $P(i)$, for each $i \in \mathcal{S}$ is a nonsingular matrix.

Notice that the infinitesimal operator emanating from the point $(\tilde{x}, i)$ at time t is given by:

$$\begin{aligned}
\mathcal{L}V(\tilde{x}(t), i) &= \dot{\tilde{x}}^\top(t)\tilde{E}^\top(i)P(i)\tilde{x}(t) + \tilde{x}^\top(t)\tilde{E}^\top(i)P(i)\dot{\tilde{x}}(t) \\
&\quad + \tilde{x}^\top(t)\left[\sum_{j=1}^{N} \lambda_{ij}E^\top(j)P(j)\right]\tilde{x}(t) \\
&= \left[\tilde{A}(i)\tilde{x}(t) + \tilde{B}(i)\omega(t)\right]^\top P(i)\tilde{x}(t) \\
&\quad + \tilde{x}^\top(t)P^\top(i)\left[\tilde{A}(i)\tilde{x}(t) + \tilde{B}(i)\omega(t)\right] \\
&\quad + \tilde{x}^\top(t)\left[\sum_{j=1}^{N} \lambda_{ij}E^\top(j)P(j)\right]\tilde{x}(t) \\
&= \tilde{x}^\top(t)\left[\tilde{A}^\top(i)P(i) + P^\top(i)\tilde{A}(i) + \sum_{j=1}^{N} \lambda_{ij}E^\top(j)P(j)\right]\tilde{x}(t) \\
&\quad + \tilde{x}^\top(t)P^\top(i)\tilde{B}(i)\omega(t) + \omega^\top(t)\tilde{B}^\top(i)P(i)\tilde{x}(t),
\end{aligned}$$

and

$$\begin{aligned}
&e^\top(t)e(t) - \gamma^2\omega^\top(t)\omega(t) \\
&= \left[\tilde{C}(i)\tilde{x}(t) + \tilde{D}(i)\omega(t)\right]^\top\left[\tilde{C}(i)\tilde{x}(t) + \tilde{D}(i)\omega(t)\right] - \gamma^2\omega^\top(t)\omega(t) \\
&= \tilde{x}^\top(t)\tilde{C}^\top(i)\tilde{C}(i)\tilde{x}(t) + \tilde{x}^\top(t)\tilde{C}^\top(i)\tilde{D}(i)\omega(t) \\
&\quad + \omega^\top(t)\tilde{D}^\top(i)\tilde{C}(i)\tilde{x}(t) + \omega^\top(t)\tilde{D}^\top(i)\tilde{D}(i)\omega(t) - \gamma^2\omega^\top(t)\omega(t).
\end{aligned}$$

Combining these two relations, we get:

$$\begin{aligned}
e^\top(t)e(t) - \gamma^2\omega^\top(t)\omega(t) + \mathcal{L}V(x(t), i) &= \tilde{x}^\top(t)\Big[\tilde{A}^\top(i)P(i) + P^\top(i)\tilde{A}(i) \\
&\quad + \sum_{j=1}^{N} \lambda_{ij}E^\top(j)P(j) + \tilde{C}^\top(i)\tilde{C}(i)\Big]\tilde{x}(t) \\
&\quad + \tilde{x}^\top(t)\left[P^\top(i)\tilde{B}(i) + \tilde{C}^\top(i)\tilde{D}(i)\right]\omega(t) \\
&\quad + \omega^\top(t)\left[\tilde{B}^\top(i)P(i) + \tilde{D}^\top(i)\tilde{C}(i)\right]\tilde{x}(t) \\
&\quad + \omega^\top(t)\left[\tilde{D}^\top(i)\tilde{D}(i) - \gamma^2\mathbb{I}\right]\omega(t).
\end{aligned}$$

which gives in matrix form:

$$\begin{aligned}
e^\top(t)e(t) &- \gamma^2\omega^\top(t)\omega(t) + \mathcal{L}V(x(t), i) \\
&= \left[\tilde{x}^\top(t) \; \omega^\top(t)\right]\Lambda_n(i)\begin{bmatrix}\tilde{x}(t) \\ \omega(t)\end{bmatrix} \\
&= \tilde{\xi}^\top(t)\Lambda_n(i)\tilde{\xi}(t), \qquad (7.13)
\end{aligned}$$

with $\tilde{\xi}(t) = \begin{bmatrix} \tilde{x}(t) \\ \omega(t) \end{bmatrix}$ and $\Lambda_n(i)$ is defined by:

$$
\Lambda_n(i) = \left[\begin{array}{c|c}
\begin{array}{c}
\tilde{A}^\top(i)P(i) \\
+P^\top(i)\tilde{A}(i) + \lambda_{ii}E^\top(i)P(i) \\
+ \sum_{j=1, j \neq i}^N \varepsilon_P \lambda_{ij} [P(j) + P^\top(j)] \\
+\tilde{C}^\top(i)\tilde{C}(i) \\
\tilde{B}^\top(i)P(i) + D^\top(i)\tilde{C}(i)
\end{array} &
\begin{array}{c}
P^\top(i)\tilde{B}(i) + \tilde{C}^\top(i)\tilde{D}(i) \\
\\
\\
\\
\tilde{D}^\top(i)\tilde{D}(i) - \gamma^2 \mathbb{I}
\end{array}
\end{array} \right].
$$

Adding and subtracting $\mathscr{L}V(x(t), i)$ to the $\mathscr{H}_\infty$ performance, (7.11), we get the following:

$$
J_T = \mathbb{E}\left[\int_0^T [e^\top(t)e(t) - \gamma^2 w^\top(t)w(t) + \mathscr{L}V(\tilde{x}(t), r_t)]dt \right]
$$
$$
- \mathbb{E}\left[\int_0^T \mathscr{L}V(\tilde{x}(t), r_t)dt \right].
$$

From the expression of the Lyapunov candidate function, we have:

$$
\mathbb{E}[V(\tilde{x}_0, r_0)] = \mathbb{E}[\tilde{x}^\top(0)\tilde{E}^\top(r_0)P(r_0)\tilde{x}(0)] . \tag{7.14}
$$

Note that $\tilde{x}(0) = \left[x^\top(0) \ x^\top(0) - \hat{x}^\top(0) \right]^\top = \left[x_0^\top, x_0^\top \right]^\top$.
In view of (7.8) and (7.14), we have:

$$
\mathbb{E}[V(\tilde{x}(0), r(0))] = \left\{ \mathbb{E}\left[x^\top(0) \begin{bmatrix} \mathbb{I} & \mathbb{I} \end{bmatrix} \tilde{E}^\top(r_0)P(r_0) \begin{bmatrix} \mathbb{I} \\ \mathbb{I} \end{bmatrix} x(0) \right] \right\}
$$
$$
\leq \gamma^2 \mathbb{E}\left[x^\top(0)Rx(0) \right].
$$

Using now Dynkin's formula, i.e.,

$$
\mathbb{E}\left[\int_0^T \mathscr{L}V(\tilde{x}(t), r_t)dt \right] = \mathbb{E}[V(\tilde{x}_T, r_T)] - \mathbb{E}[V(\tilde{x}_0, r_0)] .
$$

First of all notice that the $\mathscr{H}_\infty$ performance, J_T, can be rewritten as follows:

$$
J_T = \mathbb{E}\left[\int_0^T \tilde{\xi}_t^\top \Lambda_n(r_t)\tilde{\xi}_t dt \right] + \mathbb{E}[V(\tilde{x}(0), r_0)] - \mathbb{E}[V(\tilde{x}(T), r_T)] ,
$$

which implies:

$$
J_T \leq \mathbb{E}\left[\int_0^T \tilde{\xi}_t^\top \Lambda_n(r_t)\tilde{\xi}_t dt \right] + \mathbb{E}[V(\tilde{x}(0), r_0)] . \tag{7.15}
$$

Combining this with the fact that $\Lambda_n(i) < 0$ for all $i \in \mathscr{S}$, the following holds for all $T > 0$:

$$
J_T \leq \mathbb{E}[V(\tilde{x}(0), r_0)] \leq \gamma^2 x^\top(0)Rx(0) .
$$

Therefore, we get:

$$
J_\infty = \mathbb{E}\left[\int_0^\infty [e^\top(t)e(t) - \gamma^2 w^\top(t)w(t)]dt \right]
$$
$$
\leq \gamma^2 x^\top(0)Rx(0) .
$$

This gives in turn that:

$$\|e\|_2^2 \leq \gamma \left[\|w\|_2^2 + x^\top(0)Rx(0)\right],$$

and this ends the proof of Theorem 7.2.1.

For a given set of gains of the filter of the form (7.4), we can compute the minimum disturbance rejection by solving the following convex optimization problem:

$$
P: \begin{cases}
\min\limits_{\substack{v>0, \\ P=(P(1),\cdots,P(N))}} v \\[2ex]
\text{s.t:} \\[1ex]
\varepsilon_P \left[P(i) + P^\top(i)\right] \geq E^\top(i)P(i) = P^\top(i)E(i) \geq 0, \\[2ex]
\begin{bmatrix}
\tilde{J}_1(i) & P^\top(i)\tilde{B}(i) & \tilde{C}^\top(i) \\
\tilde{B}^\top(i)P(i) & -v\mathbb{I} & \tilde{D}^\top(i) \\
\tilde{C}(i) & \tilde{D}(i) & -\mathbb{I}
\end{bmatrix} < 0, \\[4ex]
\begin{bmatrix} \mathbb{I} & \mathbb{I} \end{bmatrix} \tilde{E}^\top(r_0)P(r_0) \begin{bmatrix} \mathbb{I} \\ \mathbb{I} \end{bmatrix} \leq vR,
\end{cases}
$$

where $v = \gamma^2$.

But since we don't have yet developed a way to compute the filter gains, this optimization problem is useless. The design of the filter's gains should be included in an optimization problem similar to this one that can help us to determine simultaneously the filter's gains and the minimum disturbance rejection.

Notice that the condition (7.7) is nonlinear in $P(i)$ and the design filter parameters. To cast the design of the $\mathscr{H}_\infty$ filter in the LMI framework, let us transform this condition in order to compute the gains $K_A(i)$, $K_B(i)$ and $K_C(i)$.

Let us first of all compute $\tilde{J}_1(i)$, $P^\top(i)\tilde{B}(i)$, $\tilde{C}^\top(i)$, and $\tilde{D}^\top(i)$ in function of $A(i)$, $B(i)$, $C_y(i)$, $D_y(i)$, $C_z(i)$ and $D_z(i)$. Using the expression of $\tilde{A}(i)$, $\tilde{B}(i)$, $\tilde{C}(i)$ and $\tilde{D}(i)$, and assuming that $P(i) = \text{diag}[X_1(i), X_2(i)]$ we get:

$$\tilde{J}_1(i) = \tilde{A}^\top(i)P(i) + P^\top(i)\tilde{A}(i) + \sum_{j=1}^{N} \lambda_{ij}E^\top(j)P(j)$$

$$= \begin{bmatrix}
\tilde{J}_{X_1}(i) & \begin{bmatrix} A^\top(i)X_2(i) \\ -C_y^\top(i)K_B^\top(i)X_2(i) \\ -K_A^\top(i)X_2(i) \end{bmatrix} \\[4ex]
\begin{bmatrix} X_2^\top(i)A(i) \\ -X_2^\top(i)K_B(i)C_y(i) \\ -X_2^\top(i)K_A(i) \end{bmatrix} & \tilde{J}_{X_2}(i)
\end{bmatrix},$$

$$
\tilde{P}^\top(i)\tilde{B}(i) = \begin{bmatrix} X_1^\top(i) & 0 \\ 0 & X_2^\top(i) \end{bmatrix} \begin{bmatrix} B(i) \\ B(i) - K_B(i)D_y(i) \end{bmatrix}
$$

$$
= \begin{bmatrix} X_1^\top(i)B(i) \\ X_2^\top(i)B(i) - X_2^\top(i)K_B(i)D_y(i) \end{bmatrix},
$$

$$\tilde{C}(i) = \begin{bmatrix} C_z(i) - K_C(i) & K_C(i) \end{bmatrix},$$

$$\tilde{D}(i) = D_z(i),$$

with

$$\tilde{J}_{X_1}(i) = A^\top(i)X_1(i) + X_1^\top(i)A(i) + \sum_{j=1}^N \lambda_{ij}E^\top(j)X_1(j),$$

$$\tilde{J}_{X_2}(i) = K_A^\top(i)X_2(i) + X_2^\top(i)K_A(i) + \sum_{j=1}^N \lambda_{ij}E^\top(j)X_2(j).$$

Using these relations, (7.7) becomes:

$$\begin{bmatrix} \tilde{J}_{X_1}(i) & \begin{bmatrix} A^\top(i)X_2(i) \\ -C_y^\top(i)K_B^\top(i)X_2(i) \\ -K_A^\top(i)X_2(i) \end{bmatrix} & X_1^\top(i)B(i) & C_z^\top(i) - K_C^\top(i) \\ \begin{bmatrix} X_2^\top(i)A(i) \\ -X_2^\top(i)K_B(i)C_y(i) \\ -X_2^\top(i)K_A(i) \end{bmatrix} & \tilde{J}_{X_2}(i) & \begin{bmatrix} X_2^\top(i)B(i) \\ -X_2^\top(i)K_B(i)D_y(i) \end{bmatrix} & K_C^\top(i) \\ B^\top(i)X_1(i) & \begin{bmatrix} B^\top(i)X_2(i) \\ -D_y^\top(i)K_B^\top(i)X_2(i) \end{bmatrix} & -\gamma^2\mathbb{I} & D_z^\top(i) \\ C_z(i) - K_C(i) & K_C(i) & D_z(i) & -\mathbb{I} \end{bmatrix} < 0.$$

Letting $Y(i) = X_2^\top(i)K_A(i)$, $Z(i) = X_2^\top(i)K_B(i)$, and $W(i) = K_C(i)$, we get:

$$\begin{bmatrix} \tilde{J}_{X_1}(i) & \begin{bmatrix} A^\top(i)X_2(i) \\ -C_y^\top(i)Z^\top(i) \\ -Y^\top(i) \end{bmatrix} & X_1^\top(i)B(i) & \begin{bmatrix} C_z^\top(i) \\ -W^\top(i) \end{bmatrix} \\ \begin{bmatrix} X_2^\top(i)A(i) \\ -Z(i)C_y(i) \\ -Y(i) \end{bmatrix} & J_{X_2}(i) & \begin{bmatrix} X_2^\top(i)B(i) \\ -Z(i)D_y(i) \end{bmatrix} & W^\top(i) \\ B^\top(i)X_1(i) & \begin{bmatrix} B^\top(i)X_2(i) \\ -D_y^\top(i)Z^\top(i) \end{bmatrix} & -\gamma^2\mathbb{I} & D_z^\top(i) \\ C_z(i) - W(i) & W(i) & D_z(i) & -\mathbb{I} \end{bmatrix} < 0,$$

with

$$J_{X_2}(i) = Y^\top(i) + Y(i) + \sum_{j=1}^N \lambda_{ij}E^\top(j)X_2(j).$$

Notice also that the condition, $\varepsilon_P[P(i) + P^\top(i)] \geq \tilde{E}^\top(i)P(i) = P^\top(i)\tilde{E}(i) \geq 0$, becomes:

$$E^\top(i)X_1(i) = X_1^\top(i)E(i) \geq 0,$$
$$E^\top(i)X_2(i) = X_2^\top(i)E(i) \geq 0.$$

For the last relation of the theorem, we have:

$$\begin{bmatrix} \mathbb{I} & \mathbb{I} \end{bmatrix}\begin{bmatrix} E^\top(r_0) & 0 \\ 0 & E^\top(r_0) \end{bmatrix}\begin{bmatrix} X_1(r_0) & 0 \\ 0 & X_2(r_0) \end{bmatrix}\begin{bmatrix} \mathbb{I} \\ \mathbb{I} \end{bmatrix}$$
$$= E^\top(r_0)X_1(r_0) + E^\top(r_0)X_2(r_0) < \gamma^2 R.$$

The following theorem gives the results for the design of the gains of the $\mathscr{H}_\infty$ filter.

Theorem 7.2.2 *Let γ and R be respectively given positive constant and a symmetric and positive-definite matrix representing the weighting of the initial conditions. If there exist sets of nonsingular matrices $X_1 = (X_1(1), \cdots, X_1(N))$, $X_1(i) \in \mathbb{R}^{n \times n}$ and $X_2 = (X_2(1), \cdots, X_2(N))$, $X_2(i) \in \mathbb{R}^{n \times n}$ and matrices $Y = (Y(1), \cdots, Y(N))$, $Y(i) \in \mathbb{R}^{n \times n}$ $Z = (Z(1), \cdots, Z(N))$, $Z(i) \in \mathbb{R}^{n \times k}$ and $W = (W(1), \cdots, W(N))$ $W(i) \in \mathbb{R}^{p \times n}$ satisfying the following set of coupled LMIs:*

$$\begin{bmatrix} J_{X_1}(i) & \begin{bmatrix} A^\top(i)X_2(i) \\ -C_y^\top(i)Z^\top(i) \\ -Y^\top(i) \end{bmatrix} & X_1^\top(i)B(i) & \begin{bmatrix} C_z^\top(i) \\ -W^\top(i) \end{bmatrix} \\ \begin{bmatrix} X_2^\top(i)A(i) \\ -Z(i)C_y(i) \\ -Y(i) \end{bmatrix} & J_{X_2}(i) & \begin{bmatrix} X_2^\top(i)B(i) \\ -Z(i)D_y(i) \end{bmatrix} & W^\top(i) \\ B^\top(i)X_1(i) & \begin{bmatrix} B^\top(i)X_2(i) \\ -D_y^\top(i)Z^\top(i) \end{bmatrix} & -\gamma^2\mathbb{I} & D_z^\top(i) \\ C_z(i) - W(i) & W(i) & D_z(i) & -\mathbb{I} \end{bmatrix} < 0, \quad (7.16)$$

$$E^\top(r_0)X_1(r_0) + E^\top(r_0)X_2(r_0) < \gamma^2 R, \quad (7.17)$$

where

$$J_{X_1}(i) = A^\top(i)X_1(i) + X_1^\top(i)A(i) + \sum_{j=1}^N \lambda_{ij}E^\top(j)X_1(j),$$

$$J_{X_2}(i) = Y^\top(i) + Y(i) + \sum_{j=1}^N \lambda_{ij}E^\top(j)X_2(j),$$

with the following constraints:

$$E^\top(i)X_1(i) = X_1^\top(i)E(i) \geq 0, \quad (7.18)$$
$$E^\top(i)X_2(i) = X_2^\top(i)E(i) \geq 0, \quad (7.19)$$

then there exists a filter of the form (7.4) such that the estimation error is piecewise regular, impulse-free and stochastically stable and bounded by:

$$\|z - \hat{z}\|_2 \leq \gamma \left[\|\omega\|_2^2 + x_0^\top R x_0 \right]^{\frac{1}{2}}. \quad (7.20)$$

The filter's gains are given by:

$$\begin{cases} K_A(i) = X_2^{-\top}(i)Y(i), \\ K_B(i) = X_2^{-\top}(i)Z(i), \\ K_C(i) = W(i). \end{cases} \quad (7.21)$$

If the initial conditions are equal to zero, the previous results becomes easier and are given by the following corollary.

Corollary 7.2.1 *Let the initial conditions of system (7.1) be zero. Let γ and ε_P be given positive constants. If there exist sets of nonsingular matrices $X_1 = (X_1(1), \cdots,$ $X_1(N))$, $X_1(i) \in \mathbb{R}^{n \times n}$ and $X_2 = (X_2(1), \cdots, X_2(N))$, $X_2(i) \in \mathbb{R}^{n \times n}$ and matrices $Y = (Y(1), \cdots, Y(N))$, $Y(i) \in \mathbb{R}^{n \times n}$ $Z = (Z(1), \cdots, Z(N))$, $Z(i) \in \mathbb{R}^{n \times k}$ and*

$W = (W(1), \cdots, W(N))$ $W(i) \in \mathbb{R}^{p \times n}$ satisfying the LMIs (7.18)-(7.16) for every $i \in \mathscr{S}$, then there exists a filter of the form (7.4) such that the estimation error is piecewise regular, impulse-free and stochastically stable and bounded by:

$$\|z - \hat{z}\|_2 \le \gamma \|\omega\|_2 .$$

The filter's gains are given by (7.21).

The minimal noise attenuation level, γ, that can be verified by the filter of the form of (7.4) can be obtained by solving the following optimization problem:

$$\mathcal{P}_0 : \begin{cases} \min\limits_{\substack{v>0, \\ X_1=(X_1(1),\cdots,X_1(N)), \\ X_2=(X_2(1),\cdots,X_2(N)), \\ Y=(Y(1),\cdots,Y(N)), \\ Z=(Z(1),\cdots,Z(N)), \\ W=(W(1),\cdots,W(N))}} v \\[2pt] s.t. \\ E^\top(i)X_1(i) = X_1^\top(i)E(i) \ge 0, \\ E^\top(i)X_2(i) = X_2^\top(i)E(i) \ge 0, \\ \Theta_v(i) < 0, \\ E^\top(r_0)X_1(r_0) + E^\top(r_0)X_2(r_0) < vR, \end{cases}$$

where $\Theta_v(i)$ is obtained from (7.16) by replacing γ^2 by v. Thus, if the convex optimization problem $\mathcal{P}_0$ has a solution, v for a given positive scalar ε_P, then by using Theorem 7.2.2, the corresponding error of the filter (7.4) is stable with noise attenuation level $\sqrt{v}$.

Notice that we can also use the bounds we gave previously. For the case of:

$$E^\top(i)P(i) \le \varepsilon_P \left[P^\top(i) + P(i) \right] ,$$

for $\varepsilon_P > 0$, the following theorem gives the results for the design of the gains of the $\mathscr{H}_\infty$ filter.

Theorem 7.2.3 *Let γ, ε_P, and R be respectively given positive constants and a symmetric and positive-definite matrix representing the weighting of the initial conditions. If there exist sets of nonsingular matrices $X_1 = (X_1(1), \cdots, X_1(N))$, $X_1(i) \in \mathbb{R}^{n \times n}$ and $X_2 = (X_2(1), \cdots, X_2(N))$, $X_2(i) \in \mathbb{R}^{n \times n}$ and matrices $Y = (Y(1), \cdots, Y(N))$, $Y(i) \in \mathbb{R}^{n \times n}$ $Z = (Z(1), \cdots, Z(N))$, $Z(i) \in \mathbb{R}^{n \times k}$ and $W = (W(1), \cdots, W(N))$ $W(i) \in \mathbb{R}^{p \times n}$ satisfying the following set of coupled LMIs:*

$$\begin{bmatrix} J_{X_1}(i) & \begin{bmatrix} A^\top(i)X_2(i) \\ -C_y^\top(i)Z^\top(i) \\ -Y^\top(i) \end{bmatrix} & X_1^\top(i)B(i) & \begin{bmatrix} C_z^\top(i) \\ -W^\top(i) \end{bmatrix} \\[14pt] \begin{bmatrix} X_2^\top(i)A(i) \\ -Z(i)C_y(i) \\ -Y(i) \end{bmatrix} & J_{X_2}(i) & \begin{bmatrix} X_2^\top(i)B(i) \\ -Z(i)D_y(i) \end{bmatrix} & W^\top(i) \\[14pt] B^\top(i)X_1(i) & \begin{bmatrix} B^\top(i)X_2(i) \\ -D_y^\top(i)Z^\top(i) \end{bmatrix} & -\gamma^2\mathbb{I} & D_z^\top(i) \\[8pt] C_z(i) - W(i) & W(i) & D_z(i) & -\mathbb{I} \end{bmatrix} < 0, \qquad (7.22)$$

$$E^\top(r_0)X_1(r_0) + E^\top(r_0)X_2(r_0) < \gamma^2 R, \qquad (7.23)$$

where

$$J_{X_1}(i) = A^\top(i)X_1(i) + X_1^\top(i)A(i) + \lambda_{ii}E^\top(i)X_1(i)$$

$$+ \sum_{j=1,j\neq i}^{N} \varepsilon_P \lambda_{ij} \left[X_1(j) + X_1^\top(j) \right],$$

$$J_{X_2}(i) = Y^\top(i) + Y(i) + \lambda_{ii}E^\top(i)X_2(i) + \sum_{j=1,j\neq i}^{N} \varepsilon_P \lambda_{ij} \left[X_2(j) + X_2^\top(j) \right],$$

with the following constraints:

$$\varepsilon_P \left[X_1(i) + X_1^\top(i) \right] \geq E^\top(i)X_1(i) = X_1^\top(i)E(i) \geq 0, \tag{7.24}$$

$$\varepsilon_P \left[X_2(i) + X_2^\top(i) \right] \geq E^\top(i)X_2(i) = X_2^\top(i)E(i) \geq 0, \tag{7.25}$$

then there exists a filter of the form (7.4) such that the estimation error is piecewise regular, impulse-free and stochastically stable and bounded by:

$$\|z - \hat{z}\|_2 \leq \gamma \left[\|\omega\|_2^2 + x_0^\top R x_0 \right]^{\frac{1}{2}}. \tag{7.26}$$

The filter's gains are given by:

$$\begin{cases} K_A(i) = X_2^{-\top}(i)Y(i), \\ K_B(i) = X_2^{-\top}(i)Z(i), \\ K_C(i) = W(i). \end{cases} \tag{7.27}$$

Now if the we assume the following to hold:

$$E^\top(i)P(i) \leq \left[\frac{1}{4}\varepsilon(i)\mathbb{I} + \varepsilon^{-1}(i)E^\top(i)P(i)P^\top(i)E(i) \right].$$

Proceeding as before we get the following theorem that summarizes the results for the design of the gains of the $\mathcal{H}_\infty$ filter.

Theorem 7.2.4 *Let γ, and R be respectively given positive constants and a symmetric and positive-definite matrix representing the weighting of the initial conditions. If there exist sets of nonsingular matrices $X_1 = (X_1(1), \cdots, X_1(N))$, $X_1(i) \in \mathbb{R}^{n\times n}$ and $X_2 = (X_2(1), \cdots, X_2(N))$, $X_2(i) \in \mathbb{R}^{n\times n}$ and matrices $Y = (Y(1), \cdots, Y(N))$, $Y(i) \in \mathbb{R}^{n\times n}$ $Z = (Z(1), \cdots, Z(N))$, $Z(i) \in \mathbb{R}^{n\times k}$, $W = (W(1), \cdots, W(N))$ $W(i) \in \mathbb{R}^{p\times n}$ and a set of positive scalars $\varepsilon = (\varepsilon(1), \cdots, \varepsilon(N))$ satisfying the following set of coupled LMIs:*

$$\begin{bmatrix} J_{X_1}(i) & \begin{matrix} A^\top(i)X_2(i) \\ -C_y^\top(i)Z^\top(i) \\ -Y^\top(i) \end{matrix} & X_1^\top(i)B(i) & \begin{bmatrix} C_z^\top(i) \\ -W^\top(i) \end{bmatrix} & S_i(X_1) & 0 \\ \begin{matrix} X_2^\top(i)A(i) \\ -Z(i)C_y(i) \\ -Y(i) \end{matrix} & J_{X_2}(i) & \begin{bmatrix} X_2^\top(i)B(i) \\ -Z(i)D_y(i) \end{bmatrix} & W^\top(i) & 0 & S_i(X_2) \\ B^\top(i)X_1(i) & \begin{bmatrix} B^\top(i)X_2(i) \\ -D_y^\top(i)Z^\top(i) \end{bmatrix} & -\gamma^2\mathbb{I} & D_z^\top(i) & 0 & 0 \\ C_z(i) - W(i) & W(i) & D_z(i) & -\mathbb{I} & 0 & 0 \\ S_i(X_1) & 0 & 0 & 0 & -\mathcal{X}_i(\varepsilon) & 0 \\ 0 & S_i(X_2) & 0 & 0 & 0 & -\mathcal{X}_i(\varepsilon) \end{bmatrix} < 0, \tag{7.28}$$

$$E^\top(r_0)X_1(r_0) + E^\top(r_0)X_2(r_0) < \gamma^2 R, \tag{7.29}$$

where

$$J_{X_1}(i) = A^\top(i)X_1(i) + X_1^\top(i)A(i) + \lambda_{ii}E^\top(i)X_1(i)$$

$$+ \sum_{j=1,j\neq i}^{N} \lambda_{ij}\frac{1}{4}\varepsilon(j)\mathbb{I},$$

$$J_{X_2}(i) = Y^\top(i) + Y(i) + \lambda_{ii}E^\top(i)X_2(i) + \sum_{j=1,j\neq i}^{N} \lambda_{ij}\frac{1}{4}\varepsilon(j)\mathbb{I},$$

$$S_i(X_1) = \left(\sqrt{\lambda_{i1}}E^\top(1)X_1(1), \cdots, \sqrt{\lambda_{ii-1}}E^\top(i-1)X_1(i-1), \right.$$

$$\left. \sqrt{\lambda_{ii+1}}E^\top(i+1)X_1(i+1), \cdots, \sqrt{\lambda_{iN}}E^\top(N)X_1(N) \right),$$

$$S_i(X_2) = \left(\sqrt{\lambda_{i1}}E^\top(1)X_2(1), \cdots, \sqrt{\lambda_{ii-1}}E^\top(i-1)X_2(i-1), \right.$$

$$\left. \sqrt{\lambda_{ii+1}}E^\top(i+1)X_2(i+1), \cdots, \sqrt{\lambda_{iN}}E^\top(N)X_2(N) \right),$$

$$\mathcal{X}_i(\varepsilon) = diag\left[\varepsilon(1)\mathbb{I}, \cdots, \varepsilon(i-1)\mathbb{I}, \varepsilon(i+1)\mathbb{I}, \cdots, \varepsilon(N)\mathbb{I}\right],$$

with the following constraints:

$$E^\top(i)X_1(i) = X_1^\top(i)E(i) \geq 0, \tag{7.30}$$

$$E^\top(i)X_2(i) = X_2^\top(i)E(i) \geq 0, \tag{7.31}$$

then there exists a filter of the form (7.4) such that the estimation error is piecewise regular, impulse-free and stochastically stable and bounded by:

$$\|z - \hat{z}\|_2 \leq \gamma\left[\|\omega\|_2^2 + x_0^\top R x_0\right]^{\frac{1}{2}}. \tag{7.32}$$

The filter's gains are given by:

$$\begin{cases} K_A(i) = X_2^{-\top}(i)Y(i), \\ K_B(i) = X_2^{-\top}(i)Z(i), \\ K_C(i) = W(i). \end{cases} \tag{7.33}$$

7.3 Robust Filtering

Let us now return back to the state equation of this chapter as given by (7.1) and consider this time the effect of the uncertainties. As we did previously, let us assume that there exists a filter of the form (7.4) and see under which conditions the estimation error will be piecewise regular, impulse-free and robust stochastically stable and it is bounded for all admissible uncertainties and all signals $\omega(t) \in \mathcal{L}_2[0, \infty)$.

Let us first of all establish the uncertain extended system in a different form. From the dynamics (7.1) notice that we have:

$$E(r_0)\dot{x}(t) = [A(r_t) + \Delta A(r_t, t)]x(t) + [B(r_t) + \Delta B(r_t, t)]\omega(t), \tag{7.34}$$

with

$$\Delta A(r_t, t) = D_A(r_t)F_A(r_t)E_A(r_t),$$

$$\Delta B(r_t, t) = D_B(r_t)F_B(r_t)E_B(r_t).$$

Also for the estimation error dynamics, $e(t) = x(t) - \hat{x}(t)$, we have:

$$E(r_t)(\dot{x}(t) - \dot{\hat{x}}(t)) = [A(r_t) + \Delta A(r_t, t)] x(t) + [B(r_t) + \Delta B(r_t, t)] \omega(t)$$
$$- K_A(r_t)\hat{x}(t) - K_B(r_t)y(t)$$
$$= \Big[A(r_t) + \Delta A(r_t, t) - K_B(r_t)\Big[C_y(r_t) + \Delta C_y(r_t, t)\Big] - K_A(r_t)\Big] x(t)$$
$$+ K_A(r_t)[x(t) - \hat{x}(t)) + \Big[B(r_t) + \Delta B(r_t, t) - K_B(r_t)D_y(r_t)\Big] \omega(t),$$

with $\Delta C_y(r_t, t) = D_{C_y}(r_t)F_{C_y}(r_t)E_{C_y}(r_t)$.

Based on these calculations, the extended state equation becomes:

$$\tilde{E}(r_t)\dot{\tilde{x}}(t) = \Big[\tilde{A}(r_t) + \Delta\tilde{A}(r_t, t)\Big] \tilde{x}(t) + \Big[\tilde{B}(r_t) + \Delta\tilde{B}(r_t, t)\Big] \omega(t), \qquad (7.35)$$

where

$$\tilde{x}(t) = \begin{bmatrix} x(t) \\ x(t) - \hat{x}(t) \end{bmatrix},$$

$$\tilde{A}(r_t) = \begin{bmatrix} A(r_t) & 0 \\ A(r_t) - K_B(r_t)C_y(r_t) - K_A(r_t) & K_A(r_t) \end{bmatrix},$$

$$\Delta\tilde{A}(r_t, t) = \begin{bmatrix} \Delta A(r_t, t) & 0 \\ \Delta A(r_t, t) - K_B(r_t)\Delta C_y(r_t, t) & 0 \end{bmatrix},$$

$$\tilde{B}(r_t) = \begin{bmatrix} B(r_t) \\ B(r_t) - K_B(r_t)D_y(r_t) \end{bmatrix},$$

$$\Delta\tilde{B}(r_t, t) = \begin{bmatrix} \Delta B(r_t, t) \\ \Delta B(r_t, t) \end{bmatrix}.$$

Notice that:

$$\Delta\tilde{A}(r_t, t) = \begin{bmatrix} \Delta A(r_t, t) & 0 \\ \Delta A(r_t, t) - K_B(r_t)\Delta C_y(r_t, t) & 0 \end{bmatrix}$$
$$= \begin{bmatrix} \Delta A(r_t, t) & 0 \\ \Delta A(r_t, t) & 0 \end{bmatrix} + \begin{bmatrix} 0 & 0 \\ -K_B(r_t)\Delta C_y(r_t, t) & 0 \end{bmatrix}$$
$$= \begin{bmatrix} D_A(r_t)F_A(r_t)E_A(r_t) & 0 \\ D_A(r_t)F_A(r_t)E_A(r_t) & 0 \end{bmatrix}$$
$$+ \begin{bmatrix} 0 & 0 \\ -K_B(r_t)D_{C_y}(r_t)F_{C_y}(r_t)E_{C_y}(r_t) & 0 \end{bmatrix}$$
$$= \begin{bmatrix} D_A(r_t) \\ D_A(r_t) \end{bmatrix} F_A(r_t) \Big[E_A(r_t)\ 0 \Big]$$
$$+ \begin{bmatrix} 0 \\ -K_B(r_t)D_{C_y}(r_t) \end{bmatrix} F_{C_y}(r_t) \Big[E_{C_y}(r_t)\ 0 \Big]$$
$$= \tilde{D}_A(r_t)F_A(r_t)\tilde{E}_A(r_t) + \tilde{D}_{C_y}(r_t)F_{C_y}(r_t)\tilde{E}_{C_y}(r_t),$$

with

$$\tilde{D}_A(r_t) = \begin{bmatrix} D_A(r_t) \\ D_A(r_t) \end{bmatrix}, \quad \tilde{E}_A(r_t) = \Big[E_A(r_t)\ 0 \Big],$$

$$\tilde{D}_{C_y}(r_t) = \begin{bmatrix} 0 \\ -K_B(r_t)D_{C_y}(r_t) \end{bmatrix}, \quad \tilde{E}_{C_y}(r_t) = \Big[E_{C_y}(r_t)\ 0 \Big],$$

and

$$\Delta\tilde{B}(r_t, t) = \begin{bmatrix} \Delta B(r_t, t) \\ \Delta B(r_t, t) \end{bmatrix} = \begin{bmatrix} D_B(r_t)F_B(r_t)E_B(r_t) \\ D_B(r_t)F_B(r_t)E_B(r_t) \end{bmatrix}$$

$$= \begin{bmatrix} D_B(r_t) \\ D_B(r_t) \end{bmatrix} F_B(r_t)E_B(r_t)$$

$$= \tilde{D}_B(r_t)F_B(r_t)\tilde{E}_B(r_t),$$

with

$$\tilde{D}_B(r_t) = \begin{bmatrix} D_B(r_t) \\ D_B(r_t) \end{bmatrix},$$

$$\tilde{E}_B(r_t) = E_B(r_t).$$

For the estimation error, we have:

$$e(t) = z(t) - \hat{z}(t) = C_z(r_t, t)x(t) + D_z(r_t)\omega(t) - K_C(r_t)\hat{x}(t)$$

$$= [C_z(r_t)x(t) + \Delta C_z(r_t, t)] x(t) - K_C(r_t)x(t)$$

$$+ K_C(r_t) [x(t) - \hat{x}(t)] + D_z(r_t)\omega(t)$$

$$= \left[\left[C_z(r_t) - K_C(r_t) \ K_C(r_t) \right] + \left[\Delta C_z(r_t, t) \ 0 \right] \right]$$

$$\times \begin{bmatrix} x(t) \\ x(t) - \hat{x}(t) \end{bmatrix} + D_z(r_t)\omega(t)$$

$$= \left[\tilde{C}(r_t) + \Delta\tilde{C}(r_t, t) \right] \tilde{x}(t) + \tilde{D}(r_t)\omega(t),$$

where

$$\tilde{C}(r_t) = \left[C_z(r_t) - K_C(r_t) \ K_C(r_t) \right],$$

$$\Delta\tilde{C}(r_t, t) = \left[\Delta C_z(r_t, t) \ 0 \right] = \left[D_C(r_t)F_C(r_t)E_C(r_t) \ 0 \right]$$

$$= D_C(r_t)F_C(r_t) \left[E_C(r_t) \ 0 \right]$$

$$= \tilde{D}_C(r_t)F(r_t)\tilde{E}_C(r_t),$$

$$\tilde{D}(r_t) = D_z(r_t).$$

Using now the second condition of Theorem 7.2.3 for the uncertain extended system, we get:

$$\varepsilon_P \left[P(i) + P^\top(i) \right] \geq \tilde{E}^\top(i)P(i) = P^\top(i)E(i) \geq 0,$$

$$\begin{bmatrix} \tilde{J}(i, t) & P^\top(i)\left[\tilde{B}(i) + \Delta\tilde{B}(i, t) \right] & \left[\tilde{C}^\top(i) + \Delta\tilde{C}^\top(i, t) \right] \\ \left[\tilde{B}^\top(i) + \Delta\tilde{B}^\top(i, t) \right] P(i) & -\gamma^2\mathbb{I} & \tilde{D}^\top(i) \\ \left[\tilde{C}(i) + \Delta\tilde{C}(i, t) \right] & \tilde{D}(i) & -\mathbb{I} \end{bmatrix} < 0,$$

with

$$\tilde{J}(i, t) = \left[\tilde{A}(i) + \Delta\tilde{A}(i, t) \right]^\top P(i) + P^\top(i)\left[\tilde{A}(i) + \Delta\tilde{A}(i, t) \right]$$

$$+ \sum_{j=1}^{N} \lambda_{ij}E^\top(j)P(j).$$

This last relation can be rewritten as follows:

$$\begin{bmatrix} \tilde{J}(i) & P^\top(i)\tilde{B}(i) & \tilde{C}^\top(i) \\ \tilde{B}^\top(i)P(i) & -\gamma^2\mathbb{I} & \tilde{D}^\top(i) \\ \tilde{C}(i) & \tilde{D}(i) & -\mathbb{I} \end{bmatrix}$$

$$+ \begin{bmatrix} \Delta\tilde{A}^\top(i,t)P(i) + P^\top(i)\Delta\tilde{A}(i,t) & P^\top(i)\Delta\tilde{B}(i,t) & \Delta\tilde{C}^\top(i,t) \\ \Delta\tilde{B}^\top(i,t)P(i) & 0 & 0 \\ \Delta\tilde{C}(i,t) & 0 & 0 \end{bmatrix} < 0,$$

with

$$\tilde{J}(i) = \tilde{A}^\top(i)P(i) + P^\top(i)\tilde{A}(i) + \lambda_{ii}E^\top(i)P(i)$$

$$+ \sum_{j=1}^N \lambda_{ij}E^\top(j)P(j).$$

This gives in turn:

$$\begin{bmatrix} \tilde{J}(i) & P^\top(i)\tilde{B}(i) & \tilde{C}^\top(i) \\ \tilde{B}^\top(i)P(i) & -\gamma^2\mathbb{I} & \tilde{D}^\top(i) \\ \tilde{C}(i) & \tilde{D}(i) & -\mathbb{I} \end{bmatrix}$$

$$+ \begin{bmatrix} P^\top(i)\Delta\tilde{A}(i,t) & P^\top(i)\Delta\tilde{B}(i,t) & \Delta\tilde{C}^\top(i,t) \\ 0 & 0 & 0 \\ 0 & 0 & 0 \end{bmatrix}$$

$$+ \begin{bmatrix} \Delta\tilde{A}^\top(i,t)P(i) & 0 & 0 \\ \Delta\tilde{B}^\top(i,t)P(i) & 0 & 0 \\ \Delta\tilde{C}(i,t) & 0 & 0 \end{bmatrix} < 0.$$

Using now the expressions of the uncertainties, we get:

$$\begin{bmatrix} \tilde{J}(i) & P^\top(i)\tilde{B}(i) & \tilde{C}^\top(i) \\ \tilde{B}^\top(i)P(i) & -\gamma^2\mathbb{I} & \tilde{D}^\top(i) \\ \tilde{C}(i) & \tilde{D}(i) & -\mathbb{I} \end{bmatrix} + \begin{bmatrix} P^\top(i)\tilde{D}_A(i)F_A(i)\tilde{E}_A(i) & 0 & 0 \\ 0 & & 0 & 0 \\ 0 & & 0 & 0 \end{bmatrix}$$

$$+ \begin{bmatrix} \tilde{E}_A^\top(i)F_A^\top(i)\tilde{D}_A^\top(i)P(i) & 0 & 0 \\ 0 & 0 & 0 \\ 0 & 0 & 0 \end{bmatrix} + \begin{bmatrix} P^\top(i)\tilde{D}_{C_y}(i)F_{C_y}(i)\tilde{E}_{C_y}(i) & 0 & 0 \\ 0 & & 0 & 0 \\ 0 & & 0 & 0 \end{bmatrix}$$

$$+ \begin{bmatrix} \tilde{E}_{C_y}^\top(i)F_{C_y}^\top(i)\tilde{D}_{C_y}^\top(i)P(i) & 0 & 0 \\ 0 & 0 & 0 \\ 0 & 0 & 0 \end{bmatrix} + \begin{bmatrix} 0 & P^\top(i)\tilde{D}_B(i)F_B(i)\tilde{E}_B(i) & 0 \\ 0 & 0 & 0 \\ 0 & 0 & 0 \end{bmatrix}$$

$$+ \begin{bmatrix} 0 & 0 & 0 \\ \tilde{E}_B^\top(i)F_B^\top(i)\tilde{D}_B^\top(i)P(i) & 0 & 0 \\ 0 & 0 & 0 \end{bmatrix} + \begin{bmatrix} 0 & 0 & \tilde{E}_C^\top(i)F_C^\top(i)\tilde{D}_C^\top(i) \\ 0 & 0 & 0 \\ 0 & 0 & 0 \end{bmatrix}$$

$$+ \begin{bmatrix} 0 & 0 & 0 \\ 0 & 0 & 0 \\ \tilde{D}_C(i)F_C(i)\tilde{E}_C(i) & 0 & 0 \end{bmatrix} < 0.$$

Noticing that:

$$\begin{bmatrix} P^{\top}(i)\tilde{D}_A(i)F_A(i)\tilde{E}_A(i)\ 0\ 0 \\ 0 \qquad\qquad 0\ 0 \\ 0 \qquad\qquad 0\ 0 \end{bmatrix} = \begin{bmatrix} P^{\top}(i)\tilde{D}_A(i) \\ 0 \\ 0 \end{bmatrix} F_A(i)\begin{bmatrix} \tilde{E}_A(i)\ 0\ 0 \end{bmatrix},$$

$$\begin{bmatrix} P^{\top}(i)\tilde{D}_{C_y}(i)F_{C_y}(i)\tilde{E}_{C_y}(i)\ 0\ 0 \\ 0 \qquad\qquad 0\ 0 \\ 0 \qquad\qquad 0\ 0 \end{bmatrix} = \begin{bmatrix} P^{\top}(i)\tilde{D}_{C_y}(i) \\ 0 \\ 0 \end{bmatrix} F_{C_y}(i)\begin{bmatrix} \tilde{E}_{C_y}(i)\ 0\ 0 \end{bmatrix},$$

$$\begin{bmatrix} 0\ P^{\top}(i)\tilde{D}_B(i)F_B(i)\tilde{E}_B(i)\ 0 \\ 0 \qquad\qquad\qquad 0 \\ 0 \qquad\qquad\qquad 0 \end{bmatrix} = \begin{bmatrix} P^{\top}(i)\tilde{D}_B(i) \\ 0 \\ 0 \end{bmatrix} F_B(i)\begin{bmatrix} 0\ \tilde{E}_B(i)\ 0 \end{bmatrix},$$

$$\begin{bmatrix} 0\ 0\ \tilde{E}_C^{\top}(i)F_C^{\top}(i)\tilde{D}_C^{\top}(i) \\ 0\ 0 \qquad\qquad 0 \\ 0\ 0 \qquad\qquad 0 \end{bmatrix} = \begin{bmatrix} \tilde{E}_C^{\top}(i) \\ 0 \\ 0 \end{bmatrix} F_C^{\top}(i)\begin{bmatrix} 0\ 0\ \tilde{D}_C^{\top}(i) \end{bmatrix},$$

and

$$\begin{bmatrix} \tilde{E}_A^{\top}(i)F_A^{\top}(i)\tilde{D}_A^{\top}(i)P(i)\ 0\ 0 \\ 0 \qquad\qquad 0\ 0 \\ 0 \qquad\qquad 0\ 0 \end{bmatrix} = \begin{bmatrix} \tilde{E}_A^{\top}(i) \\ 0 \\ 0 \end{bmatrix} F_A^{\top}(i)\begin{bmatrix} \tilde{D}_A^{\top}(i)P(i)\ 0\ 0 \end{bmatrix},$$

$$\begin{bmatrix} \tilde{E}_{C_y}^{\top}(i)F_{C_y}^{\top}(i)\tilde{D}_{C_y}^{\top}(i)P(i)\ 0\ 0 \\ 0 \qquad\qquad 0\ 0 \\ 0 \qquad\qquad 0\ 0 \end{bmatrix} = \begin{bmatrix} \tilde{E}_{C_y}^{\top}(i) \\ 0 \\ 0 \end{bmatrix} F_{C_y}^{\top}(i)\begin{bmatrix} \tilde{D}_{C_y}^{\top}(i)P(i)\ 0\ 0 \end{bmatrix},$$

$$\begin{bmatrix} 0 \qquad\qquad 0\ 0 \\ \tilde{E}_B^{\top}(i)F_B^{\top}(i)\tilde{D}_B^{\top}(i)P(i)\ 0\ 0 \\ 0 \qquad\qquad 0\ 0 \end{bmatrix} = \begin{bmatrix} 0 \\ \tilde{E}_B^{\top}(i) \\ 0 \end{bmatrix} F_B^{\top}(i)\begin{bmatrix} \tilde{D}_B^{\top}(i)P(i)\ 0\ 0 \end{bmatrix},$$

$$\begin{bmatrix} 0 \qquad\qquad 0\ 0 \\ 0 \qquad\qquad 0\ 0 \\ \tilde{D}_C(i)F_C(i)\tilde{E}_C(i)\ 0\ 0 \end{bmatrix} = \begin{bmatrix} 0 \\ 0 \\ \tilde{D}_C(i) \end{bmatrix} F_C(i)\begin{bmatrix} \tilde{E}_C(i)\ 0\ 0 \end{bmatrix}.$$

Using Lemma 1.5.1, we get:

$$\begin{bmatrix} P^{\top}(i)\tilde{D}_A(i)F_A(i)\tilde{E}_A(i)\ 0\ 0 \\ 0 \qquad\qquad 0\ 0 \\ 0 \qquad\qquad 0\ 0 \end{bmatrix} + \begin{bmatrix} \tilde{E}_A^{\top}(i)F_A^{\top}(i)\tilde{D}_A^{\top}(i)P(i)\ 0\ 0 \\ 0 \qquad\qquad 0\ 0 \\ 0 \qquad\qquad 0\ 0 \end{bmatrix}$$

$$\leq \tilde{\varepsilon}_A^{-1}(i)\begin{bmatrix} P^{\top}(i)\tilde{D}_A(i) \\ 0 \\ 0 \end{bmatrix}\begin{bmatrix} \tilde{D}_A^{\top}(i)P(i)\ 0\ 0 \end{bmatrix}$$

$$+ \tilde{\varepsilon}_A(i)\begin{bmatrix} \tilde{E}_A^{\top}(i) \\ 0 \\ 0 \end{bmatrix}\begin{bmatrix} \tilde{E}_A(i)\ 0\ 0 \end{bmatrix}$$

$$= \begin{bmatrix} \tilde{\varepsilon}_A^{-1}(i)P^{\top}(i)\tilde{D}_A(i)\tilde{D}_A^{\top}(i)P(i) + \tilde{\varepsilon}_A(i)\tilde{E}_A^{\top}(i)E_A(i)\ 0\ 0 \\ 0 \qquad\qquad\qquad 0\ 0 \\ 0 \qquad\qquad\qquad 0\ 0 \end{bmatrix},$$

for any $\tilde{\varepsilon}_A(i) > 0$,

$$\begin{bmatrix} P^\top(i)\tilde{D}_{C_y}(i)F_{C_y}(i)\tilde{E}_{C_y}(i) & 0 & 0 \\ 0 & 0 & 0 \\ 0 & 0 & 0 \end{bmatrix} + \begin{bmatrix} \tilde{E}_{C_y}^\top(i)F_{C_y}^\top(i)\tilde{D}_{C_y}^\top(i)P(i) & 0 & 0 \\ 0 & 0 & 0 \\ 0 & 0 & 0 \end{bmatrix}$$

$$\leq \tilde{\varepsilon}_{C_y}^{-1}(i)\begin{bmatrix} P^\top(i)\tilde{D}_{C_y}(i) \\ 0 \\ 0 \end{bmatrix}\begin{bmatrix} \tilde{D}_{C_y}^\top(i)P(i) & 0 & 0 \end{bmatrix}$$

$$+ \tilde{\varepsilon}_{C_y}(i)\begin{bmatrix} \tilde{E}_{C_y}^\top(i) \\ 0 \\ 0 \end{bmatrix}\begin{bmatrix} \tilde{E}_{C_y}(i) & 0 & 0 \end{bmatrix}$$

$$= \begin{bmatrix} \tilde{\varepsilon}_{C_y}^{-1}(i)P^\top(i)\tilde{D}_{C_y}(i)\tilde{D}_{C_y}^\top(i)P(i) + \tilde{\varepsilon}_{C_y}(i)\tilde{E}_{C_y}^\top(i)E_{C_y}(i) & 0 & 0 \\ 0 & 0 & 0 \\ 0 & 0 & 0 \end{bmatrix},$$

for any $\tilde{\varepsilon}_{C_y}(i) > 0$,

$$\begin{bmatrix} 0 & P^\top(i)\tilde{D}_B(i)F_B(i)\tilde{E}_B(i) & 0 \\ 0 & 0 & 0 \\ 0 & 0 & 0 \end{bmatrix} + \begin{bmatrix} 0 & 0 & 0 \\ \tilde{E}_B^\top(i)F_B^\top(i)\tilde{D}_B^\top(i)P(i) & 0 & 0 \\ 0 & 0 & 0 \end{bmatrix}$$

$$\leq \tilde{\varepsilon}_B^{-1}(i)\begin{bmatrix} P^\top(i)\tilde{D}_B(i) \\ 0 \\ 0 \end{bmatrix}\begin{bmatrix} \tilde{D}_B^\top(i)P(i) & 0 & 0 \end{bmatrix}$$

$$+ \tilde{\varepsilon}_B(i)\begin{bmatrix} 0 \\ \tilde{E}_B^\top(i) \\ 0 \end{bmatrix}\begin{bmatrix} 0 & \tilde{E}_B(i) & 0 \end{bmatrix}$$

$$= \begin{bmatrix} \tilde{\varepsilon}_B^{-1}(i)P^\top(i)\tilde{D}_B(i)\tilde{D}_B^\top(i)P(i) & 0 & 0 \\ 0 & \tilde{\varepsilon}_B(i)\tilde{E}_B^\top(i)E_B(i) & 0 \\ 0 & 0 & 0 \end{bmatrix},$$

for any $\tilde{\varepsilon}_B(i) > 0$,

$$\begin{bmatrix} 0 & 0 & \tilde{E}_C^\top(i)F_C^\top(i)\tilde{D}_C^\top(i) \\ 0 & 0 & 0 \\ 0 & 0 & 0 \end{bmatrix} + \begin{bmatrix} 0 & 0 & 0 \\ \tilde{D}_C(i)F_C(i)\tilde{E}_C(i) & 0 & 0 \\ 0 & 0 & 0 \end{bmatrix}$$

$$\leq \tilde{\varepsilon}_C^{-1}(i)\begin{bmatrix} \tilde{E}_C^\top(i) \\ 0 \\ 0 \end{bmatrix}\begin{bmatrix} \tilde{E}_C(i) & 0 & 0 \end{bmatrix}$$

$$+ \tilde{\varepsilon}_C(i)\begin{bmatrix} 0 \\ 0 \\ \tilde{D}_C(i) \end{bmatrix}\begin{bmatrix} 0 & 0 & \tilde{D}_C^\top(i) \end{bmatrix}$$

$$= \begin{bmatrix} \tilde{\varepsilon}_C^{-1}(i)\tilde{E}_C^\top(i)\tilde{E}_C(i) & 0 & 0 \\ 0 & 0 & 0 \\ 0 & 0 & \tilde{\varepsilon}_C(i)\tilde{D}_C(i)\tilde{D}_C^\top(i) \end{bmatrix}.$$

for any $\tilde{\varepsilon}_C(i) > 0$.

Based on these transformations and after using Schur complement we get the following:

$$
\left[\begin{array}{cccc}
\left[\begin{array}{c} \tilde{J}(i) \\ +\tilde{\varepsilon}_A(i)\tilde{E}_A^\top(i)\tilde{E}_A(i) \\ +\tilde{\varepsilon}_{C_y}(i)E_{C_y}^\top(i)E_{C_y}(i) \end{array}\right] & P^\top(i)\tilde{B}(i) & \tilde{C}^\top(i) & \mathcal{T}(i) \\
\tilde{B}^\top(i)P(i) & -\gamma^2\mathbb{I} + \tilde{\varepsilon}_B(i)\tilde{E}_B^\top(i)\tilde{E}_B(i) & \tilde{D}^\top(i) & 0 \\
\tilde{C}(i) & \tilde{D}(i) & -\mathbb{I} + \tilde{\varepsilon}_C(i)\tilde{D}_C(i)\tilde{D}_C^\top(i) & 0 \\
\mathcal{T}^\top(i) & 0 & 0 & -\mathcal{W}(i)
\end{array}\right] < 0,
$$

with

$$
\mathcal{T}(i) = \left[P^\top(i)\tilde{D}_A(i)\ P^\top(i)\tilde{D}_{C_y}(i)\ P^\top(i)\tilde{D}_B(i)\ \tilde{E}_C^\top(i) \right],
$$

$$
\mathcal{W}(i) = \begin{bmatrix}
\tilde{\varepsilon}_A(i)\mathbb{I} & 0 & 0 & 0 \\
0 & \tilde{\varepsilon}_{C_y}(i)\mathbb{I} & 0 & 0 \\
0 & 0 & \tilde{\varepsilon}_B(i)\mathbb{I} & 0 \\
0 & 0 & 0 & \tilde{\varepsilon}_C(i)\mathbb{I}
\end{bmatrix}.
$$

Let us now use $P(i) = \begin{bmatrix} X_1(i) & 0 \\ 0 & X_2(i) \end{bmatrix}$ with $X_1(i)$ and $X_2(i)$ are nonsingular matrices, and try to write the parameters of the extended state equation in function of the original ones, i. e.,

$$
J_1(i) = \tilde{A}^\top(i)P(i) + P^\top(i)\tilde{A}(i) + \sum_{j=1}^N \lambda_{ij}E^\top(j)P(j)
$$

$$
= \begin{bmatrix} A(i) & 0 \\ A(i) - K_B(i)C_y(i) - K_A(i) & K_A(i) \end{bmatrix}^\top \begin{bmatrix} X_1(i) & 0 \\ 0 & X_2(i) \end{bmatrix}
$$

$$
+ \begin{bmatrix} X_1^\top(i) & 0 \\ 0 & X_2^\top(i) \end{bmatrix} \begin{bmatrix} A(i) & 0 \\ A(i) - K_B(i)C_y(i) - K_A(i) & K_A(i) \end{bmatrix}
$$

$$
+ \begin{bmatrix} \sum_{j=1}^N \lambda_{ij}E^\top(j)X_1(j) & 0 \\ 0 & \sum_{j=1}^N \lambda_{ij}E^\top(j)X_2(j) \end{bmatrix}
$$

$$
= \begin{bmatrix} A^\top(i)\ A^\top(i) - C_y^\top(i)K_B^\top(i) - K_A^\top(i) \\ 0 & K_A^\top(i) \end{bmatrix} \begin{bmatrix} X_1(i) & 0 \\ 0 & X_2(i) \end{bmatrix}
$$

$$
+ \begin{bmatrix} X_1^\top(i) & 0 \\ 0 & X_2^\top(i) \end{bmatrix} \begin{bmatrix} A(i) & 0 \\ A(i) - K_B(i)C_y(i) - K_A(i) & K_A(i) \end{bmatrix}
$$

$$
+ \begin{bmatrix} \sum_{j=1}^N \lambda_{ij}E^\top(j)X_1(j) & 0 \\ 0 & \sum_{j=1}^N \lambda_{ij}E^\top(j)X_2(j) \end{bmatrix}
$$

$$
= \left[\begin{array}{c}
A^\top(i)X_1(i) + X_1^\top(i)A(i) + \sum_{j=1}^N \lambda_{ij}E^\top(j)X_1(j) \\
X_2^\top(i)A(i) - X_2^\top(i)K_B(i)C_y(i) - X_2^\top(i)K_A(i)
\end{array}\right.
$$

$$
\left.\begin{array}{c}
A^\top(i)X_2(i) - C_y^\top(i)K_B^\top(i)X_2(i) - K_A^\top(i)X_2(i) \\
K_A^\top(i)X_2(i) + X_2^\top(i)K_A(i) + \sum_{j=1}^N \lambda_{ij}E^\top(j)X_2(j)
\end{array}\right],
$$

$$P^\top(i)\tilde{B}(i) = \begin{bmatrix} X_1^\top(i) & 0 \\ 0 & X_2^\top(i) \end{bmatrix} \begin{bmatrix} B(i) \\ B(i) - K_B(i)D_y(i) \end{bmatrix}$$

$$= \begin{bmatrix} X_1^\top(i)B(i) \\ X_2^\top(i)B(i) - X_2^\top(i)K_B(i)D_y(i) \end{bmatrix},$$

$$[1pt]\tilde{C}(i) = \begin{bmatrix} C_z(i) - K_C(i) & K_C(i) \end{bmatrix},$$

$$\tilde{D}(i) = D_z(i),$$

$$\tilde{\varepsilon}_A(i)\tilde{E}_A^\top(i)\tilde{E}_A(i) = \tilde{\varepsilon}_A(i) \begin{bmatrix} E_A^\top(i) \\ 0 \end{bmatrix} \begin{bmatrix} E_A(i) & 0 \end{bmatrix}$$

$$= \begin{bmatrix} \tilde{\varepsilon}_A(i)E_A^\top(i)E_A(i) & 0 \\ 0 & 0 \end{bmatrix},$$

$$\tilde{\varepsilon}_{C_y}(i)\tilde{E}_{C_y}^\top(i)\tilde{E}_{C_y}(i) = \tilde{\varepsilon}_{C_y}(i) \begin{bmatrix} E_{C_y}^\top(i) \\ 0 \end{bmatrix} \begin{bmatrix} E_{C_y}(i) & 0 \end{bmatrix}$$

$$= \begin{bmatrix} \tilde{\varepsilon}_{C_y}(i)E_{C_y}^\top(i)E_{C_y}(i) & 0 \\ 0 & 0 \end{bmatrix},$$

$$\tilde{\varepsilon}_B(i)\tilde{E}_B^\top(i)\tilde{E}_B(i) = \tilde{\varepsilon}_B(i)E_B^\top(i)E_B(i),$$

$$\tilde{\varepsilon}_C(i)\tilde{D}_C(i)\tilde{D}_C^\top(i) = \tilde{\varepsilon}_C(i)D_C(i)D_C^\top(i),$$

$$\mathcal{T}(i) = \begin{bmatrix} P^\top(i)\tilde{D}_A(i) & P^\top(i)\tilde{D}_{C_y}(i) & P^\top(i)\tilde{D}_B(i) & \tilde{E}_C^\top(i) \end{bmatrix}$$

$$= \begin{bmatrix} \begin{bmatrix} X_1^\top(i)D_A(i) \\ X_2^\top(i)D_A(i) \end{bmatrix} & \begin{bmatrix} 0 \\ -X_2^\top(i)K_B(i)D_{C_y}(i) \end{bmatrix} \\ \begin{bmatrix} X_1^\top(i)D_B(i) \\ X_2^\top(i)D_B(i) \end{bmatrix} & \begin{bmatrix} E_C^\top(i) \\ 0 \end{bmatrix} \end{bmatrix}$$

$$= \begin{bmatrix} \mathcal{T}_1(i) \\ \mathcal{T}_2(i) \end{bmatrix}.$$

This gives us the following

$$\begin{bmatrix} J_{X_1}(i) & \mathcal{U}(i) & X_1^\top(i)B(i) & C_z^\top(i) - K_C^\top(i) & \mathcal{T}_1(i) \\ \mathcal{U}^\top(i) & J_{X_2}(i) & \begin{bmatrix} X_2^\top(i)B(i) \\ -X_2^\top(i)K_B(i)D_y(i) \end{bmatrix} & K_C^\top(i) & \mathcal{T}_2(i) \\ B^\top(i)X_1(i) & \begin{bmatrix} B^\top(i)X_2(i) \\ -D_y^\top(i)K_B^\top(i)X_2(i) \end{bmatrix} & -\gamma^2\mathbb{I} + \varepsilon_B(i)E_B^\top(i)E_B(i) & D^\top(i) & 0 \\ C_z(i) - K_C(i) & K_C(i) & D_z(i) & -\mathbb{I} + \varepsilon_C(i)D_C(i)D_C^\top(i) & 0 \\ \mathcal{T}_1^\top(i) & \mathcal{T}_2^\top(i) & 0 & 0 & -\mathcal{W}(i) \end{bmatrix} < 0,$$

where

$$J_{X_1}(i) = A^\top(i)X_1(i) + X_1^\top(i)A(i) + \sum_{j=1}^{N} \lambda_{ij}E^\top(j)X_1(j)$$

$$+ \tilde{\varepsilon}_A(i)E_A^\top(i)E_A(i) + \tilde{\varepsilon}_{C_y}(i)E_{C_y}^\top(i)E_{C_y}(i),$$

$$J_{X_2}(i) = K_A^\top(i)X_2(i) + X_2^\top(i)K_A(i) + \sum_{j=1}^{N} \lambda_{ij}E^\top(j)X_2(j),$$

$$\mathcal{U}(i) = A^\top(i)X_2(i) - C_y^\top(i)K_B^\top(i)X_2(i) - K_A^\top(i)X_2(i).$$

Letting $Y(i) = X_2(i)K_A(i)$, $Z(i) = X_2(i)K_B(i)$, and $W(i) = K_C(i)$, we get:

$$\begin{bmatrix}
J_{X_1}(i) & \mathcal{U}(i) & X_1^\top(i)B(i) \\
\mathcal{U}^\top(i) & J_{X_2}(i) & X_2^\top(i)B(i) - Z(i)D_y(i) \\
B^\top(i)X_1(i) & B^\top(i)X_2(i - D_y^\top(i)Z^\top(i) & -\gamma^2\mathbb{I} + \varepsilon_B(i)E_B^\top(i)E_B(i) \\
C_z(i) - W(i) & W(i) & D_z(i) \\
\mathcal{T}_1^\top(i) & \mathcal{T}_2^\top(i) & 0
\end{bmatrix}$$

$$\begin{matrix}
C_z^\top(i) - W^\top(i) & \mathcal{T}_1(i) \\
W^\top(i) & \mathcal{T}_2(i) \\
D^\top(i) & 0 \\
-\mathbb{I} + \varepsilon_C(i)D_C(i)D_C^\top(i) & 0 \\
0 & -\mathcal{W}(i)
\end{matrix} \Bigg] < 0,$$

where

$$J_{X_1}(i) = A^\top(i)X_1(i) + X_1^\top(i)A(i) + \sum_{j=1}^{N} \lambda_{ij}E^\top(j)X_1(j)$$

$$+ \tilde{\varepsilon}_A(i)E_A^\top(i)E_A(i) + \tilde{\varepsilon}_{C_y}(i)E_{C_y}^\top(i)E_{C_y}(i),$$

$$J_{X_2}(i) = Y^\top(i) + Y(i) + \sum_{j=1}^{N} \lambda_{ij}E^\top(j)X_2(j),$$

$$\mathcal{U}(i) = A^\top(i)X_2(i) - C_y^\top(i)Z^\top(i) - Y^\top(i).$$

The conditions $\varepsilon_P[P(i) + P^\top(i)] \geq E^\top P(i) = P^\top(i)E \geq 0$ and $\begin{bmatrix} \mathbb{I} & \mathbb{I} \end{bmatrix} E^\top P(r_0)\begin{bmatrix} \mathbb{I} \\ \mathbb{I} \end{bmatrix} <$ $\gamma^2 R$ give:

$$E(i)X_1(i) = X_2^\top(i)E^\top(i) \geq 0,$$
$$E(i)X_2(i) = X_2^\top(i)E^\top(i) \geq 0,$$
$$E^\top(r_0)X_1(r_0) + E^\top(r_0)X_2(r_0) < \gamma^2 R.$$

The following theorem gives the results for the design of the gains of the $\mathcal{H}_\infty$ filter.

Theorem 7.3.1 *Let γ, ε_P and R be respectively given positive constants and a symmetric and positive-definite matrix representing the weighting of the initial conditions. If there exist sets of nonsingular matrices $X_1 = (X_1(1), \cdots, X_1(N))$,*

$X_1(i) \in \mathbb{R}^{n \times n}$ and $X_2 = (X_2(1), \cdots, X_2(N))$, $X_2(i) \in \mathbb{R}^{n \times n}$ and sets of matrices $Y = (Y(1), \cdots, Y(N))$, $Y(i) \in \mathbb{R}^{n \times n}$, $Z = (Z(1), \cdots, Z(N))$, $Z(i) \in \mathbb{R}^{n \times k}$ and $W = (W(1), \cdots, W(N))$, $W(i) \in \mathbb{R}^{p \times n}$ and some positive constant $\varepsilon_A = (\varepsilon_A(1), \cdots, \varepsilon_A(N))$, $\varepsilon_B = (\varepsilon_B(1), \cdots, \varepsilon_C(N))$, $\varepsilon_C = (\varepsilon_C(1), \cdots, \varepsilon_C(N))$, $\varepsilon_{Cy} = (\varepsilon_{Cy}(1), \cdots, \varepsilon_{Cy}(N))$, satisfying the following set of coupled LMIs for all admissible uncertainties:

$$
\left[\begin{array}{cccc}
J_{X_1}(i) & \mathcal{U}(i) & X_1^\top(i)B(i) \\
\mathcal{U}^\top(i) & J_{X_2}(i) & X_2^\top(i)B(i) - Z(i)D_y(i) \\
B^\top(i)X_1(i) & B^\top(i)X_2(i) - D_y^\top(i)Z^\top(i) & -\gamma^2\mathbb{I} + \varepsilon_B(i)E_B^\top(i)E_B(i) \\
C_z(i) - W(i) & W(i) & D_z(i) \\
\mathcal{T}_1^\top(i) & \mathcal{T}_2^\top(i) & 0
\end{array}\right.
$$

$$
\left.\begin{array}{cc}
C_z^\top(i) - W^\top(i) & \mathcal{T}_1(i) \\
W^\top(i) & \mathcal{T}_2(i) \\
D^\top(i) & 0 \\
-\mathbb{I} + \varepsilon_C(i)D_C(i)D_C^\top(i) & 0 \\
0 & -\mathcal{W}(i)
\end{array}\right] < 0, \quad (7.36)
$$

$$
E^\top(r_0)X_1(r_0) + E^\top(r_0)X_2(r_0) < \gamma^2 R, \quad (7.37)
$$

$$
J_{X_1}(i) = A^\top(i)X_1(i) + X_1^\top(i)A(i) + \sum_{j=1}^{N} \lambda_{ij}E^\top(j)X_1(j)
$$
$$
+ \tilde{\varepsilon}_A(i)E_A^\top(i)E_A(i) + \tilde{\varepsilon}_{C_y}(i)E_{C_y}^\top(i)E_{C_y}(i),
$$

$$
J_{X_2}(i) = Y^\top(i) + Y(i) + \sum_{j=1}^{N} \lambda_{ij}E^\top(j)X_2(j),
$$

$$
\mathcal{U}(i) = A^\top(i)X_2(i) - C_y^\top(i)Z^\top(i) - Y^\top(i),
$$

$$
\mathcal{T}(i) = \left[P^\top(i)\tilde{D}_A(i) \quad P^\top(i)\tilde{D}_{C_y}(i) \quad P^\top(i)\tilde{D}_B(i) \quad \tilde{E}_C^\top(i) \right]
$$
$$
= \left[\left[\begin{array}{c} X_1^\top(i)D_A(i) \\ X_2^\top(i)D_A(i) \end{array}\right] \left[\begin{array}{c} 0 \\ -X_2^\top(i)K_B(i)D_{C_y}(i) \end{array}\right] \left[\begin{array}{c} X_1^\top(i)D_B(i) \\ X_2^\top(i)D_B(i) \end{array}\right] \left[\begin{array}{c} E_C^\top(i) \\ 0 \end{array}\right] \right]
$$
$$
= \left[\begin{array}{c} \mathcal{T}_1(i) \\ \mathcal{T}_2(i) \end{array}\right],
$$

with the following constraints:

$$
E(i)X_1(i) = X_1^\top(i)E^\top(i) \geq 0, \quad (7.38)
$$
$$
E(i)X_2(i) = X_2^\top(i)E^\top(i) \geq 0, \quad (7.39)
$$

then there exists a filter of the form (7.4) such that the estimation error is piecewise regular, impulse-free and stochastically stable and bounded by:

$$
\|z - \hat{z}\|_2 \leq \gamma \left[\|\omega\|_2^2 + x_0^\top R x_0 \right]^{\frac{1}{2}}. \quad (7.40)
$$

The filter's gains are given by:

$$
\begin{cases}
K_A(i) = X_2^{-\top}(i)Y(i), \\
K_B(i) = X_2^{-\top}(i)Z(i), \\
K_C(i) = W(i).
\end{cases} \quad (7.41)
$$

If the initial conditions are equal to zero, the previous results becomes easier and are given by the following corollary.

Corollary 7.3.1 *Let the initial conditions of system (7.1) be zero. Let γ, ε_P and R be respectively given positive constants and a symmetric and positive-definite matrix representing the weighting of the initial conditions. If there exist nonsingular matrices $X_1 = (X_1(1), \cdots, X_1(N))$, and $X_2 = (X_2(1), \cdots, X_2(N))$, and sets of matrices $Y = (Y(1), \cdots, Y(N))$, $Z = (Z(1), \cdots, Z(N))$, and $W = (W(1), \cdots, W(N))$ satisfying the set of coupled LMIs (7.38)-(7.37), then there exists a filter of the form (7.4) such that the estimation error is piecewise regular, impulse-free and stochastically stable and bounded by:*

$$\|z - \hat{z}\|_2 \leq \gamma \|\omega\|_2 \,. \tag{7.42}$$

The filter's gains are given by:

$$\begin{cases} K_A(i) = X_2^{-\top}(i)Y(i)\,, \\ K_B(i) = X_2^{-\top}(i)Z(i)\,, \\ K_C(i) = W(i)\,. \end{cases} \tag{7.43}$$

The minimal noise attenuation level, γ, that can be verified by the filter of the form of (7.4) can be obtained by solving the following optimization problem:

$$\mathcal{P}: \begin{cases} \min\limits_{\substack{v>0, \\ \varepsilon_A=(\varepsilon_A(1),\cdots,\varepsilon_A(N)), \\ \varepsilon_B=(\varepsilon_B(1),\cdots,\varepsilon_C(N)), \\ \varepsilon_C=(\varepsilon_C(1),\cdots,\varepsilon_C(N)), \\ \varepsilon_{Cy}=(\varepsilon_{Cy}(1),\cdots,\varepsilon_{Cy}(N)), \\ X_1=(X_1(1),\cdots,X_1(N)), \\ X_2=(X_2(1),\cdots,X_2(N)), \\ Y=(Y(1),\cdots,Y(N)), \\ Z=(Z(1),\cdots,Z(N)), \\ W=(W(1),\cdots,W(N))}} v \\[2pt] s.t. \\ E(i)X_1(i) = X_1^\top(i)E^\top(i) \geq 0\,, \\ E(i)X_2(i) = X_2^\top(i)E^\top(i) \geq 0\,, \\ \Psi_v(i) < 0\,, \\ E^\top(r_0)X_1(r_0) + E^\top(r_0)X_2(r_0) < vR \end{cases}$$

where $\Psi_v(i)$ is obtained from (7.36) by replacing γ^2 by v. Thus, if the convex optimization problem $\mathcal{P}$ has a solution, v, then by using Theorem 7.3.1, the corresponding error of the filter (7.4) is stable with noise attenuation level $\sqrt{v}$.

When the first bound we developed earlier, i. e.,

$$E^\top(i)P(i) \leq \varepsilon_P \left[P^\top(i) + P(i) \right]$$

we can easily developed the following theorem that gives the results for the design of the gains of the $\mathscr{H}_\infty$ filter in this case.

Theorem 7.3.2 *Let γ, ε_P and R be respectively given positive constants and a symmetric and positive-definite matrix representing the weighting of the initial conditions. If there exist sets of nonsingular matrices $X_1 = (X_1(1), \cdots, X_1(N))$,*

$X_1(i) \in \mathbb{R}^{n \times n}$ and $X_2 = (X_2(1), \cdots, X_2(N))$, $X_2(i) \in \mathbb{R}^{n \times n}$ and sets of matrices $Y = (Y(1), \cdots, Y(N))$, $Y(i) \in \mathbb{R}^{n \times n}$, $Z = (Z(1), \cdots, Z(N))$, $Z(i) \in \mathbb{R}^{n \times k}$ and $W = (W(1), \cdots, W(N))$, $W(i) \in \mathbb{R}^{p \times n}$ and some positive constant $\varepsilon_A = (\varepsilon_A(1), \cdots, \varepsilon_A(N))$, $\varepsilon_B = (\varepsilon_B(1), \cdots, \varepsilon_C(N))$, $\varepsilon_C = (\varepsilon_C(1), \cdots, \varepsilon_C(N))$, $\varepsilon_{Cy} = (\varepsilon_{Cy}(1), \cdots, \varepsilon_{Cy}(N))$, satisfying the following set of coupled LMIs for all admissible uncertainties:

$$\begin{bmatrix} J_{X_1}(i) & \mathcal{U}(i) & X_1^\top(i)B(i) \\ \mathcal{U}^\top(i) & J_{X_2}(i) & X_2^\top(i)B(i) - Z(i)D_y(i) \\ B^\top(i)X_1(i) & B^\top(i)X_2(i) - D_y^\top(i)Z^\top(i) & -\gamma^2\mathbb{I} + \varepsilon_B(i)E_B^\top(i)E_B(i) \\ C_z(i) - W(i) & W(i) & D_z(i) \\ \mathcal{T}_1^\top(i) & \mathcal{T}_2^\top(i) & 0 \end{bmatrix}$$

$$\left. \begin{matrix} C_z^\top(i) - W^\top(i) & \mathcal{T}_1(i) \\ W^\top(i) & \mathcal{T}_2(i) \\ D^\top(i) & 0 \\ -\mathbb{I} + \varepsilon_C(i)D_C(i)D_C^\top(i) & 0 \\ 0 & -W(i) \end{matrix} \right] < 0, \quad (7.44)$$

$$E^\top(r_0)X_1(r_0) + E^\top(r_0)X_2(r_0) < \gamma^2 R, \quad (7.45)$$

$$J_{X_1}(i) = A^\top(i)X_1(i) + X_1^\top(i)A(i) + \lambda_{ii}E^\top(i)X_1(i) + \sum_{j=1, j \neq i}^{N} \varepsilon_P \lambda_{ij} \left[X_1(j) + X_1^\top(j) \right]$$

$$+ \tilde{\varepsilon}_A(i)E_A^\top(i)E_A(i) + \tilde{\varepsilon}_{C_y}(i)E_{C_y}^\top(i)E_{C_y}(i),$$

$$J_{X_2}(i) = Y^\top(i) + Y(i) + \lambda_{ii}E^\top(i)X_2(i) + \sum_{j=1, j \neq i}^{N} \varepsilon_P \lambda_{ij} \left[X_2(j) + X_2^\top(j) \right],$$

$$\mathcal{U}(i) = A^\top(i)X_2(i) - C_y^\top(i)Z^\top(i) - Y^\top(i),$$

$$\mathcal{T}(i) = \begin{bmatrix} P^\top(i)\tilde{D}_A(i) & P^\top(i)\tilde{D}_{C_y}(i) & P^\top(i)\tilde{D}_B(i) & \tilde{E}_C^\top(i) \end{bmatrix}$$

$$= \begin{bmatrix} \begin{bmatrix} X_1^\top(i)D_A(i) \\ X_2^\top(i)D_A(i) \end{bmatrix} & \begin{bmatrix} 0 \\ -X_2^\top(i)K_B(i)D_{C_y}(i) \end{bmatrix} & \begin{bmatrix} X_1^\top(i)D_B(i) \\ X_2^\top(i)D_B(i) \end{bmatrix} & \begin{bmatrix} E_C^\top(i) \\ 0 \end{bmatrix} \end{bmatrix}$$

$$= \begin{bmatrix} \mathcal{T}_1(i) \\ \mathcal{T}_2(i) \end{bmatrix},$$

with the following constraints:

$$\varepsilon_P \left[X_1(i) + X_1^\top(i) \right] \geq E(i)X_1(i) = X_1^\top(i)E^\top(i) \geq 0, \quad (7.46)$$

$$\varepsilon_P \left[X_2(i) + X_2^\top(i) \right] \geq E(i)X_2(i) = X_2^\top(i)E^\top(i) \geq 0, \quad (7.47)$$

then there exists a filter of the form (7.4) such that the estimation error is piecewise regular, impulse-free and stochastically stable and bounded by:

$$\|z - \hat{z}\|_2 \leq \gamma \left[\|\omega\|_2^2 + x_0^\top R x_0 \right]^{\frac{1}{2}}. \quad (7.48)$$

The filter's gains are given by:

$$\begin{cases} K_A(i) = X_2^{-\top}(i)Y(i), \\ K_B(i) = X_2^{-\top}(i)Z(i), \\ K_C(i) = W(i). \end{cases} \quad (7.49)$$

When $E^\top(i)P(i) \le \frac{1}{4}\varepsilon(i)\mathbb{I} + \varepsilon^{-1}E^\top(i)P(i)P^\top(i)E(i)$ for any $\varepsilon(i) > 0$, the following theorem gives the results for the design of the gains of the $\mathscr{H}_\infty$ filter.

Theorem 7.3.3 *Let γ, ε_P and R be respectively given positive constants ($\varepsilon_P > 0$) and a symmetric and positive-definite matrix representing the weighting of the initial conditions. If there exist sets of nonsingular matrices $X_1 = (X_1(1), \cdots, X_1(N))$, $X_1(i) \in \mathbb{R}^{n\times n}$ and $X_2 = (X_2(1), \cdots, X_2(N))$, $X_2(i) \in \mathbb{R}^{n\times n}$ and sets of matrices $Y = (Y(1), \cdots, Y(N))$, $Y(i) \in \mathbb{R}^{n\times n}$, $Z = (Z(1), \cdots, Z(N))$, $Z(i) \in \mathbb{R}^{n\times k}$ and $W = (W(1), \cdots, W(N))$, $W(i) \in \mathbb{R}^{p\times n}$ and some positive constant $\varepsilon_A = (\varepsilon_A(1), \cdots, \varepsilon_A(N))$, $\varepsilon_B = (\varepsilon_B(1), \cdots, \varepsilon_C(N))$, $\varepsilon_C = (\varepsilon_C(1), \cdots, \varepsilon_C(N))$, $\varepsilon_{Cy} = (\varepsilon_{Cy}(1), \cdots, \varepsilon_{Cy}(N))$, satisfying the following set of coupled LMIs for all admissible uncertainties:*

$$
\left[
\begin{array}{cccc}
J_{X_1}(i) & \mathcal{U}(i) & X_1^\top(i)B(i) \\
\mathcal{U}^\top(i) & J_{X_2}(i) & X_2^\top(i)B(i) - Z(i)D_y(i) \\
B^\top(i)X_1(i) & B^\top(i)X_2(i) - D_y^\top(i)Z^\top(i) & -\gamma^2\mathbb{I} + \varepsilon_B(i)E_B^\top(i)E_B(i) \\
C_z(i) - W(i) & W(i) & D_z(i) \\
\mathcal{T}_1^\top(i) & \mathcal{T}_2^\top(i) & 0 \\
S_i(X_1) & 0 & 0 \\
0 & S_i(X_2) & 0
\end{array}
\right.
$$

$$
\left.
\begin{array}{cccccc}
C_z^\top(i) - W^\top(i) & \mathcal{T}_1(i) & S_i(X_1) & 0 \\
W^\top(i) & \mathcal{T}_2(i) & 0 & S_i(X_2) \\
D^\top(i) & 0 & 0 & 0 \\
-\mathbb{I} + \varepsilon_C(i)D_C(i)D_C^\top(i) & 0 & 0 & 0 \\
0 & -\mathcal{W}(i) & 0 & 0 \\
0 & 0 & -\mathcal{X}_i(\varepsilon) & 0 \\
0 & 0 & 0 & -\mathcal{X}_i(\varepsilon)
\end{array}
\right] < 0, \qquad (7.50)
$$

$$
E^\top(r_0)X_1(r_0) + E^\top(r_0)X_2(r_0) < \gamma^2 R, \quad (7.51)
$$

$$
J_{X_1}(i) = A^\top(i)X_1(i) + X_1^\top(i)A(i) + \lambda_{ii}E^\top(i)X_1(i) + \sum_{j=1,j\ne i}^N \varepsilon_P\lambda_{ij}\left[X_1(j) + X_1^\top(j)\right]
$$

$$
\qquad\qquad + \tilde\varepsilon_A(i)E_A^\top(i)E_A(i) + \tilde\varepsilon_{C_y}(i)E_{C_y}^\top(i)E_{C_y}(i),
$$

$$
J_{X_2}(i) = Y^\top(i) + Y(i) + \lambda_{ii}E^\top(i)X_2(i) + \sum_{j=1,j\ne i}^N \varepsilon_P\lambda_{ij}\left[X_2(j) + X_2^\top(j)\right],
$$

$$
S_i(X_1) = \left(\sqrt{\lambda_{i1}}E^\top(1)X_1(1), \cdots, \sqrt{\lambda_{ii-1}}E^\top(i-1)X_1(i-1),\right.
$$

$$
\left. \sqrt{\lambda_{ii+1}}E^\top(i+1)X_1(i+1), \cdots, \sqrt{\lambda_{iN}}E^\top(N)X_1(N)\right),
$$

$$
S_i(X_2) = \left(\sqrt{\lambda_{i1}}E^\top(1)X_2(1), \cdots, \sqrt{\lambda_{ii-1}}E^\top(i-1)X_2(i-1),\right.
$$

$$
\left. \sqrt{\lambda_{ii+1}}E^\top(i+1)X_2(i+1), \cdots, \sqrt{\lambda_{iN}}E^\top(N)X_2(N)\right),
$$

$$
\mathcal{X}_i(\varepsilon) = diag\left[\varepsilon(1)\mathbb{I}, \cdots, \varepsilon(i-1)\mathbb{I}, \varepsilon(i+1)\mathbb{I}, \cdots, \varepsilon(N)\mathbb{I}\right],
$$

$$
\mathcal{U}(i) = A^\top(i)X_2(i) - C_y^\top(i)Z^\top(i) - Y^\top(i),
$$

$$\mathcal{T}(i) = \left[P^{\mathsf{T}}(i)\tilde{D}_A(i) \quad P^{\mathsf{T}}(i)\tilde{D}_{C_y}(i) \quad P^{\mathsf{T}}(i)\tilde{D}_B(i) \quad \tilde{E}_C^{\mathsf{T}}(i) \right]$$

$$= \left[\begin{bmatrix} X_1^{\mathsf{T}}(i)D_A(i) \\ X_2^{\mathsf{T}}(i)D_A(i) \end{bmatrix} \begin{bmatrix} 0 \\ -X_2^{\mathsf{T}}(i)K_B(i)D_{C_y}(i) \end{bmatrix} \begin{bmatrix} X_1^{\mathsf{T}}(i)D_B(i) \\ X_2^{\mathsf{T}}(i)D_B(i) \end{bmatrix} \begin{bmatrix} E_C^{\mathsf{T}}(i) \\ 0 \end{bmatrix} \right]$$

$$= \begin{bmatrix} \mathcal{T}_1(i) \\ \mathcal{T}_2(i) \end{bmatrix},$$

with the following constraints:

$$E(i)X_1(i) = X_1^{\mathsf{T}}(i)E^{\mathsf{T}}(i) \geq 0, \tag{7.52}$$

$$E(i)X_2(i) = X_2^{\mathsf{T}}(i)E^{\mathsf{T}}(i) \geq 0, \tag{7.53}$$

then there exists a filter of the form (7.4) such that the estimation error is piecewise regular, impulse-free and stochastically stable and bounded by:

$$\|z - \hat{z}\|_2 \leq \gamma \left[\|\omega\|_2^2 + x_0^{\mathsf{T}} R x_0 \right]^{\frac{1}{2}}. \tag{7.54}$$

The filter's gains are given by:

$$\begin{cases} K_A(i) = X_2^{-\mathsf{T}}(i)Y(i), \\ K_B(i) = X_2^{-\mathsf{T}}(i)Z(i), \\ K_C(i) = W(i). \end{cases} \tag{7.55}$$

7.4 Numerical Examples

Example 7.4.1 *To show the usefulness of the results of this chapter, let us consider a two modes system with the following data:*

- *mode # 1:*

$$A(1) = \begin{bmatrix} -3.0 & 1.0 & 0.0 \\ 0.3 & -2.5 & 1.0 \\ -0.1 & 0.3 & -3.8 \end{bmatrix}, \ B(1) = \begin{bmatrix} 1.0 \\ 0.0 \\ 1.0 \end{bmatrix},$$

$$C_y(1) = \begin{bmatrix} 0.1 \ 1.0 \ 0.0 \end{bmatrix}, \ D_y(1) = \begin{bmatrix} 0.2 \end{bmatrix},$$

$$C_z(1) = \begin{bmatrix} 0.1 \ 1.0 \ 0.0 \end{bmatrix}, \ D_z(1) = \begin{bmatrix} 2 \end{bmatrix}.$$

- *mode # 2:*

$$A(2) = \begin{bmatrix} -4.0 & 1.0 & 0.0 \\ 0.3 & -3.0 & 1.0 \\ -0.1 & 0.3 & -4.8 \end{bmatrix}, \ B(2) = \begin{bmatrix} 1.0 \\ 0.0 \\ 2.0 \end{bmatrix},$$

$$C_y(2) = \begin{bmatrix} 0.8 \ 0.4 \ 0.0 \end{bmatrix}, \ D_y(2) = \begin{bmatrix} 0.1 \end{bmatrix},$$

$$C_z(2) = \begin{bmatrix} 0.7 \ 0.1 \ 0.0 \end{bmatrix}, \ D_z(2) = \begin{bmatrix} 3.0 \end{bmatrix}.$$

The singular matrix E is given by:

$$E = \begin{bmatrix} 1.0 & 0.0 & 0.0 \\ 0.0 & 1.0 & 0.0 \\ 0.0 & 0.0 & 0.0 \end{bmatrix}.$$

The switching between the two modes is described by the following transition rates matrix:

$$\Lambda = \begin{bmatrix} -1.0 & 1.0 \\ 1.1 & -1.1 \end{bmatrix}.$$

Solving the LMIs of the first theorem of this chapter, we get:

$$\gamma = 3,$$

$$X_1(1) = \begin{bmatrix} 3.5003 & 0.3250 & 0.0 \\ 0.3250 & 4.1628 & 0.0 \\ -0.3798 & 1.0539 & 1.2317 \end{bmatrix},$$

$$X_1(2) = \begin{bmatrix} 2.4906 & -0.0189 & 0.0 \\ -0.0189 & 7.9644 & 0.0 \\ -0.6316 & 1.4907 & 1.0078 \end{bmatrix},$$

$$X_2(1) = \begin{bmatrix} 4.9584 & -0.2900 & 0.0 \\ -0.2900 & 8.7160 & 0.0 \\ -1.5569 & -3.1224 & 1.5102 \end{bmatrix},$$

$$X_2(2) = \begin{bmatrix} 8.2968 & 1.6168 & 0.0 \\ 1.6168 & 8.5450 & 0.0 \\ -7.6142 & -2.1537 & 1.2953 \end{bmatrix},$$

$$Y(1) = \begin{bmatrix} -9.0059 & -8.1456 & 4.6406 \\ 5.8355 & -5.1498 & 19.9727 \\ -3.7035 & -19.3524 & -5.3190 \end{bmatrix},$$

$$Y(2) = \begin{bmatrix} -2.8317 & 16.0142 & 37.9277 \\ -12.0781 & -5.2683 & 18.7436 \\ -38.1334 & -19.1356 & -5.4638 \end{bmatrix},$$

$$Z(1) = \begin{bmatrix} 12.8918 \\ -16.5244 \\ 18.8044 \end{bmatrix}, \; Z(2) = \begin{bmatrix} -36.0427 \\ 5.7174 \\ 46.0611 \end{bmatrix},$$

$$W(1) = \begin{bmatrix} 0.3152 & 0.5891 & 0.4374 \end{bmatrix},$$

$$W(2) = \begin{bmatrix} 1.1091 & 1.0875 & 0.6718 \end{bmatrix},$$

which gives the following gains for the $\mathcal{H}_\infty$ filter:

$$K_A(1) = \begin{bmatrix} -2.6036 & -5.9810 & -0.1100 \\ -0.2956 & -5.3806 & 1.0261 \\ -2.4523 & -12.8146 & -3.5221 \end{bmatrix},$$

$$K_A(2) = \begin{bmatrix} -26.6191 & -11.1945 & 0.4983 \\ -3.7971 & -2.2219 & 1.0361 \\ -29.4400 & -14.7732 & -4.2182 \end{bmatrix},$$

$$K_B(1) = \begin{bmatrix} 6.6726 \\ 2.7869 \\ 12.4518 \end{bmatrix}, \quad K_B(2) = \begin{bmatrix} 27.4247 \\ 4.4429 \\ 35.5604 \end{bmatrix},$$

$$K_C(1) = \begin{bmatrix} 0.3152 & 0.5891 & 0.4374 \end{bmatrix},$$

$$K_C(2) = \begin{bmatrix} 1.1091 & 1.0875 & 0.6718 \end{bmatrix}.$$

7.5 Notes

This chapter dealt with the filtering problem of the singular class of systems with random abrupt changes. The $\mathscr{H}_\infty$ filtering and the robust $\mathscr{H}_\infty$ robust filtering problems have been considered and LMI conditions were developed. The conditions we developed in this chapter are tractable using commercial optimization tools. The content of this chapter is mainly based on the work of the author and his coauthors [17].

Part V

Singular Optimal Control

This part deals with the linear quadratic optimal control problem for the class of singular systems. In the deterministic setting this problem has attracted a lot of researchers and some interesting results has been reported in the literature. Among the contributions on the subject, we quote the works of [5, 8, 37, 61, 73, 87, 88, 97, 121, 140] and the references therein. For a complete overview on this problem we refer the reader to Mehrmann [97] and to the references therein. Most of the developed results have tried to transform the original optimal control problem of the singular systems to an equivalent one that we can solve using the existing results in the literature. For more details on this approach, we refer the reader to Dai [43] where complete details on the approach is given. More recently, the linear quadratic regular optimal control problem for time-varying singular systems have been tackled by Kurina and Marz [73].

For the class of systems we are dealing with in this volume, the only results that we know belong to the author. These results combine the ones developed for linear quadratic control problem for the singular class of systems with those developed for the jump linear quadratic regulator to solve the optimal control of the class of linear singular systems with abrupt changes. When the uncertainties are present in the dynamics, the guaranteed approach that was used either for the deterministic case or the stochastic one (see Boukas and Liu [27]) is utilized to solve the control problem of the class of uncertain linear systems with abrupt changes. The type of uncertainties that are considered in this part are of norm bounded type. Besides the uncertainties, we can have external disturbances in the dynamics that can be unfortunately modeled by a Gaussian process that is generally the cases in the linear quadratic control problem. Under the assumption that the external disturbances have finite energy or finite power, the mixed $\mathcal{H}_2/\mathcal{H}_\infty$ control problem approach is used to solve the control problem of the class of linear singular systems with abrupt changes and external disturbances.

The part is organized as follows. In Chap. 8, the uncertainties of norm bounded types are added to the class of systems and the guaranteed approach is used to solve the control problem that will consist to design a control law that makes the closed-loop system regular, impulse-free and robust stochastically stable and at the same guarantees the minimum cost. This approach is more popular since it uses a cost that is usually used in the linear quadratic control problem. In Chap. 9, the mixed $\mathcal{H}_2/\mathcal{H}_\infty$ control problem for the class of linear singular systems with abrupt changes.

8

Guaranteed Cost Control

Guaranteed cost control is one of the approaches that have been proposed in the literature to robustly stabilize dynamical uncertain systems. For normal dynamical systems, this control problem has been studied by many authors, see [27] and the references therein either for the deterministic or the stochastic cases. For the singular linear deterministic systems, only one reference has tackled this problem [53].

Our goal in this chapter is to establish conditions that permit the design of a state feedback controller that makes the closed-loop system piecewise regular, impulse-free and stochastically stable and at the same time assures the guaranteed cost. The conditions we will develop will be in the LMI setting.

The rest of this chapter is organized as follows. In Sect. 1, the problem of guaranteed cost control problem is stated and some appropriate definitions are given. In Sect. 2 some guaranteed cost bounds are computed. In Sect. 3, the guaranteed cost controller design is discussed and LMI conditions are developed to compute such controller.

8.1 Problem Statement

Consider a linear singular system with random abrupt changes that has N modes, i. e., $\mathscr{S} = \{1, 2, \ldots, N\}$. The mode switching is assumed to be governed by a continuous-time Markov process $\{r_t, t \geq 0\}$ taking values in the state space $\mathscr{S}$ and having the following infinitesimal generator

$$\Lambda = (\lambda_{ij}), i, j \in \mathscr{S},$$

where $\lambda_{ij} \geq 0, \forall j \neq i, \lambda_{ii} = -\sum_{j \neq i} \lambda_{ij}$.

The mode transition probabilities are described as follows:

$$P[r_{t+h} = j | r_t = i] = \begin{cases} \lambda_{ij} h + o(h), & j \neq i \\ 1 + \lambda_{ii} h + o(h), & j = i, \end{cases} \tag{8.1}$$

where $\lim_{h \to 0} o(h)/h = 0$.

Let the state equation of this class of systems be defined in a fundamental probability space $(\Omega, \mathcal{F}, \mathbb{P})$ and assume that its behavior is described by the following differential-algebraic equations:

$$\begin{cases} E(r_t)\dot{x}(t) = A(r_t, t)x(t) + B(r_t, t)u(t), \\ x(s) = x_0, \end{cases} \tag{8.2}$$

where $x(t) \in \mathbb{R}^n$ is the system state at time t, $u(t) \in \mathbb{R}^m$ is the control input of the system at time t, $A(r_t, t) \in \mathbb{R}^{n \times n}$ and $B(r_t, t) \in \mathbb{R}^{n \times m}$ are assumed to have uncertainties, i,e.: $A(r_t, t) = A(r_t) + D_A(r_t)F_A(r_t)E_A(r_t)$ and $B(r_t, t) = B(r_t) + D_B(r_t)F_B(r_t)E_B(r_t)$ with $A(i) \in \mathbb{R}^{n \times n}$, $D_A(i) \in \mathbb{R}^{n \times n_D}$, $E_A(i) \in \mathbb{R}^{n_E \times n}$, $B(i) \in \mathbb{R}^{n \times m}$, $D_B(i) \in \mathbb{R}^{n \times m_D}$ and $E_B(i) \in \mathbb{R}^{m_E \times n}$ are known real matrices with appropriate dimensions for each $i \in \mathcal{S}$ and $F_A(r_t) \in \mathbb{R}^{n_D \times n_E}$ and $F_B(r_t) \in \mathbb{R}^{m_D \times m_E}$ satisfy $F_A^\top(i)F_A(i) \leq \mathbb{I}$ and $F_B^\top(i)F_B(i) \leq \mathbb{I}$ for each $i \in \mathcal{S}$, the matrix $E(i)$ may be singular, and we assume $\text{rank}(E(i)) = n_r \leq n$.

Remark 8.1.1 *When the uncertainties are equal to zero the system will be referred to as nominal system. The uncertainties that satisfy the previous conditions are referred to as admissible. The uncertainties we are considering in this chapter are known in the literature as norm bounded uncertainties. When the matrix $E(i)$ is nonsingular the system (8.2) is referred to as normal system.*

Definition 8.1.1 [43]

i.   *Nominal system (8.2) is said to be regular if the characteristic polynomial, $\det(sE(i) - A(i))$ is not identically zero for each mode $i \in \mathcal{S}$.*
ii. *Nominal system (8.2) is said to be impulse free, i. e., the $\deg(\det(sE(i) - A(i))) = rank(E)$ for each mode $i \in \mathcal{S}$.*

For more details on other properties and the existence of the solution of system (8.2), we refer the reader to [122], and the references therein. In general, the regularity is often a sufficient condition for the analysis and the synthesis of singular systems.

For the system (8.2), we have the following definitions:

Definition 8.1.2 *Nominal system (8.2) is said to be stochastically stable (SS) if there exists a constant $M(x_0, r_0)$ such that*

$$\mathbb{E}\left[\int_0^\infty \|x(t)\|^2 dt \Big| x_0, r_0 \right] \leq M(x_0, r_0); \tag{8.3}$$

Definition 8.1.3 *Uncertain system (8.2) is said to be robust stochastically stable (RSS) if there exists a constant $M(r_0, x_0)$ such that (8.3) holds for all admissible uncertainties.*

Let $R_1(i)$ and $R_2(i)$, $i \in \mathcal{S}$ be two given symmetric and positive-definite matrices and consider the following cost function:

$$J(x_0, r_0) = \mathbb{E}\left[\int_0^\infty \left[x^\top(t)R_1(r_t)x(t) + u^\top(t)R_2(r_t)u(t) \right] dt \right], \tag{8.4}$$

where x_0 and r_0 are respectively the initial state and the initial mode of the system.

Definition 8.1.4 *If there exist a control law, $u(.)$ and a positive scalar ϱ representing the upper bound of the cost (8.4) such that the closed-loop system is piecewise regular, impulse-free and stochastically stable, and the cost (8.4) is bounded by ϱ, then ϱ is called the guaranteed cost, also referred to as the optimal guaranteed cost, and $u(.)$ is the associated guaranteed cost control law.*

The goal of this chapter is to design a controller of the following form:

$$u(t) = K(r_t)x(t) \tag{8.5}$$

where $K(i)$ is a design parameter that has to be determined for every $i \in \mathscr{S}$.

The aim of this chapter is to develop LMI conditions that can be used to design a state feedback controller that guarantees that the closed-loop system of the uncertain system is piecewise regular, impulse free and robust stochastically stable and the cost (8.4) is bounded for all admissible norm bounded uncertainties. Our methodology will be mainly based on the Lyapunov theory and some algebraic results. The conditions we will develop here will be in terms of the solutions of linear matrix inequalities that can be easily obtained using LMI control toolbox.

8.2 Guaranteed Cost Bound

Let us consider that the control $u(t) = 0$ for all $t \geq 0$ and see under which conditions the systems (8.2) will have a bounded cost (8.4). When the control is equal to zero for all $t \geq 0$, the cost (8.4) becomes:

$$J(x_0, r_0) = \mathbb{E}\left[\int_0^\infty \left[x^\top(t)R_1(r_t)x(t) \right] dt \right].$$

It is trivial that we should satisfy some conditions to guarantee the existence of the solution and the stochastic stability to assure that the cost function is bounded. The following theorem gives the desired results:

Theorem 8.2.1 *Let ε_P be a given positive scalar ($\varepsilon_P > 0$). If there exist a set of nonsingular matrices $P = (P(1), \cdots, P(N))$, $P(i) \in \mathbb{R}^{n \times n}$ and a set of positive scalars $\varepsilon_A = (\varepsilon_A(1), \cdots, \varepsilon_A(N))$ such that the following set of coupled LMIs holds for each $i \in \mathscr{S}$:*

$$\begin{bmatrix} J(i) & P^\top(i)D_A(i) \\ D^\top(i)P(i) & -\varepsilon_A(i)\mathbb{I} \end{bmatrix} < 0, \tag{8.6}$$

where

$$J(i) = P^\top(i)A(i) + A^\top(i)P(i) + \varepsilon_A(i)E^\top(i)E_A(i) + R_1(i)$$
$$+ \lambda_{ii}E^\top(i)P(i) + \sum_{j=1, j\neq i}^{N} \varepsilon_P \lambda_{ij}\left[P(j) + P^\top(j) \right],$$

with the following constraints:

$$\varepsilon_P\left[P(i) + P^\top(i) \right] \geq E^\top(i)P(i) = P^\top(i)E(i) \geq 0, \tag{8.7}$$

then the uncertain system with random abrupt changes (8.2) is piecewise regular, impulse-free and stochastically stable and the cost (8.4) satisfies the following for all admissible uncertainties:

$$J(x_0, r_0) \leq \varepsilon_P tr \left[\left[P(r_0) + P^\top(r_0) \right] x_0 x_0^\top \right] \tag{8.8}$$

Proof: Since $R_1(i) > 0$, for all $i \in \mathcal{S}$, from the first matrix inequality of the theorem, we get

$$\begin{bmatrix} J_1(i) & P^\top(i)D_A(i) \\ D^\top(i)P(i) & -\varepsilon_A(i)\mathbb{I} \end{bmatrix} < 0, \tag{8.9}$$

with

$$J_1(i) = P^\top(i)A(i) + A^\top(i)P(i) + \varepsilon_A(i)E^\top(i)E_A(i) + \lambda_{ii}E^\top(i)P(i)$$
$$+ \sum_{j=1, j\neq i}^{N} \varepsilon_P \lambda_{ij} \left[P(j) + P^\top(j) \right] .$$

Using this inequality and the results of Chap. 2, we know that for a given positive scalar ε_P, the uncertain singular system with random abrupt changes (8.2) is piecewise regular, impulse-free and stochastically stable if there exists a set of non-singular matrices $P = (P(1), \cdots, P(N))$, $P(i) \in \mathbb{R}^{n \times n}$ and a set of positive scalars $\varepsilon_A = (\varepsilon_A(1), \cdots, \varepsilon_A(N))$ such that the following set of coupled LMIs holds for each $i \in \mathcal{S}$:

$$\begin{cases} \varepsilon_P \left[P(i) + P^\top(i) \right] \geq E^\top(i)P(i) = P^\top(i)E(i) \geq 0 \\ \begin{bmatrix} J_1(i) & P^\top(i)D_A(i) \\ D^\top(i)P(i) & -\varepsilon_A(i)\mathbb{I} \end{bmatrix} < 0 \end{cases} \tag{8.10}$$

This implies that the system is piecewise regular, impulse-free and robust stochastically stable.

Let the (x, i) denote respectively the state of the vector state, $x(t)$ and the mode, r_t at time t and consider the following Lyapunov function:

$$V(x(t), r_t) = x^\top(t)E^\top(r_t)P(r_t)x(t) .$$

The weak infinitesimal operator, $\mathcal{L}V(.)$ emanating from (x, i) at time t is given:

$$\mathcal{L}V(x, i) = x^\top(t) \Big[P^\top(i)A(i) + A^\top(i)P(i) + P^\top(i)D_A(i)F_A(i)E_A(i)$$
$$+ E_A^\top(i)F_A^\top(i)D^\top(i)P(i)$$
$$+ \sum_{j=1}^{N} \lambda_{ij}E^\top(j)P(j) \Big] x(t) .$$

Using the fact that $E^\top(i)P(i) \leq \varepsilon_P \left[P^\top(i) + P(i) \right]$ for any $\varepsilon_P > 0$, we get:

$$\mathcal{L}V(x, i) \leq x^\top(t) \Big[P^\top(i)A(i) + A^\top(i)P(i) + P^\top(i)D_A(i)F_A(i)E_A(i)$$
$$+ E_A^\top(i)F_A^\top(i)D^\top(i)P(i) + \lambda_{ii}E^\top(i)P(i)$$
$$+ \sum_{j=1, j\neq i}^{N} \varepsilon_P \lambda_{ij} \left[P(j) + P^\top(j) \right] \Big] x(t) .$$

Based on the Lemma 1.5.1, we get:

$$\mathscr{L}V(x,i) \le x^\top(t)\Big[P^\top(i)A(i) + A^\top(i)P(i)$$
$$+ \varepsilon_A(i)E_A^\top(i)E_A(i) + \varepsilon_A^{-1}(i)P^\top(i)D_A(i)D_A^\top(i)P(i)$$
$$+ \lambda_{ii}E^\top(i)P(i) + \sum_{j=1,j\ne i}^N \varepsilon_P\lambda_{ij}\Big[P(j) + P^\top(j)\Big]\Big]x(t),$$

which gives

$$\mathscr{L}V(x,i) + x^\top(t)R_1(i)x(t) \le x^\top(t)\Big[P^\top(i)A(i) + A^\top(i)P(i) + R_1(i)$$
$$+ \varepsilon_A(i)E_A^\top(i)E_A(i) + \lambda_{ii}E^\top(i)P(i)$$
$$+ \varepsilon_A^{-1}(i)P^\top(i)D_A(i)D_A^\top(i)P(i)$$
$$+ \sum_{j=1,j\ne i}^N \varepsilon_P\lambda_{ij}\Big[P(j) + P^\top(j)\Big]\Big]x(t).$$

Combining this with the theorem conditions, the following holds for any $T > 0$:

$$\int_0^T \mathscr{L}V(x(t), r_t)dt + \int_0^T x^\top(t)R_1(r_t)x(t)dt < 0.$$

Using now the Dynkin formula and the fact that the system is piecewise regular, impulse-free and stochastically stable, we get:

$$\mathbb{E}\left[\int_0^\infty x^\top(t)R_1(r_t)x(t)dt\right] \le \mathbb{E}\left[x_0^\top E^\top(r_0)P(r_0)x_0\right]$$
$$\le \varepsilon_P\mathbb{E}\left[x_0^\top \Big[P(r_0) + P^\top(r_0)\Big]x_0\right],$$

which implies that the cost function is bounded and this ends the proof of the theorem.

In a similar manner, if we assume that the following holds:

$$E^\top(i)P(i) \le \varepsilon(i)\Big[P^\top(i)P(i)\Big], \forall \varepsilon(i) > 0,$$

we can establish the following result.

Theorem 8.2.2 *If there exist a set of nonsingular matrices* $P = (P(1), \cdots, P(N))$, $P(i) \in \mathbb{R}^{n\times n}$ *and sets of positive scalars* $\varepsilon_A = (\varepsilon_A(1), \cdots, \varepsilon_A(N))$ *and* $\varepsilon = (\varepsilon(1), \cdots, \varepsilon(N))$ *such that the following set of coupled matrix inequalities holds for each* $i \in \mathscr{S}$:

$$\begin{bmatrix} J(i) & P^\top(i)D_A(i) \\ D^\top(i)P(i) & -\varepsilon_A(i)\mathbb{I} \end{bmatrix} < 0, \tag{8.11}$$

where

$$J(i) = P^\top(i)A(i) + A^\top(i)P(i) + \varepsilon_A(i)E^\top(i)E_A(i) + R_1(i)$$
$$+ \lambda_{ii}E^\top(i)P(i) + \sum_{j=1,j\ne i}^N \varepsilon(i)\lambda_{ij}\Big[P^\top(i)P(j)\Big],$$

with the following constraints:

$$\varepsilon(i)\left[P^\top(i)P(i)\right] \geq E^\top(i)P(i) = P^\top(i)E(i) \geq 0, \tag{8.12}$$

then the uncertain system with random abrupt changes (8.2) is piecewise regular, impulse-free and stochastically stable and the cost (8.4) satisfies the following for all admissible uncertainties:

$$J(x_0, r_0) \leq \varepsilon(r_0)tr\left[\left[P^\top(r_0)P(r_0)\right]x_0 x_0^\top\right]. \tag{8.13}$$

Similarly, if we assume that the following holds:

$$E^\top(i)P(i) \leq \left[\frac{1}{4}\varepsilon^{-1}(i)\mathbb{I} + \varepsilon(i)E^\top(i)P(i)P^\top(i)E(i)\right], \forall \varepsilon(i) > 0,$$

we can establish the following result.

Theorem 8.2.3 *If there exist a set of nonsingular matrices $P = (P(1), \cdots, P(N))$, $P(i) \in \mathbb{R}^{n \times n}$ and a set of positive scalars $\varepsilon_A = (\varepsilon_A(1), \cdots, \varepsilon_A(N))$ and $\varepsilon = (\varepsilon(1), \cdots, \varepsilon(N))$ such that the following set of coupled matrix inequalities holds for each $i \in \mathscr{S}$:*

$$\begin{bmatrix} J(i) & P^\top(i)D_A(i) \\ D^\top(i)P(i) & -\varepsilon_A(i)\mathbb{I} \end{bmatrix} < 0, \tag{8.14}$$

where

$$J(i) = P^\top(i)A(i) + A^\top(i)P(i) + \varepsilon_A(i)E^\top(i)E_A(i) + R_1(i)$$
$$+ \lambda_{ii}E^\top(i)P(i) + \sum_{j=1, j \neq i}^{N} \lambda_{ij}\left[\frac{1}{4}\varepsilon^{-1}(i)\mathbb{I} + \varepsilon(i)E^\top(i)P(i)P^\top(i)E(i)\right],$$

with the following constraints:

$$E^\top(i)P(i) = P^\top(i)E(i) \geq 0, \tag{8.15}$$

then the uncertain system with random abrupt changes (8.2) is piecewise regular, impulse-free and stochastically stable and the cost (8.4) satisfies the following for all admissible uncertainties:

$$J(x_0, r_0) \leq tr\left[\left[\frac{1}{4}\varepsilon^{-1}(r_0)\mathbb{I} + \varepsilon(r_0)E^\top(r_0)P(r_0)P^\top(r_0)E(r_0)\right]x_0 x_0^\top\right]. \tag{8.16}$$

In the next section, we will deal with the design of the state feedback controller that will guarantee that the closed-loop system is piecewise regular, impulse-free and stochastically stable and the at same time assures that cost function is bounded.

8.3 Guaranteed Cost Control Design

Let us now focus on the design on the control law of the form (8.5) that will assure that the closed-loop system is piecewise regular, impulse-free and stochastically stable and at the same time gives a guaranteed cost control. For this purpose, notice that the cost function with the control law (8.5) becomes:

$$J(x_0, r_0) = \mathbb{E}\left[\int_0^\infty \left[x^\top(t)\widetilde{R}_1(r_t)x(t)\right]dt\right],$$

with $\widetilde{R}_1(i) = R_1(i) + K^\top(i)R_2(i)K(i)$ for all $i \in \mathscr{S}$.

The closed-loop state equation is given by:

$$E(r_t)\dot{x}(t) = [A(r_t, t) + B(r_t, t)K(r_t)] x(t), \ x(0) = x_0$$
$$= A_{cl}(r_t, t)x(t).$$

Based on the results of Theorem 8.2.1, the closed-loop system will be piecewise regular, impulse-free and stochastically stable and the cost function is bounded if there exist a set of nonsingular matrices $P = (P(1), \cdots, P(N)), P(i) \in \mathbb{R}^{n \times n}$ and a set of positive scalars $\varepsilon_A = (\varepsilon_A(1), \cdots, \varepsilon_A(N))$ such that the following hold for a given $\varepsilon_P > 0$:

$$\varepsilon_P \left[P(i) + P^\top(i) \right] \geq E^\top(i)P(i) = P^\top(i)E(i) \geq 0$$
$$A_{cl}^\top(i, t)P(i) + P^\top(i)A_{cl}(i, t) + \widetilde{R}_1(i) + \lambda_{ii}E^\top(i)P(i)$$
$$+ \sum_{j=1, j\neq i}^{N} \varepsilon_P \lambda_{ij} \left[P(j) + P^\top(j) \right] < 0,$$

for a given positive scalar ε_P.

Using the expression of $A_{cl}(i, t)$ and Lemma 1.5.1, we get:

$$\begin{cases} \varepsilon_P \left[P(i) + P^\top(i) \right] \geq E^\top(i)P(i) = P^\top(i)E(i) \geq 0 \\ A^\top(i)P(i) + P^\top(i)A(i) + K^\top(i)B^\top(i)P(i) + P^\top(i)B(i)K(i) \\ + \varepsilon_A(i)P^\top(i)D_A(i)D_A^\top(i)P(i) + \varepsilon_B(i)P^\top(i)D_B(i)D_B^\top(i)P(i) \\ + \varepsilon_A^{-1}(i)E_A^\top(i)E_A(i) + \varepsilon_B^{-1}(i)K^\top(i)E_B^\top(i)E_B(i)K(i) + \widetilde{R}_1(i) \\ + \lambda_{ii}E^\top(i)P(i) + \sum_{j=1, j\neq i}^{N} \varepsilon_P \lambda_{ij} \left[P(j) + P^\top(j) \right] < 0. \end{cases}$$

The second inequality matrix is nonlinear in the decision variables $P(i)$ and $K(i)$. To put it in the LMI setting, let $X(i) = P^{-1}(i)$ and pre- and post-multiplying this inequality respectively by $X^\top(i)$ and $X(i)$, we get:

$$X^\top(i)A^\top(i) + A(i)X(i) + X^\top(i)K^\top(i)B^\top(i) + B(i)K(i)X(i)$$
$$+ \varepsilon_A(i)D_A(i)D_A^\top(i) + \varepsilon_B(i)D_B(i)D_B^\top(i) + X^\top(i)\widetilde{R}_1(i)X(i)$$
$$+ \varepsilon_A^{-1}(i)X^\top(i)E_A^\top(i)E_A(i)X(i) + \lambda_{ii}X^\top(i)E^\top(i)$$
$$+ \varepsilon_B^{-1}(i)X^\top(i)K^\top(i)E_B^\top(i)E_B(i)K(i)X(i)$$
$$+ \sum_{j=1, j\neq i}^{N} \varepsilon_P \lambda_{ij} X^\top(i) \left[X^{-1}(j) + X^{-\top}(j) \right] X(i) < 0.$$

Using the fact that $X^{-1}(j) + X^{-\top}(j) \leq \mathbb{I} + [X^\top(j)X(j)]^{-1}$ and defining $Z_i(X)$ and $S_i(X)$ as follows:

$$Z_i(X) = \text{diag} \left[X^\top(1)X(1), \cdots, X^\top(i-1)X(i-1), \right.$$
$$\left. X^\top(i+1)X(i+1), \cdots, X^\top(N)X(N) \right]$$
$$S_i(X) = \left[\sqrt{\varepsilon_P \lambda_{i1}}X^\top(i), \cdots, \sqrt{\varepsilon_P \lambda_{ii-1}}X^\top(i), \sqrt{\varepsilon_P \lambda_{ii+1}}X^\top(i), \right.$$
$$\left. \cdots, \sqrt{\varepsilon_P \lambda_{iN}}X^\top(i) \right],$$

we have:

$$\sum_{j=1,j\neq i}^{N} \varepsilon_P \lambda_{ij} X^{\top}(i)\left[X^{-1}(j) + X^{-\top}(j)\right] X(i) \leq S_i(X)S_i^{\top}(X) + S_i(X)Z_i^{-1}(X)S_i^{\top}(X).$$

For the constraints $\varepsilon_P\left[P^{\top}(i) + P(i)\right] \geq E^{\top}(i)P(i) \, P^{\top}(i)E(i)$, pre- and post-multi-plying it respectively by $X^{\top}(i)$ and $X(i)$ give:

$$\varepsilon_P\left[X(i) + X^{\top}(i)\right] \geq X^{\top}(i)E^{\top}(i) = E(i)X(i) \geq 0.$$

Notice that

$$X^{\top}(i)\widetilde{R}_1(i)X(i) = X^{\top}(i)R_1(i)X(i) + X^{\top}(i)K^{\top}(i)R_2(i)K(i)X(i).$$

Letting $Y(i) = K(i)X(i)$, $S_1(i) = R_1^{\frac{1}{2}}$ and $S_2(i) = R_2^{\frac{1}{2}}$ and using now again the fact that $X^{\top}(j)X(j) \geq X^{\top}(j) + X(j) - \mathbb{I}$, and the Schur complement, we get the following results:

Theorem 8.3.1 *Let ε_P be a given positive scalar. There exists a state feedback con-troller of the form (8.5) such that the closed-loop state equation of the nominal sys-tem (8.2) is piecewise regular, impulse-free and stochastically stable and moreover the cost function (8.4) is bounded if there exist a set of nonsingular matrices $X = (X(1), \cdots, X(N))$, $X(i) \in \mathbb{R}^{n\times n}$ a set of matrices $Y = (Y(1), \cdots, Y(N))$, $Y(i) \in \mathbb{R}^{m\times n}$ and positive scalars $\varepsilon_A = (\varepsilon_A(1), \cdots, \varepsilon_A(N))$ and $\varepsilon_B = (\varepsilon_B(1), \cdots, \varepsilon_B(N))$ such that the following set of coupled LMIs holds for each $i \in \mathscr{S}$:*

$$\begin{bmatrix} \widehat{J}(i) & X^{\top}(i)E_A^{\top}(i) & Y^{\top}(i)E_B^{\top}(i) & X^{\top}(i)S_1^{\top}(i) & Y^{\top}(i)S_2^{\top}(i) & S_i(X) & S_i(X) \\ E_A(i)X(i) & -\varepsilon_A(i)\mathbb{I} & 0 & 0 & 0 & 0 & 0 \\ E_B(i)Y(i) & 0 & -\varepsilon_B(i)\mathbb{I} & 0 & 0 & 0 & 0 \\ S_1(i)X(i) & 0 & 0 & -\mathbb{I} & 0 & 0 & 0 \\ S_2(i)Y(i) & 0 & 0 & 0 & -\mathbb{I} & 0 & 0 \\ S_i^{\top}(X) & 0 & 0 & 0 & 0 & -\mathbb{I} & 0 \\ S_i^{\top}(X) & 0 & 0 & 0 & 0 & 0 & -\mathcal{X}_i(X) \end{bmatrix} < 0,$$

(8.17)

where

$$\widehat{J}(i) = A(i)X(i) + X^{\top}(i)A^{\top}(i) + B(i)Y(i) + B^{\top}(i)Y^{\top}(i) + \lambda_{ii}X^{\top}(i)E^{\top}(i)$$
$$+\varepsilon_A(i)D_A(i)D_A^{\top}(i) + \varepsilon_B(i)D_B(i)D_B^{\top}(i)$$
$$\mathcal{X}_i(X) = \text{diag}\left[X^{\top}(1) + X(1) - \mathbb{I}, \cdots, X^{\top}(i-1) + X(i-1) - \mathbb{I},\right.$$
$$\left. X^{\top}(i+1) + X(i+1) - \mathbb{I}, \cdots, X^{\top}(N) + X(N) - \mathbb{I}\right]$$
$$S_i(X) = \left[\sqrt{\varepsilon_P\lambda_{i1}}X^{\top}(i), \cdots, \sqrt{\varepsilon_P\lambda_{ii-1}}X^{\top}(i), \sqrt{\varepsilon_P\lambda_{ii+1}}X^{\top}(i),\right.$$
$$\left. \cdots, \sqrt{\varepsilon_P\lambda_{iN}}X^{\top}(i)\right],$$

with the following constraints:

$$\varepsilon_P\left[X(i) + X^{\top}(i)\right] \geq X^{\top}(i)E^{\top}(i) = E(i)X(i) \geq 0. \qquad (8.18)$$

The stabilizing controller gain is given by $K(i) = Y(i)X^{-1}(i)$, $i \in \mathcal{S}$. *Moreover the cost (8.4) satisfies the following for all admissible uncertainties:*

$$J(x_0, r_0) \leq \varepsilon_P tr\left[\left(X^{-\top}(r_0) + X^{-1}(r_0)\right) x_0 x_0^\top\right]. \tag{8.19}$$

From the practical point of view, it is of interest to determine a controller that assures the minimum guaranteed cost.

Notice that (8.19) can be bounded by: $\varepsilon_P \left[x_0^\top \left(\mathbb{I} + (X(r_0)X^\top(r_0))^{-1}\right) x_0\right] < \varrho$ for every $i \in \mathcal{S}$ which can be written as follows:

$$\begin{bmatrix} -\frac{\varrho}{\varepsilon_P} x_0^\top & x_0^\top \\ x_0 & -\mathbb{I} & 0 \\ x_0 & 0 & -(X^\top(r_0) + X(r_0) - \mathbb{I}) \end{bmatrix} < 0.$$

The following optimization problem can determine the controller that assures the minimum cost:

$$P: \begin{cases} \min\limits_{\substack{\varrho > 0, \\ X = (X(1), \cdots, X(N)), \\ Y = (Y(1), \cdots, Y(N)), \\ \varepsilon_A = (\varepsilon_A(1), \cdots, \varepsilon_A(N)), \\ \varepsilon_B = (\varepsilon_B(1), \cdots, \varepsilon_B(N))}} \varrho \\ s.t: (8.17), (8.18), (8.19). \end{cases}$$

The following corollary gives the results on the design of the controller that assures that the closed-loop system is piecewise regular, impulse-free and stochastically stable and simultaneously guarantees the smallest guaranteed cost.

Corollary 8.3.1 *Let* $\varrho > 0$, $X = (X(1), \cdots, X(N))$, $Y = (Y(1), \cdots, Y(N))$, $\varepsilon_A = (\varepsilon_A(1), \cdots, \varepsilon_A(N))$, *and* $\varepsilon_B = (\varepsilon_B(1), \cdots, \varepsilon_B(N))$ *be the solution of the optimization problem P for a given positive scalar* ε_P. *Then, the controller (8.5) with* $K(i) = Y(i)X^{-1}(i)$ *assures that the closed-loop system is piecewise regular, impulse-free and stochastically stable and simultaneously guarantees the smallest guaranteed cost.*

If we use the results of Theorem 8.2.2, we can easily, by following similar steps as we did for the previous results, establish the results of the following theorem.

Theorem 8.3.2 *There exists a state feedback controller of the form (8.5) such that the closed-loop state equation of the nominal system (8.2) is piecewise regular, impulse-free and stochastically stable and moreover the cost function (8.4) is bounded if there exist a set of nonsingular matrices* $X = (X(1), \cdots, X(N))$, $X(i) \in \mathbb{R}^{n \times n}$ *a set of matrices* $Y = (Y(1), \cdots, Y(N))$, $Y(i) \in \mathbb{R}^{m \times n}$ *and positive scalars* $\varepsilon_A = (\varepsilon_A(1), \cdots, \varepsilon_A(N))$, $\varepsilon_B = (\varepsilon_B(1), \cdots, \varepsilon_B(N))$ *and* $\varepsilon = (\varepsilon(1), \cdots, \varepsilon(N))$ *such that the following set of coupled LMIs holds for each* $i \in \mathcal{S}$:

$$\begin{bmatrix} \widehat{J}(i) & X^\top(i)E_A^\top(i) & Y^\top(i)E_B^\top(i) & X^\top(i)S_1^\top(i) & Y^\top(i)S_2^\top(i) & S_i(X) \\ E_A(i)X(i) & -\varepsilon_A(i)\mathbb{I} & 0 & 0 & 0 & 0 \\ E_B(i)Y(i) & 0 & -\varepsilon_B(i)\mathbb{I} & 0 & 0 & 0 \\ S_1(i)X(i) & 0 & 0 & -\mathbb{I} & 0 & 0 \\ S_2(i)Y(i) & 0 & 0 & 0 & -\mathbb{I} & 0 \\ S_i^\top(X) & 0 & 0 & 0 & 0 & -\mathcal{X}_i(X) \end{bmatrix} < 0, \tag{8.20}$$

where

$$\widehat{J}(i) = A(i)X(i) + X^{\top}(i)A^{\top}(i) + B(i)Y(i) + B^{\top}(i)Y^{\top}(i) + \lambda_{ii}X^{\top}(i)E^{\top}(i)$$
$$+\varepsilon_A(i)D_A(i)D_A^{\top}(i) + \varepsilon_B(i)D_B(i)D_B^{\top}(i)$$

$$X_i(X) = \text{diag}\left[X^{\top}(1) + X(1) - \varepsilon(i)\mathbb{I}, \cdots, X^{\top}(i-1) + X(i-1) - \varepsilon(i-1)\mathbb{I},\right.$$
$$X^{\top}(i+1) + X(i+1) - \varepsilon(i+1)\mathbb{I}, \cdots, X^{\top}(N) + X(N) - \varepsilon(N)\mathbb{I}\Big]$$

$$S_i(X) = \left[\sqrt{\lambda_{i1}}X^{\top}(i), \cdots, \sqrt{\lambda_{ii-1}}X^{\top}(i), \sqrt{\lambda_{ii+1}}X^{\top}(i),\right.$$
$$\left. \cdots, \sqrt{\lambda_{iN}}X^{\top}(i)\right],$$

with the following constraints:

$$\varepsilon(i)\mathbb{I} \geq X^{\top}(i)E^{\top}(i) = E(i)X(i) \geq 0. \tag{8.21}$$

The stabilizing controller gain is given by $K(i) = Y(i)X^{-1}(i)$, $i \in \mathscr{S}$. *Moreover the cost (8.4) satisfies the following for all admissible uncertainties:*

$$J(x_0, r_0) \leq tr\left[\left(\left(\varepsilon^{-1}(r_0)X(r_0)X^{\top}(r_0)\right)^{-1}\right)x_0 x_0^{\top}\right]. \tag{8.22}$$

Remark 8.3.1 *For the cost we can establish similar results. In fact notice that:*

$$x_0^{\top}\left(\varepsilon^{-1}(r_0)X(r_0)X^{\top}(r_0)\right)^{-1} x_0 \leq \varrho$$

can be rewritten as follows:

$$\begin{bmatrix} -\varrho & x_0^{\top} \\ x_0 & -\varepsilon^{-1}(r_0)X(r_0)X^{\top}(r_0) \end{bmatrix}.$$

Using now the fact that $\varepsilon^{-1}(r_0)X(r_0)X^{\top}(r_0) \geq X(r_0) + X^{\top}(r_0) - \varepsilon(r_0)\mathbb{I}$, *we get the desired results:*

$$\begin{bmatrix} -\varrho & x_0^{\top} \\ x_0 & -(X^{\top}(r_0) + X(r_0) - \varepsilon(r_0)\mathbb{I}) \end{bmatrix}.$$

The following optimization problem can determine the controller that assures the minimum cost:

$$P: \begin{cases} \min\limits_{\substack{\varrho>0, \\ X=(X(1),\cdots,X(N)), \\ Y=(Y(1),\cdots,Y(N)), \\ \varepsilon_A=(\varepsilon_A(1),\cdots,\varepsilon_A(N)), \\ \varepsilon_B=(\varepsilon_B(1),\cdots,\varepsilon_B(N)), \\ \varepsilon=(\varepsilon(1),\cdots,\varepsilon(N))}} \varrho \\ s.t : (8.20), (8.21), (8.22) \end{cases}$$

The following corollary gives the results on the design of the controller that assures that the closed-loop system is piecewise regular, impulse-free and stochastically stable and simultaneously guarantees the smallest guaranteed cost.

Corollary 8.3.2 *Let* $\varrho > 0$, $X = (X(1), \cdots, X(N))$, $Y = (Y(1), \cdots, Y(N))$, $\varepsilon_A = (\varepsilon_A(1), \cdots, \varepsilon_A(N))$, $\varepsilon_B = (\varepsilon_B(1), \cdots, \varepsilon_B(N))$ *and* $\varepsilon = (\varepsilon(1), \cdots, \varepsilon(N))$ *be the solution of the optimization problem P. Then, the controller (8.5) with* $K(i) = Y(i)X^{-1}(i)$ *assures that the closed-loop system is piecewise regular, impulse-free and stochastically stable and simultaneously guarantees the smallest guaranteed cost.*

If we use the results of Theorem 8.2.3, we can easily establish the results of the following theorem.

Theorem 8.3.3 *There exists a state feedback controller of the form (8.5) such that the closed-loop state equation of the nominal system (8.2) is piecewise regular, impulse-free and stochastically stable and moreover the cost function (8.4) is bounded if there exist a set of nonsingular matrices* $X = (X(1), \cdots, X(N))$, $X(i) \in \mathbb{R}^{n \times n}$ *a set of matrices* $Y = (Y(1), \cdots, Y(N))$, $Y(i) \in \mathbb{R}^{m \times n}$ *and positive scalars* $\varepsilon_A = (\varepsilon_A(1), \cdots, \varepsilon_A(N))$, $\varepsilon_B = (\varepsilon_B(1), \cdots, \varepsilon_B(N))$ *and* $\varepsilon = (\varepsilon(1), \cdots, \varepsilon(N))$ *such that the following set of coupled LMIs holds for each* $i \in \mathscr{S}$:

$$
\begin{bmatrix}
\widehat{J}(i) & X^\top(i)E_A^\top(i) & Y^\top(i)E_B^\top(i) & X^\top(i)S_1^\top(i) & Y^\top(i)S_2^\top(i) & \mathcal{Z}_i(X) & \mathcal{S}_i(X) \\
E_A(i)X(i) & -\varepsilon_A(i)\mathbb{I} & 0 & 0 & 0 & 0 & 0 \\
E_B(i)Y(i) & 0 & -\varepsilon_B(i)\mathbb{I} & 0 & 0 & 0 & 0 \\
S_1(i)X(i) & 0 & 0 & -\mathbb{I} & 0 & 0 & 0 \\
S_2(i)Y(i) & 0 & 0 & 0 & -\mathbb{I} & 0 & 0 \\
\mathcal{Z}_i^\top(X) & 0 & 0 & 0 & 0 & -\mathcal{X}_i(\varepsilon) & 0 \\
\mathcal{S}_i^\top(X) & 0 & 0 & 0 & 0 & 0 & -\mathcal{X}_i(X)
\end{bmatrix} < 0,
$$

$$(8.23)$$

where

$$
\begin{aligned}
\widehat{J}(i) &= A(i)X(i) + X^\top(i)A^\top(i) + B(i)Y(i) + B^\top(i)Y^\top(i) + \lambda_{ii}X^\top(i)E^\top(i) \\
&\quad + \varepsilon_A(i)D_A(i)D_A^\top(i) + \varepsilon_B(i)D_B(i)D_B^\top(i) \\
\mathcal{X}_i(\varepsilon) &= \mathrm{diag}\left[4\varepsilon(1)\mathbb{I}, \cdots, 4\varepsilon(i-1)\mathbb{I}, 4\varepsilon(i+1)\mathbb{I}, \cdots, 4\varepsilon(N)\mathbb{I}\right] \\
\mathcal{Z}_i(X) &= \left[\sqrt{\lambda_{i1}}X^\top(i), \cdots, \sqrt{\lambda_{ii-1}}X^\top(i), \sqrt{\lambda_{ii+1}}X^\top(i), \cdots, \sqrt{\lambda_{iN}}X^\top(i)\right], \\
\mathcal{X}_i(X) &= \mathrm{diag}\left[X^\top(1) + X(i) - \varepsilon(1)\mathbb{I}, \cdots, X^\top(i-1) + X(i-1) - \varepsilon(i-1)\mathbb{I}, \right. \\
&\quad \left. X^\top(i+1) + X(i+1) - \varepsilon(i+1)\mathbb{I}, \cdots, X^\top(N) + X(N) - \varepsilon(N)\mathbb{I}\right] \\
\mathcal{S}_i(X) &= \left[\sqrt{\lambda_{i1}}X^\top(i)E^\top(1), \cdots, \sqrt{\lambda_{ii-1}}X^\top(i)E^\top(i-1), \sqrt{\lambda_{ii+1}}X^\top(i)E^\top(i+1), \right. \\
&\quad \left. \cdots, \sqrt{\lambda_{iN}}X^\top(i)E^\top(N)\right],
\end{aligned}
$$

with the following constraints:

$$X^\top(i)E^\top(i) = E(i)X(i) \geq 0. \tag{8.24}$$

The stabilizing controller gain is given by $K(i) = Y(i)X^{-1}(i)$, $i \in \mathscr{S}$. *Moreover the cost (8.4) satisfies the following for all admissible uncertainties:*

$$J(x_0, r_0) \leq \mathrm{tr}\left[\left(\frac{1}{4}\varepsilon^{-1}(r_0)\mathbb{I} + E^\top(r_0)\left(\varepsilon^{-1}(r_0)X^\top(r_0)X(r_0)\right)^{-1}E(r_0)\right)x_0 x_0^\top\right]. \tag{8.25}$$

Remark 8.3.2 *For the cost we can establish similar results. In fact notice that:*

$$x_0^\top\left(\frac{1}{4}\varepsilon^{-1}(r_0)\mathbb{I} + E^\top(r_0)\left(\varepsilon^{-1}(r_0)X^\top(r_0)X(r_0)\right)^{-1}E(r_0)\right)x_0 \leq \varrho$$

can be rewritten as follows:

$$\begin{bmatrix} -\varrho & x_0^\top & x_0^\top E^\top(r_0) \\ x_0 & -4\varepsilon(r_0)\mathbb{I} & 0 \\ E(r_0)x_0 & 0 & -\varepsilon^{-1}(r_0)X^\top(r_0)X(r_0) \end{bmatrix} \leq 0 .$$

Using now the fact that $\varepsilon^{-1}(r_0)X(r_0)X^\top(r_0) \geq X(r_0) + X^\top(r_0) - \varepsilon(r_0)\mathbb{I}$, *we get the desired results:*

$$\begin{bmatrix} -\varrho & x_0^\top & x_0^\top E^\top(r_0) \\ x_0 & -4\varepsilon(r_0)\mathbb{I} & 0 \\ E(r_0)x_0 & 0 & -(X(r_0) + X^\top(r_0) - \varepsilon(r_0)\mathbb{I}) \end{bmatrix} \leq 0 .$$

The following optimization problem can determine the controller that assures the minimum cost:

$$\text{P} : \begin{cases} \min\limits_{\substack{\varrho > 0, \\ X = (X(1),\cdots,X(N)), \\ Y = (Y(1),\cdots,Y(N)), \\ \varepsilon_A = (\varepsilon_A(1),\cdots,\varepsilon_A(N)), \\ \varepsilon_B = (\varepsilon_B(1),\cdots,\varepsilon_B(N)), \\ \varepsilon = (\varepsilon(1),\cdots,\varepsilon(N))}} \varrho \\ s.t : (8.23), (8.24), (8.25) . \end{cases}$$

8.4 Numerical Example

To illustrate the effectiveness of the proposed results, let us consider a dynamical singular system with random abrupt changes with two modes with the following data:

- mode # 1:

$$A(1) = \begin{bmatrix} -1.0 & 0.0 & 1.0 \\ 0.0 & 0.0 & 1.0 \\ 0.0 & -1.0 & -1.0 \end{bmatrix} ,$$

$$B(1) = \begin{bmatrix} 0.3 & 0.0 \\ 0.0 & 0.1 \\ 0.2 & 1.0 \end{bmatrix} ,$$

$$D_A(1) = \begin{bmatrix} 0.1 \\ 0.2 \\ 0.3 \end{bmatrix} , \quad E_A(1) = \begin{bmatrix} 0.3 & 0.2 & 0.1 \end{bmatrix} ,$$

$$D_B(1) = \begin{bmatrix} 0.2 \\ 0.3 \\ 0.1 \end{bmatrix} , \quad E_B(1) = \begin{bmatrix} 0.2 & 0.1 \end{bmatrix} ,$$

$$R_1(1) = \begin{bmatrix} 1.0 & 0.0 & 0.0 \\ 0.0 & 1.0 & 0.0 \\ 0.0 & 0.0 & 1.0 \end{bmatrix} ,$$

$$R_2(1) = \begin{bmatrix} 2.0 & 0.0 \\ 0.0 & 4.0 \end{bmatrix} ;$$

- mode # 2:

$$A(2) = \begin{bmatrix} 1.0 & 0.0 & 1.0 \\ 0.0 & 0.0 & 1.0 \\ 0.0 & 1.0 & -1.0 \end{bmatrix},$$

$$B(2) = \begin{bmatrix} 0.1 & 0.0 \\ 0.0 & 0.0 \\ 0.1 & 0.2 \end{bmatrix},$$

$$D_A(2) = \begin{bmatrix} 0.2 \\ 0.1 \\ 0.3 \end{bmatrix}, \quad E_A(2) = \begin{bmatrix} 0.3 & 0.2 & 0.1 \end{bmatrix},$$

$$D_B(2) = \begin{bmatrix} 0.3 \\ 0.1 \\ 0.2 \end{bmatrix}, \quad E_B(2) = \begin{bmatrix} 0.1 & 0.2 \end{bmatrix},$$

$$R_1(2) = \begin{bmatrix} 1.0 & 0.0 & 0.0 \\ 0.0 & 1.0 & 0.0 \\ 0.0 & 0.0 & 1.0 \end{bmatrix}, \quad R_2(2) = \begin{bmatrix} 4.0 & 0.0 \\ 0.0 & 2.0 \end{bmatrix}.$$

Solving the LMIs (8.23)-(8.24) of Theorem 8.3.3, we get:

$$\varepsilon_A(1) = \varepsilon_A(2) = 0.0037,$$
$$\varepsilon_B(1) = \varepsilon_B(2) = 0.0219,$$
$$\varepsilon(1) = 0.0019,$$
$$\varepsilon(2) = 0.0014,$$

$$X(1) = \begin{bmatrix} 0.0076 & 0.0009 & 0.0 \\ 0.0009 & 0.0026 & 0.0 \\ -0.0064 & -0.0028 & 0.0343 \end{bmatrix},$$

$$X(2) = \begin{bmatrix} 0.0024 & 0.0006 & 0.0 \\ 0.0006 & 0.0018 & 0.0 \\ -0.0016 & -0.0018 & 0.0025 \end{bmatrix},$$

$$Y(1) = \begin{bmatrix} -0.0533 & 0.0048 & 0.0090 \\ 0.0067 & -0.0234 & -0.1812 \end{bmatrix},$$

$$Y(2) = \begin{bmatrix} -0.0429 & 0.0008 & -0.0307 \\ 0.0070 & 0.0007 & -0.0306 \end{bmatrix},$$

which give the following gains:

$$K(1) = \begin{bmatrix} -7.4133 & 4.7730 & 0.2626 \\ -1.8693 & -14.0204 & -5.2849 \end{bmatrix},$$

$$K(2) = \begin{bmatrix} -25.3618 & -2.8919 & -12.1526 \\ -2.6612 & -10.5641 & -12.0991 \end{bmatrix}.$$

8.5 Notes

This chapter dealt with the guaranteed cost control problem of the singular class of systems with random abrupt changes. This approach of stabilization is one of the most popular one since it uses the cost function that is usually in the jump linear quadratic regulator. A state feedback controller that assures that closed-loop state equation either for the nominal system or the uncertain is piecewise regular, impulse-free and stochastically stable is designed in the LMI setting. Three approaches were developed for this purpose. The conditions we developed in this chapter are tractable using commercial optimization tools. The content of this chapter is mainly based on the work of the author and his coauthors [16].

9

Mixed $\mathcal{H}_2/\mathcal{H}_\infty$ Control Problem

In the previous chapters we covered separately the $\mathcal{H}_2$ and the $\mathcal{H}_\infty$ control problems. The guaranteed cost control has also been treated. In this chapter, we will combine the $\mathcal{H}_2$ and $\mathcal{H}_\infty$ techniques to get what it is known in the literature as the mixed $\mathcal{H}_2/\mathcal{H}_\infty$ control problem. This approach consists of determining a controller that makes the closed-loop state equation of the class of system we are considering piecewise regular, impulse-free and stochastically stable and at the same time guarantees the disturbance rejection with a certain given level $\gamma > 0$ and assures a guaranteed cost.

The mixed $\mathcal{H}_2/\mathcal{H}_\infty$ control problem in the deterministic setting has been treated by many authors among them we quote the works of [10, 139, 67, 77] where nonlinear Riccati equations have been used. The authors of [103] used the LMI techniques to solve the mixed $\mathcal{H}_2/\mathcal{H}_\infty$ control problem. For the class of normal systems we are dealing with, some results have been reported in the literature among them we quote the works of Boukas [13] where a state feedback controller has been designed to stabilize the class of Markovian jump linear systems with time-delay, and de Farias et al. [44] where an output feedback controller has been designed using the LMI setting for the nominal systems of Markovian jump linear systems. To the best of our knowledge, the mixed $\mathcal{H}_2/\mathcal{H}_\infty$ control problem has never been tackled for the class of systems we are treating in this volume and this work is the first extension to this class of systems.

Our goal in this chapter is to develop LMI conditions that allow the state feedback controller that will guarantee what it is required in the previous design specifications. The nominal and the uncertain cases are covered. The rest of the chapter is organized as follows. In Sect. 2, the problem is stated and some appropriate definitions are given. In Sect. 3, the nominal case is covered and the LMI conditions to design the state feedback controller are established. In Sect. 4, the uncertain case is tackled and the robust state feedback controller that assures the desired requirements is designed using LMI setting. In Sect. 5, a case study is presented to show the effectiveness of the developed results.

9.1 Problem Statement

Consider a singular system with random abrupt changes that has N modes, i.e., $\mathscr{S} = \{1, 2, \ldots, N\}$. The mode switching is assumed to be governed by a continuous-time Markov process $\{r_t, t \geq 0\}$ taking values in the state space $\mathscr{S}$ and having the following infinitesimal generator

$$\Lambda = (\lambda_{ij}), i, j \in \mathscr{S},$$

where $\lambda_{ij} \geq 0, \forall j \neq i, \lambda_{ii} = -\sum_{j \neq i} \lambda_{ij}$.

The mode transition probabilities are described as follows:

$$P[r_{t+h} = j | r_t = i] = \begin{cases} \lambda_{ij} h + o(h), & j \neq i, \\ 1 + \lambda_{ii} h + o(h), & j = i, \end{cases} \tag{9.1}$$

where $\lim_{h \to 0} o(h)/h = 0$.

Let the state equation of this class of systems be defined in a probability space $(\Omega, \mathcal{F}, \mathbb{P})$ and assume that its behavior is described by the following differential-algebraic equations:

$$\begin{cases} E(r_t)\dot{x}(t) = A(r_t, t)x(t) + B(r_t, t)u(t) + B_w(r_t)w(t), x(0) = x_0, \\ y(t) = C_y(r_t, t)x(t) + D_y(r_t, t)u(t), \\ z(t) = C_z(r_t, t)x(t) + D_z(r_t, t)u(t), \end{cases} \tag{9.2}$$

where $x(t) \in \mathbb{R}^n$ is the system state at time t, $u(t) \in \mathbb{R}^m$ is the control input of the system at time t, $w(t) \in \mathbb{R}^k$ is the external disturbance of the system at time t, $y(t) \in \mathbb{R}^p$ and $z(t) \in \mathbb{R}^l$ are the controlled output of the system at time t, $A(r_t, t)$, $B(r_t, t)$, $C_y(r_t, t)$, and $C_z(r_t, t)$ are assumed to have uncertainties, i.e.: $A(r_t, t) = A(r_t) + D_A(r_t)F_A(r_t)E_A(r_t)$, $B(r_t, t) = B(r_t) + D_B(r_t)F_B(r_t)E_B(r_t)$, $C_y(r_t, t) = C_y(r_t) + D_{C_y}(r_t)F_{C_y}(r_t)E_{C_y}(r_t)$ $C_z(r_t, t) = C_z(r_t) + D_{C_z}(r_t)F_{C_z}(r_t)E_{C_z}(r_t)$, $D_y(r_t, t) = D_y(r_t) + D_{D_y}(r_t)F_{D_y}(r_t)E_{D_y}(r_t)$ and $D_z(r_t, t) = D_z(r_t) + D_{D_z}(r_t)F_{D_z}(r_t)E_{D_z}(r_t) E_{C_z}(r_t)$ with $A(i) \in \mathbb{R}^{n \times n}$, $D_A(i) \in \mathbb{R}^{n \times n_D}$, $E_A(i) \in \mathbb{R}^{n_E \times n}$, $B(i) \in \mathbb{R}^{n \times m}$, $D_B(i) \in \mathbb{R}^{n \times m_D}$, $E_B(i \in \mathbb{R}^{m_E \times m}$, $B_w(i) \in \mathbb{R}^{n \times k}$, $D_{B_w}(i) \in \mathbb{R}^{n \times k_D}$, $E_{B_w}(i) \in \mathbb{R}^{k_E \times n}$, $C_y(i) \in \mathbb{R}^{p \times n}$, $D_y(i) \in \mathbb{R}^{p \times m}$, $D_{C_y}(i) \in \mathbb{R}^{p \times n_C}$, $E_{C_y}(i) \in \mathbb{R}^{m_C \times n}$, $C_z(i) \in \mathbb{R}^{l \times n}$, $D_z(i) \in \mathbb{R}^{l \times m}$, $D_{C_z}(i) \in \mathbb{R}^{l \times m_D}$, and $E_{C_z}(i) \in \mathbb{R}^{m_D \times l}$ are known real matrices with appropriate dimensions for each $i \in \mathscr{S}$ and $F_A(r_t) \in \mathbb{R}^{n_D \times n_E}$, $F_B(r_t) \in \mathbb{R}^{m_D \times m_E}$, $F_{B_w}(r_t) \in \mathbb{R}^{k_D \times k_E}$, $F_{C_y}(i) \in \mathbb{R}^{n_C \times m_C}$, and $F_{C_z}(i) \in \mathbb{R}^{m_D \times m_C}$ satisfy $F_A^\top(i)F_A(i) \leq I$, $F_B^\top(i)F_B(i) \leq I$, $F_{B_w}^\top(i)F_{B_w}(i) \leq I$, $F_{C_y}^\top(i)F_{C_y}(i) \leq I$ and $F_{C_z}^\top(i)F_{C_z}(i) \leq I$ for each $i \in \mathscr{S}$, the matrix $E(i)$ may be singular, and we assume $0 \leqslant \text{rank}(E(i)) = n_r \leq n$.

Remark 9.1.1 *When the uncertainties are equal to zero the system will be referred to as nominal system. The uncertainties that satisfy the previous conditions are referred to as admissible. The uncertainties we are considering in this chapter are known in the literature as norm bounded uncertainties. When the matrix $E(i)$ for each mode $i \in \mathscr{S}$ is nonsingular, the system (9.2) is referred to as normal system.*

Definition 9.1.1 [43]

   i. Nominal system (9.2) is said to be regular if the characteristic polynomial, $\det(sE(i) - A(i))$ is not identically zero for each mode $i \in \mathscr{S}$.

   ii. Nominal system (9.2) is said to be impulse free, i.e., the $\deg(\det(sE(i) - A(i))) = \text{rank}(E(i))$ for each mode $i \in \mathscr{S}$.

For more details on other properties and the existence of the solution of system (9.2), we refer the reader to [122], and the references therein. In general, the regularity is often a sufficient condition for the analysis and the synthesis of singular systems.

For the system (9.2), we have the following definitions:

Definition 9.1.2 *Nominal system (9.2) is said to be stochastically stable (SS) if there exists a constant $T(r_0, x_0)$ such that*

$$\mathbb{E}\left[\int_0^\infty \|x(t)\|^2 dt \Big| x_0, r_0\right] \leq T(r_0, x_0). \tag{9.3}$$

Definition 9.1.3 *Uncertain system (9.2) is said to be robust stochastically stable (RSS) if there exists a constant $T(r_0, x_0)$ such that (9.3) holds for all admissible uncertainties.*

Definition 9.1.4 *Let $\gamma > 0$ be a given positive constant. System (9.2) with $u(t) \equiv 0$ is said to be stochastically stable with γ−disturbance attenuation if there exists a constant $M(x_0, r_0)$ with $M(0, r_0) = 0$, for all $r_0 \in \mathscr{S}$, such that the following holds:*

$$\|z\|_2 \triangleq \left[\mathbb{E}\int_0^\infty z^\top(t)z(t)dt|(x_0, r_0)\right]^{1/2} \leq \gamma\left[\|w\|_2^2 + M(x_0, r_0)\right]^{\frac{1}{2}}. \tag{9.4}$$

Definition 9.1.5 *System (9.2) with $u(t) \equiv 0$ is said to be internally stochastically stable if there exists a set of nonsingular matrices $P = (P(1), \cdots, P(N))$, satisfying the following for every $i \in \mathscr{S}$:*

$$P^\top(i)A(i, t) + A^\top(i, t)P(i) + \sum_{j=1}^N \lambda_{ij}E^\top(j)P(j) < 0, \tag{9.5}$$

with the following constraints:

$$E^\top P(i) = P^\top(i)E \geq 0. \tag{9.6}$$

By virtue of Definition 2.1.1, it is obvious that internal stochastically stable means that system (9.2) is stochastically stable in case of $w(t) \equiv 0$, i. e., system (9.2) being free of input disturbance. Likewise, we can give the following definitions:

Definition 9.1.6 *System (9.2) with $u(t) \equiv 0$ is said to be internally stochastically stable if it is stochastically stable in case of $w(t) \equiv 0$.*

Definition 9.1.7 *System (9.2) is said to be stabilizable with γ-disturbance in the stochastic sense if there exists a control law such that the closed-loop system under this control law is piecewise regular, impulse-free and stochastically stable and satisfies (9.4).*

For the uncertain system we will have similar definitions that we can summarize as follows:

Definition 9.1.8 *Let $\gamma > 0$ be a given positive constant. System (9.2) with $u(t) \equiv 0$ is said to be robust stochastically stable with γ-disturbance attenuation if there exists a constant $M(x_0, r_0)$ with $M(0, r_0) = 0$, for all $r_0 \in \mathscr{S}$, such that (9.4) holds for all admissible uncertainties.*

Definition 9.1.9 *System (9.2) with $u(t) \equiv 0$ is said to be internally robust stochastically stable if there exists a set of nonsingular matrices $P = (P(1), \cdots, P(N))$, satisfying the following for every $i \in \mathscr{S}$ and for all admissible uncertainties:*

$$P^\top(i)A(i, t) + A^\top(i, t)P(i) + \sum_{j=1}^{N} \lambda_{ij} E^\top(j)P(j) < 0, \tag{9.7}$$

with the following constraints:

$$E^\top P(i) = P^\top(i)E \geq 0. \tag{9.8}$$

By virtue of Definition 2.1.1, it is obvious that internal robust stochastically stable means that system (9.2) is robust stochastically stable in case of $w(t) \equiv 0$, i. e., system (9.2) being free of input disturbance. Likewise, we can give the following definitions:

Definition 9.1.10 *System (9.2) with $u(t) \equiv 0$ is said to be internally robust stochastically stable if it is robust stochastically stable in case of $w(t) \equiv 0$.*

Definition 9.1.11 *System (9.2) is said to be robust stabilizable with γ-disturbance in the robust stochastically stable sense if there exists a control law such that the closed-loop system under this control law is robust stochastically stable and satisfies (9.4).*

Consider the following cost function:

$$J(x_0, r_0) = \mathbb{E}\left[\int_0^\infty z^\top(t)z(t)dt \right], \tag{9.9}$$

where x_0 and r_0 are respectively the initial state and the initial mode of the system.

Definition 9.1.12 *If there exist a control law, $u(.)$ and a positive scalar ϱ representing the upper bound of the cost (9.9) such that the closed-loop system is piecewise regular, impulse-free and stochastically stable, and the cost (9.9) is bounded by ϱ, then ϱ is called the guaranteed cost, also referred to as the optimal guaranteed cost, and $u(.)$ is the associated guaranteed cost control.*

Remark 9.1.2 *In the rest of this paper we use the upper bounds we developed earlier to establish the results of this chapter. The definitions will be adapted in consequence.*

The goal of this chapter is to design a controller of the following form:

$$u(t) = K(r_t)x(t), \tag{9.10}$$

where $K(i)$ is a design parameter that has to be determined for every $i \in \mathcal{S}$, that makes the closed-loop state equation of the nominal or the uncertain systems piecewise regular, impulse-free and stochastically stable and at the same time

- minimizes the upper bound of the worst case $\mathcal{H}_2$ performance.
- and guarantees (9.4) when the external disturbance is viewed as an square integrable disturbance signal (finite energy or finite power).

The aim of this chapter is to develop LMI conditions that can be used to design a state feedback controller that guarantees the previous specifications. The conditions we will develop here can be solved using LMI control toolbox.

9.2 Mixed $\mathcal{H}_2/\mathcal{H}_\infty$ Control: Nominal Case

Let us introduce the following performance measures:

$$J_{\mathcal{H}_2} = \mathbb{E}\left[\int_0^\infty y^\top(t)y(t)dt\right] \tag{9.11}$$

$$J_{\mathcal{H}_\infty} = \mathbb{E}\left[\int_0^\infty \left[z^\top(t)z(t) - \gamma^2 w^\top(t)w(t)\right]dt\right] \tag{9.12}$$

The objective of this section is to synthesize the state feedback controller that guarantees that the closed-loop system is piecewise regular, impulse-free and stochastically stable and at the same time minimizes the $\mathcal{H}_2$ performance (9.11) and assures the disturbance rejection of level $\gamma > 0$.

Let us consider the nominal case first and then extend the results for the uncertain case. Combining the dynamics and the controller expressions we get:

$$\begin{cases} E(r_t)\dot{x}(t) = \bar{A}(r_t)x(t) + B_w(r_t, t)w(t), x(0) = x_0, \\ y(t) = \bar{C}_y(r_t, t)x(t), \\ z(t) = \bar{C}_z(r_t, t)x(t), \end{cases}$$

where

$$\bar{A}(i) = A(i) + B(i)K(i)$$
$$\bar{C}_y(i) = C(i) + D_y(i)K(i)$$
$$\bar{C}_z(i) = C_z(i) + D_z(i)K(i).$$

For the $\mathcal{H}_2$ performance measure we have the following result.

Theorem 9.2.1 *If there exists a nonsingular set of matrices $P = (P(1), \cdots, P(N))$, $P(i) \in \mathbb{R}^{n \times n}$ when $w(t) = 0$ for all $t \geq 0$ such that the following holds:*

$$\bar{A}^\top(i)P(i) + P^\top(i)\bar{A}(i) + \sum_{j=1}^{N} \lambda_{ij}E^\top(j)P(j) + \bar{C}_y^\top(i)\bar{C}_y(i) < 0, \qquad (9.13)$$

with the following constraints:

$$E^\top(i)P(i) = P^\top(i)E(i) \geq 0, \qquad (9.14)$$

then the controller $u(t) = K(r_t)x(t)$, with given set of gains $K = (K(1), \cdots, K(N))$, is an $\mathcal{H}_2$ optimal controller satisfying the minimization of the $\mathcal{H}_2$ performance measure with $J_{\mathcal{H}_2} \leq x^\top(0)E^\top(r_0)P(r_0)x(0)$.

Proof: Suppose that (9.13)-(9.14) hold for a set of nonsingular matrices $P(i)$, $i = 1, \cdots, N$, then it is easy to conclude that the following hold:

$$\bar{A}^\top(i)P(i) + P^\top(i)\bar{A}(i) + \sum_{j=1}^{N} \lambda_{ij}E^\top(j)P(j) < 0,$$

with the following constraints:

$$E^\top(i)P(i) = P^\top(i)E(i) \geq 0.$$

This implies that the closed-loop system is piecewise regular, impulse-free and stochastically stable.

Let us now consider the following Lyapunov candidate function:

$$V(x(t), r_t) = x^\top(t)E^\top(r_t)P(r_t)x(t).$$

Then, if at time t, $x(t) = x$ and $r_t = i$, $i \in \mathcal{S}$, the infinitesimal operator emanating from the point (x, i) at time t is given by:

$$\mathcal{L}V(x(t), i) = x^\top(t)\left[\bar{A}^\top(i)P(i) + P^\top(i)\bar{A}(i) + \sum_{j=1}^{N} \lambda_{ij}E^\top(j)P(j)\right]x(t).$$

Using the condition (9.13), we get:

$$x^\top(t)\left[\bar{A}^\top(i)P(i) + P^\top(i)\bar{A}(i) + \sum_{j=1}^{N} \lambda_{ij}E^\top(j)P(j) + \bar{C}_y^\top(i)\bar{C}_y(i)\right]x(t) < 0,$$

which is equivalent to:

$$\mathcal{L}V(x(t), i) + x^\top(t)\left[\bar{C}_y^\top(i)\bar{C}_y(i)\right]x(t) < 0.$$

Integrating both sides from 0 to ∞ and using Dynkin formula and the fact that the closed-system is stable, we get:

$$J_{\mathcal{H}_2} \leq V(x(0), r_0) = x^\top(0)E^\top(r_0)P(r_0)x(0).$$

This ends the proof of the theorem.

For the $\mathcal{H}_\infty$ performance measure we have the following result.

Theorem 9.2.2 *Let γ be a given positive scalar. If there exists a nonsingular set of matrices $P = (P(1), \cdots, P(N))$, $P(i) \in \mathbb{R}^{n \times n}$ such that the following holds:*

$$\begin{bmatrix} \bar{A}^\top(i)P(i) + P^\top(i)\bar{A}(i) + \sum_{j=1}^N \lambda_{ij}E^\top(j)P(j) + \bar{C}_z^\top(i)\bar{C}_z(i) & \star \\ B_w^\top(i)P(i) & -\gamma^2\mathbb{I} \end{bmatrix} < 0, \qquad (9.15)$$

with the following constraints:

$$E^\top(i)P(i) = P^\top(i)E(i) \geq 0, \qquad (9.16)$$

then the controller $u(t) = K(r_t)x(t)$, with a given set of gains $K = (K(1), \cdots, K(N))$, guarantees the disturbance rejection of level γ.

Proof: Suppose that (9.15)-(9.16) hold for a set of nonsingular matrices $P(i)$, $i = 1, \cdots, N$, then it is easy to conclude that the closed-loop system is piecewise regular, impulse-free and stochastically stable when external disturbance is equal to zero.

Now if define $\eta(t) = (x^\top(t), w^\top(t))^\top$, then for any nonzero external disturbance belonging to the set $\mathscr{L}_2[0, \infty)$, notice that we have:

$$z^\top(t)z(t) - \gamma^2 w^\top(t)w(t) = x^\top(t)\bar{C}_z^\top(i)\bar{C}_z(i)x(t) - \gamma^2 w^\top(t)w(t),$$

and

$$\mathscr{L}V(x(t), i) = x^\top(t)\left[\bar{A}^\top(i)P(i) + P^\top(i)\bar{A}(i) + \sum_{j=1}^N \lambda_{ij}E^\top(j)P(j)\right]x(t)$$
$$+ 2x^\top(t)P^\top(i)B_w(i)w(t).$$

This implies that:

$$z^\top(t)z(t) - \gamma^2 w^\top(t)w(t) + \mathscr{L}V(x(t), r_t) = \eta^\top(t)\Theta(r_t)\eta(t),$$

with

$$\Theta(i) = \begin{bmatrix} \bar{A}^\top(i)P(i) + P^\top(i)\bar{A}(i) + \sum_{j=1}^N \lambda_{ij}E^\top(j)P(j) + \bar{C}_z^\top(i)\bar{C}_z(i) & \star \\ B_w^\top(i)P(i) & -\gamma^2\mathbb{I} \end{bmatrix}.$$

Since the closed-loop system is stochastically stable it results that:

$$\mathbb{E}\left[\int_0^\infty \left[z^\top(t)z(t) - \gamma^2 w^\top(t)w(t) + \mathscr{L}V(x(t), r_t)\right]dt\right] - \mathbb{E}\left[\int_0^\infty \mathscr{L}V(x(t), r_t)dt\right]$$
$$\leq \mathbb{E}\left[\int_0^\infty \eta^\top(t)\Theta(r_t)\eta(t)dt\right] + V(x(0), r_0).$$

Using now the fact that $\Theta(i) < 0$ for each $i \in \mathscr{S}$, we get:

$$\mathbb{E}\left[\int_0^\infty \left[z^\top(t)z(t) - \gamma^2 w^\top(t)w(t)\right]dt\right] \leq x^\top(0)E^\top(r_0)P(r_0)x(0),$$

which yields

$$\|z\|_2^2 \leq \gamma^2 \|w\|_2^2 + x^\top(0)E^\top(r_0)P(r_0)x(0).$$

This ends the proof of the theorem.

The matrix inequalities (9.13) and (9.15) are nonlinear in the decision variables. To transform them in the LMI setting, we can proceed as usual. First of all notice that:

$$\sum_{j=1}^N \lambda_{ij}E^\top(j)P(j) = \lambda_{ii}E^\top(i)P(i) + \sum_{j=1,j\neq i}^N \lambda_{ij}E^\top(j)P(j).$$

Using the fact that $E^\top(j)P(j) \leq \varepsilon_P\left[P^\top(j) + P(j)\right]$ for any given $\varepsilon_P > 0$, the conditions (9.13)-(9.14) become after pre- and post-multiplying respectively by $X^\top(i) = P^{-\top}(i)$:

$$X^\top(i)\bar{A}^\top(i) + \bar{A}(i)X(i) + \lambda_{ii}X^\top(i)E^\top(i) + \sum_{j=1,j\neq i}^N \varepsilon_P\lambda_{ij}X^\top(i)\left[X^{-\top}(j) + X^{-1}(j)\right]X(i)$$

$$+ \bar{X}^\top(i)\bar{C}_y^\top(i)\bar{C}_y(i)X(i) < 0$$

$$\varepsilon_P\left[X^\top(i) + X(i)\right] \geq P^\top(i)E^\top(i) = E(i)X(i) \geq 0.$$

Based on the fact that

$$\left[X^\top(i)X(i)\right]^{-1} \geq X^{-\top}(i) + X^{-1}(i) - \mathbb{I},$$

and letting

$$Y(i) = K(i)X(i)$$

$$S_i(X) = \left[\sqrt{\varepsilon_P\lambda_{i1}}X^\top(i), \cdots, \sqrt{\varepsilon_P\lambda_{ii-1}}X^\top(i), \sqrt{\varepsilon_P\lambda_{ii+1}}X^\top(i), \cdots, \sqrt{\varepsilon_P\lambda_{iN}}X^\top(i)\right]$$

$$X_i(X) = \mathrm{diag}\left[X^\top(1) + X(1) - \mathbb{I}, \cdots, X^\top(i-1) + X(i-1) - \mathbb{I},\right.$$

$$\left. X^\top(i+1) + X(i+1) - \mathbb{I}, \cdots, X^\top(N) + X(N) - \mathbb{I}\right],$$

we get the following:

$$\begin{bmatrix} J_1(X) & X^\top(i)C_y^\top(i) + Y^\top(i)D_y^\top(i) & S_i(X) & S_i(X) \\ C_y(i)X(i) + D_y(i)Y(i) & -\mathbb{I} & 0 & \\ S_i^\top(X) & 0 & -\mathbb{I} & 0 \\ S_i^\top(X) & 0 & 0 & -X_i(X) \end{bmatrix} < 0$$

$$\varepsilon_P\left[X^\top(i) + X(i)\right] \geq P^\top(i)E^\top(i) = E(i)X(i) \geq 0.$$

with

$$J_1(X) = A(i)X(i) + X^\top(i)A^\top(i) + B(i)Y(i) + Y^\top(i)B^\top(i) + \lambda_{ii}X^\top(i)E^\top(i)$$

For the conditions (9.15)-(9.16), proceeding similarly and by pre- and post-multiplying respectively by $\mathrm{diag}\,[X^\top(i),\mathbb{I}]$ and $\mathrm{diag}\,[X(i),\mathbb{I}]$, we get:

$$\begin{bmatrix} J_2(X) & X^\top(i)C_z^\top(i)+Y^\top(i)D_z^\top(i) & B_w(i)X(i) & S_i(X) & S_i(X) \\ C_z(i)X(i)+D_z(i)Y(i) & -\mathbb{I} & 0 & 0 & 0 \\ X^\top(i)B_w^\top(i) & 0 & -\gamma^2\mathbb{I} & 0 & 0 \\ S_i^\top(X) & 0 & 0 & -\mathbb{I} & 0 \\ S_i^\top(X) & 0 & 0 & 0 & -X_i(X) \end{bmatrix} < 0$$

$$\varepsilon_P\left[X^\top(i)+X(i)\right] \geq P^\top(i)E^\top(i) = E(i)X(i) \geq 0,$$

with

$$J_2(X) = A(i)X(i)+X^\top(i)A^\top(i)+B(i)Y(i)+Y^\top(i)B^\top(i)+\lambda_{ii}X^\top(i)E^\top(i).$$

For the performance measure we can show the following LMI:

$$\begin{bmatrix} -\frac{\alpha}{\varepsilon_P} & x^\top(0) & x^\top(0) \\ x(0) & -\mathbb{I} & 0 \\ x(0) & 0 & -(X^\top(r_0)+X(r_0)-\mathbb{I}) \end{bmatrix} \leq 0.$$

The results of this development are summarized by the following theorem.

Theorem 9.2.3 *Let γ and ε_P be given positive scalars. If there exist a set of non-singular matrices $X = (X(1),\cdots,X(N))$, $X(i) \in \mathbb{R}^{n\times n}$, and a set of matrices $Y = (Y(1),\cdots,Y(N))$, $Y(i) \in \mathbb{R}^{m\times n}$, such that the following optimization problem is feasible*

$$\left\{\begin{array}{l} \min \alpha \\ s.t.: \\ \begin{bmatrix} J_1(X) & X^\top(i)C_y^\top(i)+Y^\top(i)D_y^\top(i) & S_i(X) & S_i(X) \\ C_y(i)X(i)+D_y(i)Y(i) & -\mathbb{I} & 0 & \\ S_i^\top(X) & 0 & -\mathbb{I} & 0 \\ S_i^\top(X) & 0 & 0 & -X_i(X) \end{bmatrix} < 0 \\ \begin{bmatrix} J_2(X) & X^\top(i)C_z^\top(i)+Y^\top(i)D_z^\top(i) & B_w(i)X(i) & S_i(X) & S_i(X) \\ C_z(i)X(i)+D_z(i)Y(i) & -\mathbb{I} & 0 & 0 & 0 \\ X^\top(i)B_w^\top(i) & 0 & -\gamma^2\mathbb{I} & 0 & 0 \\ S_i^\top(X) & 0 & 0 & -\mathbb{I} & 0 \\ S_i^\top(X) & 0 & 0 & 0 & -X_i(X) \end{bmatrix} < 0 \\ \begin{bmatrix} -\frac{\alpha}{\varepsilon_P} & x^\top(0) & x^\top(0) \\ x(0) & -\mathbb{I} & 0 \\ x(0) & 0 & -(X^\top(r_0)+X(r_0)-\mathbb{I}) \end{bmatrix} \leq 0 \end{array}\right.$$

where

$$J_1(X) = A(i)X(i) + X^\top(i)A^\top(i) + B(i)Y(i) + Y^\top(i)B^\top(i) + \lambda_{ii}X^\top(i)E^\top(i),$$

$$J_2(X) = A(i)X(i) + X^\top(i)A^\top(i) + B(i)Y(i) + Y^\top(i)B^\top(i) + \lambda_{ii}X^\top(i)E^\top(i),$$

$$S_i(X) = \left[\sqrt{\varepsilon_P \lambda_{i1}}X^\top(i), \cdots, \sqrt{\varepsilon_P \lambda_{ii-1}}X^\top(i), \sqrt{\varepsilon_P \lambda_{ii+1}}X^\top(i), \cdots, \sqrt{\varepsilon_P \lambda_{iN}}X^\top(i) \right],$$

$$X_i(X) = diag \left[X^\top(1) + X(1) - \mathbb{I}, \cdots, X^\top(i-1) + X(i-1) - \mathbb{I}, \right.$$
$$\left. X^\top(i+1) + X(i+1) - \mathbb{I}, \cdots, X^\top(N) + X(N) - \mathbb{I} \right],$$

with the following constraints $\varepsilon_P \left[X^\top(i) + X(i) \right] \geq P^\top(i)E^\top(i) = E(i)X(i) \geq 0$, *then the controller* $u(t) = K(r_t)x(t)$ *with* $K(i) = Y(i)X^{-1}(i)$ *is a mixed* $\mathscr{H}_2/\mathscr{H}_\infty$ *controller that satisfies the desired performance.*

If we use the fact that $E^\top(i)P(i) \leq \varepsilon(i)P^\top(i)P(i)$ for $\varepsilon(i) > 0$, and letting

$$S_i(X) = \left[\sqrt{\lambda_{i1}}X^\top(i), \cdots, \sqrt{\lambda_{ii-1}}X^\top(i), \sqrt{\lambda_{ii+1}}X^\top(i), \cdots, \sqrt{\lambda_{iN}}X^\top(i) \right]$$

$$X_i(X) = diag \left[X^\top(1) + X(1) - \varepsilon(1)\mathbb{I}, \cdots, X^\top(i-1) + X(i-1) - \varepsilon(i-1)\mathbb{I}, \right.$$
$$\left. X^\top(i+1) + X(i+1) - \varepsilon(i+1)\mathbb{I}, \cdots, X^\top(N) + X(N) - \varepsilon(N)\mathbb{I} \right],$$

we get the results of the following theorem.

Theorem 9.2.4 *Let* γ *and* ε_P *be given positive scalars. If there exist a set of nonsingular matrices* $X = (X(1), \cdots, X(N))$, $X(i) \in \mathbb{R}^{n \times n}$, *and a set of matrices* $Y = (Y(1), \cdots, Y(N))$, $Y(i) \in \mathbb{R}^{m \times n}$ *such that the following optimization problem is feasible*

$$\begin{cases} \min \alpha \\ s.t. : \\ \begin{bmatrix} J_1(X) & \star & \star \\ C_y(i)X(i) + D_y(i)Y(i) & -\mathbb{I} & \star \\ S_i^\top(X) & 0 & -X_i(X) \end{bmatrix} < 0 \\ \begin{bmatrix} J_2(X) & \star & \star & \star \\ C_z(i)X(i) + D_z(i)Y(i) & -\mathbb{I} & \star & \star \\ X^\top(i)B_w^\top(i) & 0 & -\gamma^2\mathbb{I} & \star \\ S_i^\top(X) & 0 & 0 & -X_i(X) \end{bmatrix} < 0 \\ \begin{bmatrix} -\alpha & x^\top(0) \\ x(0) & -(X^\top(r_0) + X(r_0) - \varepsilon(r_0)\mathbb{I}) \end{bmatrix} \leq 0, \end{cases}$$

where

$$J_1(X) = A(i)X(i) + X^\top(i)A^\top(i) + B(i)Y(i) + Y^\top(i)B^\top(i) + \lambda_{ii}X^\top(i)E^\top(i)$$

$$J_2(X) = A(i)X(i) + X^\top(i)A^\top(i) + B(i)Y(i) + Y^\top(i)B^\top(i) + \lambda_{ii}X^\top(i)E^\top(i)$$

$$S_i(X) = \left[\sqrt{\lambda_{i1}}X^\top(i), \cdots, \sqrt{\lambda_{ii-1}}X^\top(i), \sqrt{\lambda_{ii+1}}X^\top(i), \cdots, \sqrt{\lambda_{iN}}X^\top(i) \right]$$

$$X_i(X) = diag \left[X^\top(1) + X(1) - \varepsilon(1)\mathbb{I}, \cdots, X^\top(i-1) + X(i-1) - \varepsilon(i-1)\mathbb{I}, \right.$$
$$\left. X^\top(i+1) + X(i+1) - \varepsilon(i+1)\mathbb{I}, \cdots, X^\top(N) + X(N) - \varepsilon(N)\mathbb{I} \right],$$

with the following constraints $\varepsilon(i)\mathbb{I} \geq P^\top(i)E^\top(i) = E(i)X(i) \geq 0$, *then the controller* $u(t) = K(r_t)x(t)$ *with* $K(i) = Y(i)X^{-1}(i)$ *is a mixed* $\mathcal{H}_2/\mathcal{H}_\infty$ *controller that satisfies the desired performance.*

If we use the fact that $E^\top(i)P(i) \leq \frac{1}{4}\varepsilon^{-1}(i)\mathbb{I} + \varepsilon(i)E^\top(i)P(i)P^\top(i)E(i)$ for $\varepsilon(i) > 0$, and letting

$$Y(i) = K(i)X(i)$$

$$S_i(X) = \left[\sqrt{\lambda_{i1}}X^\top(i)E^\top(1), \cdots, \sqrt{\lambda_{ii-1}}X^\top(i)E^\top(i-1), \right.$$
$$\left. \sqrt{\lambda_{ii+1}}X^\top(i)E^\top(i+1), \cdots, \sqrt{\lambda_{iN}}X^\top(i)E^\top(N) \right],$$

$$\mathcal{X}_i(X) = \mathrm{diag}\left[X^\top(1) + X(1) - \varepsilon(1)\mathbb{I}, \cdots, X^\top(i-1) + X(i-1) - \varepsilon(i-1)\mathbb{I}, \right.$$
$$\left. X^\top(i+1) + X(i+1) - \varepsilon(i+1)\mathbb{I}, \cdots, X^\top(N) + X(N) - \varepsilon(N)\mathbb{I} \right],$$

$$\mathcal{Z}_i(X) = \left[\sqrt{\lambda_{i1}}X^\top(i), \cdots, \sqrt{\lambda_{ii-1}}X^\top(i), \sqrt{\lambda_{ii+1}}X^\top(i), \cdots, \sqrt{\lambda_{iN}}X^\top(i) \right],$$

$$\mathcal{X}_i(\varepsilon) = \mathrm{diag}\left[4\varepsilon(1)\mathbb{I}, \cdots, 4\varepsilon(i-1)\mathbb{I}, 4\varepsilon(i+1)\mathbb{I}, \cdots, 4\varepsilon(N)\mathbb{I} \right],$$

we get the results of the following theorem.

Theorem 9.2.5 *Let* γ *and* ε_P *be given positive scalars. If there exists a set of nonsingular matrces* $X = (X(1), \cdots, X(N))$, $X(i) \in \mathbb{R}^{n \times n}$, *and a set of matrices* $Y = (Y(1), \cdots, Y(N))$, $Y(i) \in \mathbb{R}^{m \times n}$, *such that the following optimization problem is feasible*

$$\begin{cases} \min \alpha \\ s.t. : \\ \begin{bmatrix} J_1(X) & \star & \star \\ C_y(i)X(i) + D_y(i)Y(i) & -\mathbb{I} & \star \\ S_i^\top(X) & 0 & -\mathcal{X}_i(X) \end{bmatrix} < 0 \\ \begin{bmatrix} J_2(X) & \star & \star & \star & \star \\ C_z(i)X(i) + D_z(i)Y(i) & -\mathbb{I} & \star & \star & \star \\ X^\top(i)B_w^\top(i) & 0 & -\gamma^2\mathbb{I} & \star & \star \\ \mathcal{Z}_i^\top(X) & 0 & 0 & -\mathcal{X}_i(\varepsilon) & \star \\ S_i^\top(X) & 0 & 0 & 0 & -\mathcal{X}_i(X) \end{bmatrix} < 0 \\ \begin{bmatrix} -\alpha & x^\top(0) & x^\top(0)E^\top(r_0) \\ x(0) & -4\varepsilon(r_0)\mathbb{I} & 0 \\ E(r_0)x(0) & 0 & -(X^\top(r_0) + X(r_0) - \varepsilon(r_0)\mathbb{I}) \end{bmatrix} \leq 0, \end{cases}$$

where

$$J_1(X) = A(i)X(i) + X^\top(i)A^\top(i) + B(i)Y(i) + Y^\top(i)B^\top(i) + \lambda_{ii}X^\top(i)E^\top(i)$$

$$J_2(X) = A(i)X(i) + X^\top(i)A^\top(i) + B(i)Y(i) + Y^\top(i)B^\top(i) + \lambda_{ii}X^\top(i)E^\top(i)$$

$$S_i(X) = \left[\sqrt{\lambda_{i1}}X^\top(i)E^\top(1), \cdots, \sqrt{\lambda_{ii-1}}X^\top(i)E^\top(i-1), \right.$$
$$\left. \sqrt{\lambda_{ii+1}}X^\top(i)E^\top(i+1), \cdots, \sqrt{\lambda_{iN}}X^\top(i)E^\top(N) \right]$$

$$\mathcal{X}_i(X) = diag\left[X^\top(1) + X(1) - \varepsilon(1)\mathbb{I}, \cdots, X^\top(\iota - 1) + X(i - 1) - \varepsilon(i - 1)\mathbb{I},\right.$$
$$\left. X^\top(i + 1) + X(i + 1) - \varepsilon(i + 1)\mathbb{I}, \cdots, X^\top(N) + X(N) - \varepsilon(N)\mathbb{I}\right]$$
$$\mathcal{Z}_i(X) = \left[\sqrt{\lambda_{i1}}X^\top(i), \cdots, \sqrt{\lambda_{ii-1}}X^\top(i), \sqrt{\lambda_{ii+1}}X^\top(i), \cdots, \sqrt{\lambda_{iN}}X^\top(i)\right]$$
$$\mathcal{X}_i(\varepsilon) = diag\left[4\varepsilon(1)\mathbb{I}, \cdots, 4\varepsilon(i - 1)\mathbb{I}, 4\varepsilon(i + 1)\mathbb{I}, \cdots, 4\varepsilon(N)\mathbb{I}\right],$$

with the following constraints $P^\top(i)E^\top(i) = E(i)X(i) \geq 0$, then the controller $u(t) = K(r_t)x(t)$ with $K(i) = Y(i)X^{-1}(i)$ is a mixed $\mathscr{H}_2/\mathscr{H}_\infty$ controller that satisfies the desired performance.

9.3 Mixed $\mathscr{H}_2/\mathscr{H}_\infty$ Control: Uncertain Case

Let us now consider the uncertain system. The closed-loop dynamics are given by:

$$\begin{cases} E(r_t)\dot{x}(t) = [A(r_t) + D_A(r_t)F_A(r_t)E_A(r_t) \\ \qquad + B(r_t)K(r_t) + D_B(r_t)F_B(r_t)E_B(r_t)K(r_t)]\, x(t) + B_w(r_t)w(t) \\ y(t) = \left[C_y(r_t) + D_{C_y}(r_t)F_{C_y}(r_t)E_{C_y}(r_t) \right. \\ \qquad \left. + D_y(r_t)K(r_t) + D_{D_y}(r_t)F_{D_y}(r_t)E_{D_y}(r_t)K(r_t)\right] x(t) \\ z(t) = \left[C_z(r_t) + D_{C_z}(r_t)F_{C_z}(r_t)E_{C_z}(r_t) \right. \\ \qquad \left. + D_z(r_t)K(r_t) + D_{D_z}(r_t)F_{D_z}(r_t)E_{D_z}(r_t)K(r_t)\right] x(t). \end{cases}$$

Notice that

$$\begin{bmatrix} D_A(i)F_A(i)E_A(i)X(i)\ 0\ 0\ 0 \\ 0 \qquad\qquad 0\ 0\ 0 \\ 0 \qquad\qquad 0\ 0\ 0 \\ 0 \qquad\qquad 0\ 0\ 0 \end{bmatrix} = \begin{bmatrix} D_A(i) \\ 0 \\ 0 \\ 0 \end{bmatrix} F_A(i)\left[E_A(i)X(i)\ 0\ 0\ 0\right],$$

$$\begin{bmatrix} D_B(i)F_B(i)E_B(i)Y(i)\ 0\ 0\ 0 \\ 0 \qquad\qquad 0\ 0\ 0 \\ 0 \qquad\qquad 0\ 0\ 0 \\ 0 \qquad\qquad 0\ 0\ 0 \end{bmatrix} = \begin{bmatrix} D_B(i) \\ 0 \\ 0 \\ 0 \end{bmatrix} F_B(i)\left[E_B(i)Y(i)\ 0\ 0\ 0\right],$$

$$\begin{bmatrix} 0 \qquad\qquad 0\ 0\ 0 \\ D_{C_y}(i)F_{C_y}(i)E_{C_y}(i)X(i)\ 0\ 0\ 0 \\ 0 \qquad\qquad 0\ 0\ 0 \\ 0 \qquad\qquad 0\ 0\ 0 \end{bmatrix} = \begin{bmatrix} 0 \\ D_{C_y}(i) \\ 0 \\ 0 \end{bmatrix} F_{C_y}(i)\left[E_{C_y}(i)X(i)\ 0\ 0\ 0\right],$$

$$\begin{bmatrix} 0 \qquad\qquad 0\ 0\ 0 \\ D_{D_y}(i)F_{D_y}(i)E_{D_y}(i)Y(i)\ 0\ 0\ 0 \\ 0 \qquad\qquad 0\ 0\ 0 \\ 0 \qquad\qquad 0\ 0\ 0 \end{bmatrix} = \begin{bmatrix} 0 \\ D_{D_y}(i) \\ 0 \\ 0 \end{bmatrix} F_{D_y}(i)\left[E_{D_y}(i)Y(i)\ 0\ 0\ 0\right],$$

$$
\begin{bmatrix}
0 & 0\,0\,0 \\
D_{C_z}(i)F_{C_z}(i)E_{C_z}(i)X(i) & 0\,0\,0 \\
0 & 0\,0\,0 \\
0 & 0\,0\,0
\end{bmatrix}
=
\begin{bmatrix}
0 \\
D_{C_z}(i) \\
0 \\
0
\end{bmatrix}
F_{C_z}(i)\begin{bmatrix} E_{C_z}(i)X(i) & 0\ 0\ 0 \end{bmatrix},
$$

$$
\begin{bmatrix}
0 & 0\,0\,0 \\
D_{D_z}(i)F_{D_z}(i)E_{D_z}(i)Y(i) & 0\,0\,0 \\
0 & 0\,0\,0 \\
0 & 0\,0\,0
\end{bmatrix}
=
\begin{bmatrix}
0 \\
D_{D_z}(i) \\
0 \\
0
\end{bmatrix}
F_{D_z}(i)\begin{bmatrix} E_{D_z}(i)Y(i) & 0\ 0\ 0 \end{bmatrix},
$$

and using Lemma 1.5.1, we get

$$
\begin{bmatrix}
D_A(i)F_A(i)E_A(i)X(i) & 0\,0\,0 \\
0 & 0\,0\,0 \\
0 & 0\,0\,0 \\
0 & 0\,0\,0
\end{bmatrix}
+
\begin{bmatrix}
D_A(i)F_A(i)E_A(i)X(i) & 0\,0\,0 \\
0 & 0\,0\,0 \\
0 & 0\,0\,0 \\
0 & 0\,0\,0
\end{bmatrix}^{\top}
$$
$$
\leq
\begin{bmatrix}
\varepsilon_A(i)D_A(i)D_A^{\top}(i) + \varepsilon_A^{-1}X^{\top}(i)E_A^{\top}(i)E_A(i)X(i) & 0\,0\,0 \\
0 & 0\,0\,0 \\
0 & 0\,0\,0 \\
0 & 0\,0\,0
\end{bmatrix},
$$

$$
\begin{bmatrix}
D_B(i)F_B(i)E_B(i)Y(i) & 0\,0\,0 \\
0 & 0\,0\,0 \\
0 & 0\,0\,0 \\
0 & 0\,0\,0
\end{bmatrix}
+
\begin{bmatrix}
D_B(i)F_B(i)E_B(i)Y(i) & 0\,0\,0 \\
0 & 0\,0\,0 \\
0 & 0\,0\,0 \\
0 & 0\,0\,0
\end{bmatrix}^{\top}
$$
$$
\leq
\begin{bmatrix}
\varepsilon_B(i)D_B(i)D_B^{\top}(i) + \varepsilon_B^{-1}X^{\top}(i)E_B^{\top}(i)E_B(i)X(i) & 0\,0\,0 \\
0 & 0\,0\,0 \\
0 & 0\,0\,0 \\
0 & 0\,0\,0
\end{bmatrix},
$$

$$
\begin{bmatrix}
0 & 0\,0\,0 \\
D_{C_y}(i)F_{C_y}(i)E_{C_y}(i)X(i) & 0\,0\,0 \\
0 & 0\,0\,0 \\
0 & 0\,0\,0
\end{bmatrix}
+
\begin{bmatrix}
0 & 0\,0\,0 \\
D_{C_y}(i)F_{C_y}(i)E_{C_y}(i)X(i) & 0\,0\,0 \\
0 & 0\,0\,0 \\
0 & 0\,0\,0
\end{bmatrix}^{\top}
$$
$$
\leq
\begin{bmatrix}
\varepsilon_{C_y}^{-1}X^{\top}(i)E_{C_y}^{\top}(i)E_{C_y}(i)X(i) & 0 & 0\,0 \\
0 & \varepsilon_{C_y}(i)D_{C_y}(i)D_{C_y}^{\top}(i) & 0\,0 \\
0 & 0 & 0\,0 \\
0 & 0 & 0\,0
\end{bmatrix},
$$

$$
\begin{bmatrix}
0 & 0\,0\,0 \\
D_{D_y}(i)F_{D_y}(i)E_{D_y}(i)Y(i) & 0\,0\,0 \\
0 & 0\,0\,0 \\
0 & 0\,0\,0
\end{bmatrix}
+
\begin{bmatrix}
0 & 0\,0\,0 \\
D_{D_y}(i)F_{D_y}(i)E_{D_y}(i)Y(i) & 0\,0\,0 \\
0 & 0\,0\,0 \\
0 & 0\,0\,0
\end{bmatrix}^{\top}
$$
$$
\leq
\begin{bmatrix}
\varepsilon_{D_y}^{-1}Y^{\top}(i)E_{D_y}^{\top}(i)E_{D_y}(i)Y(i) & 0 & 0\,0 \\
0 & \varepsilon_{D_y}(i)D_{D_y}(i)D_{D_y}^{\top}(i) & 0\,0 \\
0 & 0 & 0\,0 \\
0 & 0 & 0\,0
\end{bmatrix},
$$

$$\begin{bmatrix} 0 & 0\,0\,0 \\ D_{C_z}(i)F_{C_z}(i)E_{C_z}(i)X(i) & 0\,0\,0 \\ 0 & 0\,0\,0 \\ 0 & 0\,0\,0 \end{bmatrix} + \begin{bmatrix} 0 & 0\,0\,0 \\ D_{C_z}(i)F_{C_z}(i)E_{C_z}(i)X(i) & 0\,0\,0 \\ 0 & 0\,0\,0 \\ 0 & 0\,0\,0 \end{bmatrix}^\top$$

$$\leq \begin{bmatrix} \varepsilon_{C_z}^{-1}X^\top(i)E_{C_z}^\top(i)E_{C_z}(i)X(i) & 0 & 0\,0 \\ 0 & \varepsilon_{C_z}(i)D_{C_z}(i)D_{C_z}^\top(i) & 0\,0 \\ 0 & 0 & 0\,0 \\ 0 & 0 & 0\,0 \end{bmatrix},$$

$$\begin{bmatrix} 0 & 0\,0\,0 \\ D_{D_z}(i)F_{D_z}(i)E_{D_z}(i)Y(i) & 0\,0\,0 \\ 0 & 0\,0\,0 \\ 0 & 0\,0\,0 \end{bmatrix} + \begin{bmatrix} 0 & 0\,0\,0 \\ D_{D_z}(i)F_{D_z}(i)E_{D_z}(i)Y(i) & 0\,0\,0 \\ 0 & 0\,0\,0 \\ 0 & 0\,0\,0 \end{bmatrix}^\top$$

$$\leq \begin{bmatrix} \varepsilon_{D_z}^{-1}Y^\top(i)E_{D_z}^\top(i)E_{D_z}(i)Y(i) & 0 & 0\,0 \\ 0 & \varepsilon_{D_z}(i)D_{D_z}(i)D_{D_z}^\top(i) & 0\,0 \\ 0 & 0 & 0\,0 \\ 0 & 0 & 0\,0 \end{bmatrix}$$

for $\varepsilon_A(i) > 0$, $\varepsilon_B(i) > 0$, $\varepsilon_{C_y}(i) > 0$, $\varepsilon_{D_y}(i) > 0$, $\varepsilon_{C_z}(i) > 0$ and $\varepsilon_{D_z}(i) > 0$, $\forall i \in \mathcal{S}$.
Let $\mathcal{X}_1(X)$, $\mathcal{Y}_1(X)$, $\mathcal{Z}_1(X)$, $\mathcal{X}_2(X)$, $\mathcal{Y}_2(X)$ and $\mathcal{Z}_2(X)$ be defined as follows:

$$\mathcal{X}_1(X) = \left[X^\top(i)E_A^\top(i), X^\top(i)E_B^\top(i), X^\top(i)E_{C_y}^\top(i), X^\top(i)E_{D_y}^\top(i) \right],$$

$$\mathcal{X}_2(X) = \left[X^\top(i)E_A^\top(i), X^\top(i)E_B^\top(i), X^\top(i)E_{C_z}^\top(i), X^\top(i)E_{D_z}^\top(i) \right],$$

$$\mathcal{Y}_1(X) = \mathrm{diag}\left[\varepsilon_A(i), \varepsilon_B(i), \varepsilon_{C_y}(i), \varepsilon_{D_y}(i) \right],$$

$$\mathcal{Y}_2(X) = \mathrm{diag}\left[\varepsilon_A(i), \varepsilon_B(i), \varepsilon_{C_z}(i), \varepsilon_{D_z}(i) \right],$$

$$\mathcal{Z}_1(X) = \mathbb{I} - \varepsilon_{C_y}(i)D_{C_y}(i)D_{C_y}^\top(i) - \varepsilon_{D_y}(i)D_{D_y}(i)D_{D_y}^\top(i),$$

$$\mathcal{Z}_2(X) = \mathbb{I} - \varepsilon_{C_z}(i)D_{C_z}(i)D_{C_z}^\top(i) - \varepsilon_{D_z}(i)D_{D_z}(i)D_{D_z}^\top(i).$$

Using now Theorem 9.2.3 and these relations we get the following the results.

Theorem 9.3.1 *Let γ and ε_P be given positive scalars. If there are a set of non-singular matrices $X = (X(1), \cdots, X(N))$, $X(i) \in \mathbb{R}^{n \times n}$, a set of matrices $Y = (Y(1), \cdots, Y(N))$, $Y(i) \in \mathbb{R}^{m \times n}$, and sets of positive scalars $\varepsilon_A = (\varepsilon_A(1), \cdots, \varepsilon_A(N))$, $\varepsilon_B = (\varepsilon_B(1), \cdots, \varepsilon_B(N))$, $\varepsilon_{C_y} = \left(\varepsilon_{C_y}(1), \cdots, \varepsilon_{C_y}(N)\right)$, $\varepsilon_{C_z} = \left(\varepsilon_{C_z}(1), \cdots, \varepsilon_{C_z}(N)\right)$, $\varepsilon_{D_y} = \left(\varepsilon_{D_y}(1), \cdots, \varepsilon_{D_y}(N)\right)$ and $\varepsilon_{D_z} = \left(\varepsilon_{D_z}(1), \cdots, \varepsilon_{D_z}(N)\right)$ such that the following optimization problem is feasible*

$$
\begin{cases}
\min \alpha \\
s.t. : \\
\begin{bmatrix}
J_1(X) & \star & \star & \star & \star \\
C_y(i)X(i) + D_y(i)Y(i) - \mathscr{Z}_1(X) & \star & \star & \star & \star \\
\mathscr{X}_1^\top(X) & 0 & -\mathscr{Y}_1(X) & \star & \star \\
S_i^\top(X) & 0 & 0 & -\mathbb{I} & \star \\
S_i^\top(X) & 0 & 0 & 0 & -\mathcal{X}_i(X)
\end{bmatrix} < 0 \\
\begin{bmatrix}
J_2(X) & \star & \star & \star & \star & \star \\
C_z(i)X(i) + D_z(i)Y(i) - \mathscr{Z}_2(X) & \star & \star & \star & \star & \star \\
\mathscr{X}_2^\top(X) & 0 & -\mathscr{Y}_2(X) & \star & \star & \star \\
X^\top(i)B_w^\top(i) & 0 & & -\gamma^2\mathbb{I} & \star & \star \\
S_i^\top(X) & 0 & & 0 & -\mathbb{I} & \star \\
S_i^\top(X) & 0 & & 0 & 0 & -\mathcal{X}_i(X)
\end{bmatrix} < 0 \\
\begin{bmatrix}
-\frac{\alpha}{\varepsilon_P} & x^\top(0) & x^\top(0) \\
x(0) & -\mathbb{I} & 0 \\
x(0) & 0 & -(X^\top(r_0) + X(r_0) - \mathbb{I})
\end{bmatrix} \le 0,
\end{cases}
$$

where

$$
\begin{aligned}
J_1(X) &= A(i)X(i) + X^\top(i)A^\top(i) + B(i)Y(i) + Y^\top(i)B^\top(i) + \lambda_{ii}X^\top(i)E^\top(i) \\
&\quad + \varepsilon_A(i)D_A(i)D_A^\top(i) + \varepsilon_B(i)D_B(i)D_B^\top(i), \\
J_2(X) &= A(i)X(i) + X^\top(i)A^\top(i) + B(i)Y(i) + Y^\top(i)B^\top(i) + \lambda_{ii}X^\top(i)E^\top(i) \\
&\quad + \varepsilon_A(i)D_A(i)D_A^\top(i) + \varepsilon_B(i)D_B(i)D_B^\top(i), \\
S_i(X) &= \left[\sqrt{\varepsilon_P\lambda_{i1}}X^\top(i), \cdots, \sqrt{\varepsilon_P\lambda_{ii-1}}X^\top(i), \sqrt{\varepsilon_P\lambda_{ii+1}}X^\top(i), \cdots, \sqrt{\varepsilon_P\lambda_{iN}}X^\top(i) \right], \\
\mathcal{X}_t(X) &= diag\left[X^\top(1) + X(1) - \mathbb{I}, \cdots, X^\top(i-1) + X(i-1) - \mathbb{I}, \right. \\
&\qquad \left. X^\top(i+1) + X(i+1) - \mathbb{I}, \cdots, X^\top(N) + X(N) - \mathbb{I} \right], \\
\mathscr{X}_1(X) &= \left[X^\top(i)E_A^\top(i), X^\top(i)E_B^\top(i), X^\top(i)E_{C_y}^\top(i), X^\top(i)E_{D_y}^\top(i) \right], \\
\mathscr{X}_2(X) &= \left[X^\top(i)E_A^\top(i), X^\top(i)E_B^\top(i), X^\top(i)E_{C_z}^\top(i), X^\top(i)E_{D_z}^\top(i) \right], \\
\mathscr{Y}_1(X) &= diag\left[\varepsilon_A(i), \varepsilon_B(i), \varepsilon_{C_y}(i), \varepsilon_{D_y}(i) \right], \\
\mathscr{Y}_2(X) &= diag\left[\varepsilon_A(i), \varepsilon_B(i), \varepsilon_{C_z}(i), \varepsilon_{D_z}(i) \right], \\
\mathscr{Z}_1(X) &= \mathbb{I} - \varepsilon_{C_y}(i)D_{C_y}(i)D_{C_y}^\top(i) - \varepsilon_{D_y}(i)D_{D_y}(i)D_{D_y}^\top(i), \\
\mathscr{Z}_2(X) &= \mathbb{I} - \varepsilon_{C_z}(i)D_{C_z}(i)D_{C_z}^\top(i) - \varepsilon_{D_z}(i)D_{D_z}(i)D_{D_z}^\top(i).
\end{aligned}
$$

with the following constraints $\varepsilon_P\left[X^\top(i) + X(i)\right] \geq P^\top(i)E^\top(i) = E(i)X(i) \geq 0$, *then the controller* $u(t) = K(r_t)x(t)$ *with* $K(i) = Y(i)X^{-1}(i)$ *is a mixed* $\mathcal{H}_2/\mathcal{H}_\infty$ *controller that satisfies the desired performance.*

If we use the fact that $E^\top(i)P(i) \leq \varepsilon(i)P^\top(i)P(i)$ for $\varepsilon(i) > 0$, and the results of Theorem 9.2.4, we get the following theorem.

Theorem 9.3.2 *Let* γ *and* ε_P *be given positive scalars. If there exist a set of nonsingular matrices* $X = (X(1), \cdots, X(N))$, $X(i) \in \mathbb{R}^{n \times n}$, *a set of matrices, and sets of positive scalars* $\varepsilon_A = (\varepsilon_A(1), \cdots, \varepsilon_A(N))$, $\varepsilon_B = (\varepsilon_B(1), \cdots, \varepsilon_B(N))$, $\varepsilon_{C_y} = \left(\varepsilon_{C_y}(1), \cdots, \varepsilon_{C_y}(N)\right)$, $\varepsilon_{C_z} = \left(\varepsilon_{C_z}(1), \cdots, \varepsilon_{C_z}(N)\right)$, $\varepsilon_{D_y} = \left(\varepsilon_{D_y}(1), \cdots, \varepsilon_{D_y}(N)\right)$ *and* $\varepsilon_{D_z} = \left(\varepsilon_{D_z}(1), \cdots, \varepsilon_{D_z}(N)\right)$, $Y = (Y(1), \cdots, Y(N))$, $Y(i) \in \mathbb{R}^{m \times n}$ *such that the following optimization problem is feasible*

$$
\begin{cases}
\min \alpha \\
s.t.: \\
\begin{bmatrix}
J_1(X) & \star & \star & \star \\
C_y(i)X(i) + D_y(i)Y(i) - \mathscr{Z}_1(X) & \star & \star \\
\mathscr{X}_1^\top(X) & 0 & -\mathscr{Y}_1(X) & \star \\
S_i^\top(X) & 0 & 0 & -\mathcal{X}_i(X)
\end{bmatrix} < 0 \\
\begin{bmatrix}
J_2(X) & \star & \star & \star & \star \\
C_z(i)X(i) + D_z(i)Y(i) - \mathscr{Z}_2(X) & \star & \star & \star \\
\mathscr{X}_2^\top(X) & 0 & -\mathscr{Y}_2(X) & \star & \star \\
X^\top(i)B_w^\top(i) & 0 & 0 & -\gamma^2\mathbb{I} & \star \\
S_i^\top(X) & 0 & 0 & 0 & -\mathcal{X}_i(X)
\end{bmatrix} < 0 \\
\begin{bmatrix}
-\alpha & x^\top(0) \\
x(0) - (X^\top(r_0) + X(r_0) - \varepsilon(r_0)\mathbb{I})
\end{bmatrix} \leq 0,
\end{cases}
$$

where

$$J_1(X) = A(i)X(i) + X^\top(i)A^\top(i) + B(i)Y(i) + Y^\top(i)B^\top(i) + \lambda_{ii}X^\top(i)E^\top(i)$$
$$+ \varepsilon_A(i)D_A(i)D_A^\top(i) + \varepsilon_B(i)D_B(i)D_B^\top(i),$$

$$J_2(X) = A(i)X(i) + X^\top(i)A^\top(i) + B(i)Y(i) + Y^\top(i)B^\top(i) + \lambda_{ii}X^\top(i)E^\top(i)$$
$$+ \varepsilon_A(i)D_A(i)D_A^\top(i) + \varepsilon_B(i)D_B(i)D_B^\top(i),$$

$$S_i(X) = \left[\sqrt{\lambda_{i1}}X^\top(i), \cdots, \sqrt{\lambda_{ii-1}}X^\top(i), \sqrt{\lambda_{ii+1}}X^\top(i), \cdots, \sqrt{\lambda_{iN}}X^\top(i)\right],$$

$$\mathcal{X}_i(X) = diag\left[X^\top(1) + X(1) - \varepsilon(1)\mathbb{I}, \cdots, X^\top(i-1) + X(i-1) - \varepsilon(i-1)\mathbb{I}, \right.$$
$$\left. X^\top(i+1) + X(i+1) - \varepsilon(i+1)\mathbb{I}, \cdots, X^\top(N) + X(N) - \varepsilon(N)\mathbb{I}\right],$$

$$\mathscr{X}_1(X) = \left[X^\top(i)E_A^\top(i), X^\top(i)E_B^\top(i), X^\top(i)E_{C_y}^\top(i), X^\top(i)E_{D_y}^\top(i)\right],$$

$$\mathscr{X}_2(X) = \left[X^\top(i)E_A^\top(i), X^\top(i)E_B^\top(i), X^\top(i)E_{C_z}^\top(i), X^\top(i)E_{D_z}^\top(i)\right],$$

$$\mathscr{Y}_1(X) = diag\left[\varepsilon_A(i), \varepsilon_B(i), \varepsilon_{C_y}(i), \varepsilon_{D_y}(i)\right],$$

$$\mathscr{Y}_2(X) = diag\left[\varepsilon_A(i), \varepsilon_B(i), \varepsilon_{C_z}(i), \varepsilon_{D_z}(i)\right],$$

$$\mathscr{Z}_1(X) = \mathbb{I} - \varepsilon_{C_y}(i)D_{C_y}(i)D_{C_y}^\top(i) - \varepsilon_{D_y}(i)D_{D_y}(i)D_{D_y}^\top(i),$$

$$\mathscr{Z}_2(X) = \mathbb{I} - \varepsilon_{C_z}(i)D_{C_z}(i)D_{C_z}^\top(i) - \varepsilon_{D_z}(i)D_{D_z}(i)D_{D_z}^\top(i).$$

with the following constraints $\varepsilon(i)\mathbb{I} \geq P^\top(i)E^\top(i) = E(i)X(i) \geq 0$, then the controller $u(t) = K(r_t)x(t)$ with $K(i) = Y(i)X^{-1}(i)$ is a mixed $\mathscr{H}_2/\mathscr{H}_\infty$ controller that satisfies the desired performance.

If we use the fact that $E^\top(i)P(i) \leq \frac{1}{4}\varepsilon^{-1}(i)\mathbb{I} + \varepsilon(i)E^\top(i)P(i)P^\top(i)E(i)$ for $\varepsilon(i) > 0$, and the results of Theorem 9.2.5, we get the following theorem.

Theorem 9.3.3 *Let γ and ε_P be given positive scalars. If there exist a set of non-singular matrices $X = (X(1), \cdots, X(N))$, $X(i) \in \mathbb{R}^{n\times n}$, a set of matrices $Y = (Y(1), \cdots, Y(N))$, $Y(i) \in \mathbb{R}^{m\times n}$, and sets of positive scalars $\varepsilon_A = (\varepsilon_A(1), \cdots, \varepsilon_A(N))$, $\varepsilon_B = (\varepsilon_B(1), \cdots, \varepsilon_B(N))$, $\varepsilon_{C_y} = \left(\varepsilon_{C_y}(1), \cdots, \varepsilon_{C_y}(N)\right)$, $\varepsilon_{C_z} = \left(\varepsilon_{C_z}(1), \cdots, \varepsilon_{C_z}(N)\right)$, $\varepsilon_{D_y} = \left(\varepsilon_{D_y}(1), \cdots, \varepsilon_{D_y}(N)\right)$ and $\varepsilon_{D_z} = \left(\varepsilon_{D_z}(1), \cdots, \varepsilon_{D_z}(N)\right)$ such that the following optimization problem is feasible*

$$\begin{cases} \min \alpha \\ s.t. : \\ \begin{bmatrix} J_1(X) & \star & \star & \star \\ C_y(i)X(i) + D_y(i)Y(i) - \mathscr{Z}_1(X) & \star & \star \\ \mathscr{X}_1^\top(X) & 0 & -\mathscr{Y}_1(X) & \star \\ S_i^\top(X) & 0 & 0 & -\mathcal{X}_i(X) \end{bmatrix} < 0 \\ \begin{bmatrix} J_2(X) & \star & \star & \star & \star & \star \\ C_z(i)X(i) + D_z(i)Y(i) - \mathscr{Z}_2(X) & \star & \star & \star & \star \\ \mathscr{X}_2^\top(X) & 0 & -\mathscr{Y}_2(X) & \star & \star & \star \\ X^\top(i)B_w^\top(i) & 0 & 0 & -\gamma^2\mathbb{I} & \star & \star \\ \mathcal{Z}_i^\top(X) & 0 & 0 & 0 & -\mathcal{X}_i(\varepsilon) & \star \\ S_i^\top(X) & 0 & 0 & 0 & 0 & -\mathcal{X}_i(X) \end{bmatrix} < 0 \\ \begin{bmatrix} -\alpha & x^\top(0) & x^\top(0)E^\top(r_0) \\ x(0) & -4\varepsilon(r_0)\mathbb{I} & 0 \\ E(r_0)x(0) & 0 & -\left(X^\top(r_0) + X(r_0) - \varepsilon(r_0)\mathbb{I}\right) \end{bmatrix} \leq 0, \end{cases}$$

where

$$J_1(X) = A(i)X(i) + X^\top(i)A^\top(i) + B(i)Y(i) + Y^\top(i)B^\top(i) + \lambda_{ii}X^\top(i)E^\top(i)$$
$$+ \varepsilon_A(i)D_A(i)D_A^\top(i) + \varepsilon_B(i)D_B(i)D_B^\top(i),$$

$$J_2(X) = A(i)X(i) + X^\top(i)A^\top(i) + B(i)Y(i) + Y^\top(i)B^\top(i) + \lambda_{ii}X^\top(i)E^\top(i)$$
$$+ \varepsilon_A(i)D_A(i)D_A^\top(i) + \varepsilon_B(i)D_B(i)D_B^\top(i),$$

$$S_i(X) = \left[\sqrt{\lambda_{i1}}X^\top(i)E^\top(1), \cdots, \sqrt{\lambda_{ii-1}}X^\top(i)E^\top(i-1),\right.$$
$$\left. \sqrt{\lambda_{ii+1}}X^\top(i)E^\top(i+1), \cdots, \sqrt{\lambda_{iN}}X^\top(i)E^\top(N)\right],$$

$$X_i(X) = diag\left[X^\top(1) + X(1) - \varepsilon(1)\mathbb{I}, \cdots, X^\top(i-1) + X(i-1) - \varepsilon(i-1)\mathbb{I},\right.$$
$$\left. X^\top(i+1) + X(i+1) - \varepsilon(i+1)\mathbb{I}, \cdots, X^\top(N) + X(N) - \varepsilon(N)\mathbb{I}\right],$$
$$Z_i(X) = \left[\sqrt{\lambda_{i1}}X^\top(i), \cdots, \sqrt{\lambda_{ii-1}}X^\top(i), \sqrt{\lambda_{ii+1}}X^\top(i), \cdots, \sqrt{\lambda_{iN}}X^\top(i)\right]$$
$$X_i(\varepsilon) = diag\left[4\varepsilon(1)\mathbb{I}, \cdots, 4\varepsilon(i-1)\mathbb{I}, 4\varepsilon(i+1)\mathbb{I}, \cdots, 4\varepsilon(N)\mathbb{I}\right],$$
$$\mathcal{X}_1(X) = \left[X^\top(i)E_A^\top(i), X^\top(i)E_B^\top(i), X^\top(i)E_{C_y}^\top(i), X^\top(i)E_{D_y}^\top(i)\right],$$
$$\mathcal{X}_2(X) = \left[X^\top(i)E_A^\top(i), X^\top(i)E_B^\top(i), X^\top(i)E_{C_z}^\top(i), X^\top(i)E_{D_z}^\top(i)\right],$$
$$\mathcal{Y}_1(X) = diag\left[\varepsilon_A(i), \varepsilon_B(i), \varepsilon_{C_y}(i), \varepsilon_{D_y}(i)\right],$$
$$\mathcal{Y}_2(X) = diag\left[\varepsilon_A(i), \varepsilon_B(i), \varepsilon_{C_z}(i), \varepsilon_{D_z}(i)\right],$$
$$\mathcal{Z}_1(X) = \mathbb{I} - \varepsilon_{C_y}(i)D_{C_y}(i)D_{C_y}^\top(i) - \varepsilon_{D_y}(i)D_{D_y}(i)D_{D_y}^\top(i),$$
$$\mathcal{Z}_2(X) = \mathbb{I} - \varepsilon_{C_z}(i)D_{C_z}(i)D_{C_z}^\top(i) - \varepsilon_{D_z}(i)D_{D_z}(i)D_{D_z}^\top(i).$$

with the following constraints $P^\top(i)E^\top(i) = E(i)X(i) \geq 0$, then the controller $u(t) = K(r_t)x(t)$ with $K(i) = Y(i)X^{-1}(i)$ is a mixed $\mathcal{H}_2/\mathcal{H}_\infty$ controller that satisfies the desired performance.

9.4 Numerical Example

To illustrate the effectiveness of the proposed results, let us consider a dynamical singular system with random abrupt changes with two modes with the following data:

- mode # 1:

$$A(1) = \begin{bmatrix} -1.0 & 0.0 & 1.0 \\ 0.0 & 0.0 & 1.0 \\ 0.0 & -1.0 & -1.0 \end{bmatrix},$$

$$B(1) = \begin{bmatrix} 0.3 & 0.0 \\ 0.0 & 0.1 \\ 0.2 & 1.0 \end{bmatrix},$$

$$C_y(1) = \begin{bmatrix} 1 & 0 & 0 \\ 0 & 1 & 0 \end{bmatrix}$$

$$D_y(1) = \begin{bmatrix} 1 & 0 \\ 0 & 1 \end{bmatrix}$$

$$D_A(1) = \begin{bmatrix} 0.1 \\ 0.2 \\ 0.3 \end{bmatrix}, E_A(1) = \begin{bmatrix} 0.3 & 0.2 & 0.1 \end{bmatrix},$$

$$D_B(1) = \begin{bmatrix} 0.2 \\ 0.3 \\ 0.1 \end{bmatrix}, E_B(1) = \begin{bmatrix} 0.2 & 0.1 \end{bmatrix},$$

$$R_1(1) = \begin{bmatrix} 1.0 & 0.0 & 0.0 \\ 0.0 & 1.0 & 0.0 \\ 0.0 & 0.0 & 1.0 \end{bmatrix},$$

$$R_2(1) = \begin{bmatrix} 2.0 & 0.0 \\ 0.0 & 4.0 \end{bmatrix};$$

- mode # 2:

$$A(2) = \begin{bmatrix} 1.0 & 0.0 & 1.0 \\ 0.0 & 0.0 & 1.0 \\ 0.0 & 1.0 & -1.0 \end{bmatrix},$$

$$B(2) = \begin{bmatrix} 0.1 & 0.0 \\ 0.0 & 0.0 \\ 0.1 & 0.2 \end{bmatrix},$$

$$C_y(2) = \begin{bmatrix} 1.1 & 0 & 0 \\ 0 & 1.1 & 0 \end{bmatrix}$$

$$D_y(2) = \begin{bmatrix} 1 & 0 \\ 0 & 1 \end{bmatrix}$$

$$D_A(2) = \begin{bmatrix} 0.2 \\ 0.1 \\ 0.3 \end{bmatrix}, E_A(2) = \begin{bmatrix} 0.3 & 0.2 & 0.1 \end{bmatrix},$$

$$D_B(2) = \begin{bmatrix} 0.3 \\ 0.1 \\ 0.2 \end{bmatrix}, E_B(2) = \begin{bmatrix} 0.1 & 0.2 \end{bmatrix},$$

$$R_1(2) = \begin{bmatrix} 1.0 & 0.0 & 0.0 \\ 0.0 & 1.0 & 0.0 \\ 0.0 & 0.0 & 1.0 \end{bmatrix}, R_2(2) = \begin{bmatrix} 4.0 & 0.0 \\ 0.0 & 2.0 \end{bmatrix}.$$

Solving the LMIs of Theorem 9.2.5, we get:

$$X(1) = \begin{bmatrix} 0.0168 & 0.0048 & 0.0 \\ 0.0048 & 0.0145 & 0.0 \\ -0.0065 & -0.0063 & 0.0433 \end{bmatrix},$$

$$X(2) = \begin{bmatrix} 0.0109 & 0.0086 & 0.0 \\ 0.0086 & 0.0191 & 0.0 \\ -0.0192 & -0.0205 & 0.0313 \end{bmatrix},$$

$$Y(1) = \begin{bmatrix} -0.1495 & 0.0285 & -0.0408 \\ 0.0169 & -0.0227 & -0.1114 \end{bmatrix},$$

$$Y(2) = \begin{bmatrix} -0.0625 & -0.0161 & -0.0503 \\ -0.0716 & -0.1239 & -0.0669 \end{bmatrix},$$

which give the following gains:

$$K(1) = \begin{bmatrix} -10.7268 & 5.1338 & -0.9415 \\ 0.8621 & -2.9626 & -2.5718 \end{bmatrix},$$

$$K(2) = \begin{bmatrix} -10.0781 & 1.9540 & -1.6034 \\ -5.3069 & -6.3866 & -2.1361 \end{bmatrix}.$$

9.5 Notes

This chapter dealt with the guaranteed cost control problem of the singular class of systems with random abrupt changes. This approach of stabilization is one of the most popular one since it uses the cost function that usually in the jump linear quadratic regulator. A state feedback controller that assures that closed-loop state equation either for the nominal system or the uncertain is piecewise regular, impulse-free and stochastically stable is designed in the LMI setting. The conditions we developed in this chapter are tractable using commercial optimization tools. The content of this chapter is mainly based on the work of the author and his coauthors [19].

10

Computation Tools

Most of the conditions we presented in this book are in the LMI setting with some equality conditions. Both conditions are linear and therefore the existing tools like the LMI-toolbox of MATLAB[1] may be used to solve them after some transformation. The LMI-toolbox doesn't handle the equality conditions. We can still transform the problem with these equality constraints to an equivalent optimization problem that can be handled by the LMI-toolbox of MATLAB.

Another tool that may be appropriate for our case is the combination YALMIP[2] and SeDuMi[3]. YALMIP is a MATLAB toolbox for rapid prototyping of optimization problems. The package was initially designed to solve semi-definite programming, but the latest release extends this scope significantly. YALMIP can now be used for solving problems among them we quote:

- convex linear, quadratic, second order cone and semi-definite programming
- non-convex quadratic and semi-definite programming (local and global)
- mixed integer conic programming
- multiparametric programming
- geometric programming

SeDuMi is a great piece of software for optimization over symmetric cones. It was developed by J.F. Sturm. The Advanced Optimization Lab at McMaster University is now in charge of the development and maintenance of this software.

Both YALMIP and SeDuMi are developed to run under MATLAB.

10.1 Transformation

In this section we will see how to transform the feasibility problem for the class of singular systems with random abrupt changes that has equality conditions to an

---

[1] MATLAB is a product of The Mathworks, Inc., Natick, MA, USA.

[2] http://control.ee.ethz.ch/joloef/yalmip.php

[3] http://sedumi.mcmaster.ca/

equivalent one that may be solved using the LMI-toolbox of MATLAB. For this pur-
posc let us consider the stability problem. As we have seen at Chap. 2, the conditions
in this case are:

$$E^{\top}(i)P(i) = E^{\top}(i)E(i) \geq 0 \tag{10.1}$$

$$A^{\top}(i)P(i) + P^{\top}(i)A(i) + \sum_{j=1}^{N} \lambda_{ij}E^{\top}(j)P(j) < 0, \tag{10.2}$$

where $P(i)$, $i \in \mathscr{S}$ is a decision variable that we have to determine and that should
be nonsingular matrix.

The constraint $E^{\top}(i)P(i) = P^{\top}(i)E(i)$ may be difficult to solve with some commercial LMI toolboxes like LMI-Toolbox of MATLAB. To overcome this we can
use the following LMI condition that may approximate this constraint:

$$\left[E^{\top}(i)P(i) - P^{\top}(i)E(i)\right]^{\top}\left[E^{\top}(i)P(i) - P^{\top}(i)E(i)\right] \leq \beta\mathbb{I},$$

that gives the following LMI:

$$\begin{bmatrix} -\beta\mathbb{I} & [E^{\top}(i)P(i) - P^{\top}(i)E(i)]^{\top} \\ [E^{\top}(i)P(i) - P^{\top}(i)E(i)] & -\mathbb{I} \end{bmatrix} \leq 0. \tag{10.3}$$

Therefore the solution of our problem is brought to the minimization of β subject
to the LMIs (10.2)-(10.3) and $\beta \geq 0$ that we should minimize.

As it is shown now the conditions are in the LMI setting and therefore the LMI-
toolbox of MATLAB can be used for this purpose.

10.2 YALMIP and SeDuMi

As a second alternate we can recourse to YALMIP and SeDuMi. YALMIP is a powerful tool that offers an easy way to enter the conditions to be solved. SeDuMi is
a powerful solver that can handle both inequalities and equalities. Both of these tools
are distributed free of charge but they require MATLAB.

To sow how these tools can be used we consider a two modes system with the
following data:

- mode 1:

$$A(1) = \begin{bmatrix} 0 & 1 & 0 & 0 \\ 0 & 0 & 1 & 0 \\ 0 & 0 & 0 & 1 \\ -1 & -2 & -3 & -4 \end{bmatrix},$$

- mode 2:

$$A(1) = \begin{bmatrix} 0 & 1 & 0 & 0 \\ 0 & 0 & 1 & 0 \\ 0 & 0 & 0 & 1 \\ -2 & -3 & -4 & -1 \end{bmatrix},$$

Let us assume that the switching between these two modes is described by a continuous-time Markov process with the following data:

$$\Lambda = \begin{bmatrix} -1 & 1 \\ 1.1 & -1.1 \end{bmatrix},$$

To check the stability of such system, we can use the appropriate results for that and then write a program for this purpose. The readers who are familiar with MATLAB programming are invited to read some online tutorial that can be found on the Internet. There are also good book that were written for this purpose. The following one was written to run under MATLAB and uses YALMIP and SeDuMi to solve the feasibility problem of the stability for the system we are considering:

```
%
% Singular systems: Stability problem
%
clear all;
yalmip('clear')

global n A L

lambda=[-1 1; 1.1 -1.1];

E1=[1 0 0 0; 0 1 0 0; 0 0 0 0; 0 0 0 0];
E2=[1 0 0 0; 0 1 0 0; 0 0 0 0; 0 0 0 0];

% mode 1
A1=[0 1 0 0; 0 0 1 0; 0 0 0 1; -1 -2 -3 -4];

% mode 2
A2=[0 1 0 0; 0 0 1 0; 0 0 0 1; -2 -3 -4 -1];

m=rank(E1);
n=size(A1,1);

%%%%%%%%%%%%%%%%%
%  Stability %
%%%%%%%%%%%%%%%%%

%
%  Theorem 2.2.1
%

% Matrices to extract the components of the matrix Pi that
% has the following form:
%          | F1i 0 |
%    P = | ----- |
%          |  F3i   |
```

```
%
% Pi = a*F1i*a'+b*F3i

a=[eye(m); zeros(n-m,m)];
b=[zeros(m,n-m); eye(n-m)];

%
%    Standard conditions
%

% mode 1
F11=sdpvar(m,m,'symmetric');
F31=sdpvar(n-m,n,'full');

% mode 2
F12=sdpvar(m,m,'symmetric');
F32=sdpvar(n-m,n,'full');

% LMI 1-2
F=set(F11>0)
F=F+set(F12>0)

P1=a*F11*a'+b*F31;
P2=a*F12*a'+b*F32;

% LMI 3-5
JP1=A1'*P1+P1'*A1+lambda(1,1)*E1'*P1+lambda(1,2)*E2'*P2;
F=F+set(JP1 < 0)
F=F+set(E1'*P1 >= 0)

% LMI 6-8
JP2=A2'*P2+P2'*A2+lambda(2,2)*E2'*P2+lambda(2,1)*E1'*P1;
F=F+set(JP2 < 0)
F=F+set(E2'*P2 >= 0)

Sol=solvesdp(F)                    % Solve the problem

F11=double(F11)                    % Extract the decision variable F11
F12=double(F12)

F31=double(F31)
F32=double(F32)

P1=a*F11*a'+b*F31                  % Compute the decision variable P1
P2=a*F12*a'+b*F32

checkset(F)                        % Check if the problem is feasible
```

Running this program gives the following solution:

- mode 1:

$$P(1) = \begin{bmatrix} 0.9587 & 0.4762 & 0.0 & 0.0 \\ 0.4762 & 1.5184 & 0.0 & 0.0 \\ 0.8901 & 2.0083 & 0.9492 & 0.0000 \\ 0.1913 & 0.4396 & 0.1435 & 0.1250 \end{bmatrix};$$

- mode 2:

$$P(1) = \begin{bmatrix} 0.9331 & 0.4054 & 0.0 & 0.0 \\ 0.4054 & 1.4226 & 0.0 & 0.0 \\ 1.1243 & 1.8034 & 2.0987 & -0.0000 \\ 0.1243 & 0.3034 & 0.0987 & 0.5000 \end{bmatrix}.$$

Notice that the two matrices $P(1)$ and $P(2)$ are both nonsingular, that can be checked by computing the eigenvalues of these matrices or by computing directly the inverse of the two matrices. We have also $E^{\top}(1)P(1) \geq 0$ and $E^{\top}(2)P(2) \geq 0$ since their eigenvalues are respectively:

mode	eigenvalues
1	0 0 0.6862 1.7908
2	0 0 0.7043 1.6514

Table 10.1. Eigenvalues of the matrix $E^{\top}(i)P(i)$

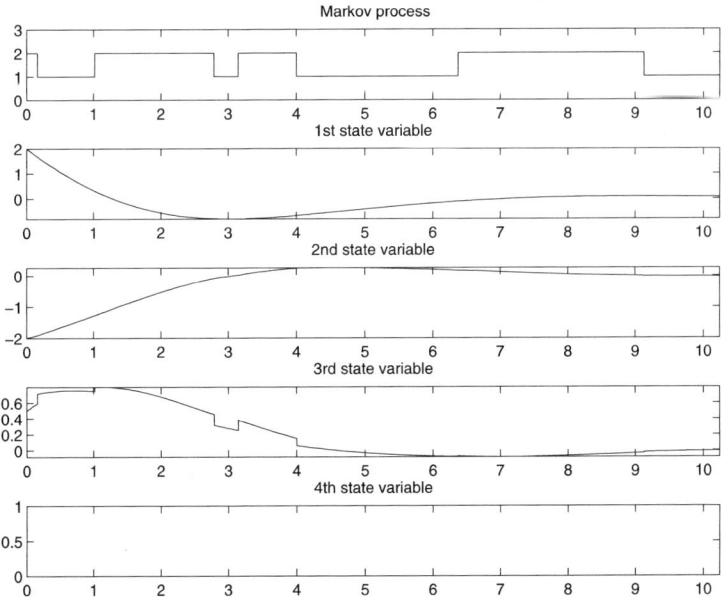

Fig. 10.1. The behavior of the system states as a function of time t

Based on the results of the appropriate theorem, we conclude that the singular system with random abrupt changes is piecewise regular, impulse-free and stochastically stable. Simulation results of this system as illustrated at Fig. 10.1 confirms the results since all the state converges to zero once time goes infinity. The initial conditions vector used for this simulation is:

$$x_0 = \begin{bmatrix} 2 \\ -2 \\ 3 \\ 0 \end{bmatrix}.$$

As it is shown in the figure, the fast states have discontinuities when the mode jumps from one value to another.

Programs for stabilization (state feedback, static output feedback, $\mathcal{H}_\infty$ state feedback, etc.) can be rewritten following the same idea of this one. Readers interested by these programs can contact the author at el-kebir.boukas@polymtl.ca.

References

1. J. Raouf, and E. K. Boukas, "Stabilization of Discontnuous Singular Systems with Markovian Switching and saturating inputs, " in *Proc. of the 33rd IEEE America Control Conference*, pp. 2245–2250, July 2007.
2. E. K. Boukas, "On Stability and Stabilizability of Singular Stochastic Systems with Delays", vol. 127, pp. 249–262, 2005.
3. A. B. H. Adamou-Mitiche, L. Mitiche, and V. Sima, "Model reduction for descriptor systems," in *First Int. Symp. Control, Communications and Signal Processing*, March 2004, pp. 827–830.
4. A. Ailon, "An approach for pole assignment in singular systems," *IEEE Trans. Automatic Control*, vol. 34, no. 8, pp. 889–893, August 1989.
5. D. Arzelier, J. Bernussou, and G. Garcia, "A convex problem formulation to solve the linear quadratic problem for singular systems," in *Proc. 32nd IEEE Conf. Decision and Control*, vol. 4, December 1993, pp. 3300–3303.
6. V. Bajic, "Algebraic conditions for stability of linear singular systems," in *IEEE Int. Symp. Circuits and Systems*, vol. 2, June 1991, pp. 1089–1092.
7. A. Bassong-Onana, M. Zasadzinski, and M. Darouach, "Application of Kalman filtering techniques in singular systems," in *Proc. 32nd IEEE Conf. Decision and Control*, vol. 4, December 1993, pp. 3308–3310.
8. D. Bender and A. L. Laub, "The linear quadratic regulator for descriptor systems," *IEEE Trans. Automatic Control*, vol. 32, no. 3, pp. 672–688, March 1987.
9. D. Bernstein, "The optimal projection equations for static and dynamic output feedback: The singular case," *IEEE Trans. Automatic Control*, vol. 32, no. 12, pp. 1139–1143, December 1987.
10. D. Bernstein and W. M. Haddad, "Lqg control with $\mathcal{H}_\infty$ performance bound: A Riccati approach," *IEEE Trans. Automatic Control*, vol. 34, no. 3, pp. 295–305, March 1989.
11. S. Bittanti, P. Colaneri, and M. Mongiovi, "From singular to nonsingular filtering of periodic systems: Filling the gap with the spectral interactor matrix," *IEEE Trans. Automatic Control*, vol. 44, no. 1, pp. 222–227, January 1999.
12. E. K. Boukas, *Systèmes Asservis.* Montréal: Éditions de l'École Polytechnique de Montréal, 1995.
13. E. K. Boukas, "Stochastic mixed $\mathcal{H}_2/\mathcal{H}_\infty$ control of time-varying delay systems," *Nonlinear Dynam. Syst. Theor.*, vol. 3, no. 2, pp. 119–137, 2003.
14. E. K. Boukas, *Stochastic Switching Systems: Analysis and Design.* Boston: Birkhauser, 2005.

15. E. K. Boukas, "Manufacturing systems: LMI approach," *IEEE Trans. Automatic Control*, vol. 51, no. 6, pp. 1014–1018, 2006.

16. E. K. Boukas, "On guaranteed cost control of singular systems with random abrupt changes," *Submitted for publications*, 2007.

17. E. K. Boukas, "On $\mathcal{H}_\infty$ filtering of singular systems with random abrupt changes," *Submitted for publication*, 2007.

18. E. K. Boukas, "On $\mathcal{H}_\infty$ state feedback stabilization of singular systems with random abrupt changes," *Submitted for publication*, 2007.

19. E. K. Boukas, "On mixed $\mathcal{H}_2/\mathcal{H}_\infty$ control of singular systems with random abrupt changes," *Submitted for publication*, 2007.

20. E. K. Boukas, "On observer-based stabilization of singular systems with random abrupt changes," *Submitted for publication*, 2007.

21. E. K. Boukas, "On robust stability of singular systems with random abrupt changes," *Submitted for publication*, 2007.

22. E. K. Boukas, "On state feedback stabilization of singular systems with random abrupt changes," *Submitted for publication*, 2007.

23. E. K. Boukas, "On static output feedback stabilization of singular systems with random abrupt changes," *Submitted for publication*, 2007.

24. E. K. Boukas and Z. K. Liu, "Robust $\mathcal{H}_\infty$ control of discrete-time Markovian jump linear systems with mode-dependent time-delay," *IEEE Trans. Automatic Control*, vol. 46, pp. 1918–1924, 2001.

25. E. K. Boukas and Z. K. Liu, "Robust stability and stability of Markov jump linear uncertain systems with mode-dependent time delays," *J. Opt. Theor. Appl.*, vol. 209, pp. 587–600, 2001.

26. E. K. Boukas and Z. K. Liu, "Suboptimal design of regulators for jump linear system with time-multiplied quadratic cost," *IEEE Trans. Automatic Control*, vol. 46, no. 1, pp. 131–136, 2001.

27. E. K. Boukas and Z. K. Liu, *Deterministic and Stochastic Time-Delay Systems*. Boston: Birkhauser, 2002.

28. E. K. Boukas and Z. K. Liu, "Delay-dependent stabilization of singularly perturbed jump linear systems," *Int. J. Control*, vol. 77, no. 3, pp. 310–319, 2004.

29. E. K. Boukas and H. Yang, "Exponential stability of stochastic systems with Markovian jumping parameters," *Automatica*, vol. 35, pp. 1437–1441, 1999.

30. M. Boutayeb, M. Darouach, H. Rafaralahy, and G. Krzakala, "Asymptotic observers for a class of nonlinear singular systems," in *Proc. American Control Conf.*, vol. 2, June 1994, pp. 1440–1441.

31. Y. Y. Cao and J. Lam, "Robust $\mathcal{H}_\infty$ control of uncertain Markovian jump systems with time-delay," *IEEE Trans. Automatic Control*, vol. 45, no. 1, 2000.

32. C. Chen and Y. Liu, "Lyapunov stability analysis of linear singular dynamical systems," in *IEEE Int. Conf. Intelligent Processing Systems, ICIPS '97*, vol. 1, October 1997, pp. 635–639.

33. S.-J. Chen and J.-L. Lin, "Robust stability of discrete time-delay uncertain singular systems," *IEE Proc. Control Theory and Applications*, vol. 151, no. 1, pp. 45–52, January 2004.

34. S. Chen and J.-H. Chou, "Stability robustness of linear discrete singular time-delay systems with structured parameter uncertainties," *IEE Proc. Control Theory and Applications*, vol. 150, no. 3, pp. 295–302, May 2003.

35. S. Chen and J. Lin, "Robust stability analysis of uncertain singular systems," in *Proc. 36th SICE Annual Conference*, July 1997, pp. 1001–1006.

36. Z. Chen and J. Huang, "Solution of output regulation of singular nonlinear systems by normal output feedback," *IEEE Trans. Automatic Control*, vol. 47, no. 5, pp. 808–813, May 2002.

37. Z. Cheng, J. Yan, and K. Zhao, "The linear-quadratic optimal regulator of time-invariant discrete singular systems," in *Proc. 31st IEEE Conference on Decision and Control*, vol. 1, , December 1992, pp. 989–990.

38. D. Chu, "A case study for the open question: Disturbance decoupling problem for singular systems by output feedback," *IEEE Trans. Automatic Control*, vol. 46, no. 12, pp. 1924–1930, December 2001.

39. J. Cobb, "A unified theory of full-order and low-order observers based on singular system theory," *IEEE Transactions on Automatic Control*, vol. 39, no. 12, pp. 2497–2502, December 1994.

40. L. Dai, "An $\mathscr{H}_\infty$ method for decentralized stabilization in large-scale singular systems," in *Proc. IEEE Int. Conf. Systems, Man, and Cybernetics*, vol. 1, , August 1988, pp. 722–725.

41. L. Dai, "Observers for discrete singular systems," *IEEE Trans. Automatic Control*, vol. 33, no. 2, pp. 187–191, February 1988.

42. L. Dai, "Filtering and lqg problems for discrete-time stochastic singular systems," *IEEE Trans. Automatic Control*, vol. 34, no. 10, pp. 1105–1108, October 1989.

43. L. Dai, *Singular Control Systems, Volume 118 of Lecture Notes in Control and Information Sciences*.   New York: Springer, 1989.

44. D. P. de Farias, J. C. Geromel, J. B. R. D. Val, and O. L. V. Costa, "Output feedback control of Markov jump linear systems in continuous-time," *IEEE Trans. Automatic Control*, vol. 45, no. 5, pp. 944–949, May 2000.

45. C. E. de Souza and M. D. Fragoso, "Robust $\mathscr{H}_\infty$ filtering for Markovian jump linear systems," in *Proc. 35th IEEE Conf. Decision and Control*, , Kobe, Japan, December 1996.

46. X. Dong and Q. Zhang, "Robust $\mathscr{H}_\infty$ control for singular systems with state delay and parameter uncertainty," in *Fifth World Congr. Intelligent Control and Automation*, vol. 2, June 2004, pp. 1035–1039.

47. Q. Fang, "LMI based state feedback $\mathscr{H}_2$ control of linear singular systems," in *Fifth World Congr. Intelligent Control and Automation*, vol. 2, June 2004, pp. 988–992.

48. J. Feng and A. Cheng, "Robust $\mathscr{H}_\infty$ control of uncertain singular systems with delay in state," in *The 2002 Int. Conf. Control and Automation*, June 2002, p. 224.

49. X. Feng, K. A. Loparo, Y. Ji, and H. J. Chizeck, "Stochastic stability properties of jump linear systems," *IEEE Trans. Automatic Control*, vol. 37, pp. 38–53, 1992.

50. L. R. Fletcher, "Pole assignment and controllability subspaces in descriptor systems," *Int. J. Control*, vol. 66, pp. 677–709, 1997.

51. E. Fridman, "A Lyapunov-based approach to stability of descriptor systems with delay," in *Proc. 40th IEEE Conf. Control and Decision*, Orlando, FL, USA, December 2001, pp. 2850–2855.

52. E. Fridman and U. Shaked, "An improved stabilization method for linear time-delay systems," *IEEE Trans. Automatic on Control*, vol. 47, no. 11, pp. 1931–1937, 2002.

53. H. Gang, L. Cenfeng, and L. Xu, "Resilient guaranteed cost control to tolerate perturbations for uncertain singular systems," in *Fifth World Congr. Intelligent Control and Automation*, vol. 2, June 2004, pp. 1128–1131.

54. Z. Gao, X. Wang, J. Wang, and G. Li, "Internal properness and stability in singular decentralized control systems," in *Proc. 1997 American Control Conf.*, vol. 4, June 1997, pp. 2520–2521.

55. A. Germani, C. Manes, and P. Palumbo, "Optimal linear filtering for stochastic non-Gaussian descriptor systems," in *Proc. 40th IEEE Conf. Decision and Control*, vol. 3, December 2001, pp. 2514–2519.

56. A. Germani, C. Manes, and P. Palumbo, "Filtering of switching systems via a singular minimax approach," in *Proc. 41st IEEE Conf. Decision and Control*, vol. 3, December 2002, pp. 2600–2605.

57. A. Germani, C. Manes, and P. Palumbo, "Polynomial filtering for stochastic non-Gaussian descriptor systems," *IEEE Trans. Circuits and Systems I: Regular Papers*, vol. 51, no. 8, pp. 1561–1576, August 2004.

58. Z. H. Guan, Y. Q. Liu, and X. C. Wen, "Decentralized stabilization of singular and time-delay large-scale control systems with impulsive solutions," *IEEE Trans. Automatic Control*, vol. 40, no. 8, pp. 1437–1441, August 1995.

59. J. Ishihara and M. Terra, "On the Lyapunov theorem for singular systems," *IEEE Trans. Automatic Control*, vol. 47, no. 11, pp. 1926–1930, November 2002.

60. Y. Ji and H. J. Chizeck, "Controllability, stabilizability, and continuous-time Markovian jump linear quadratic control," *IEEE Trans. Automatic Control*, vol. 35, no. 7, pp. 777–788, 1990.

61. Y. Jinxi, C. Zhaolin, and Y. Yi, "The linear-quadratic optimal regulation for continuous time-varying singular systems," in *Proc. 34th IEEE Conf. Decision and Control*, vol. 4, December 1995, pp. 3920–3921.

62. N. A. Kablar, "Singularly impulsive or generalized impulsive dynamical systems: Lyapunov and asymptotic stability," in *Proc. 42nd IEEE Conf. Decision and Control*, vol. 1, December 2003, pp. 173–175.

63. N. A. Kablar and D. Debeukovic, "Finite-time stability of time-varying linear singular systems," in *Proc. 37th IEEE Conf. Decision and Control*, vol. 4, December 1998, pp. 3831–3836.

64. N. A. Kablar and D. Debeukovic, "Finite-time instability of time-varying linear singular systems," in *Proc. 1999 American Control Conf.*, vol. 3, June 1999, pp. 1796–1800.

65. I. Y. Kats and A. A. Martynyuk, *Stability and Stabilization of Nonlinear Systems with Random Structures*.   New York: Taylor and Francis, 2002.

66. H. Khalil, *Nonlinear Systems*.   Englewood Cliffs, NJ: Prentice-Hall, 2002.

67. P. P. Khargonekar and M. A. Rotea, "Mixed $\mathcal{H}_2/\mathcal{H}_\infty$ control: A convex optimization approach," *IEEE Trans. Automatic Control*, vol. 36, no. 7, pp. 824–837, July 1991.

68. J. Kim, J. Lee, and H. Park, "Robust $\mathcal{H}_\infty$ control of singular systems with time delays and uncertainties via LMI approach," in *Proc. American Control Conf.*, vol. 1, May 2002, pp. 620–621.

69. N. N. Krasovskii and E. A. Lidskii, "Analysis design of controller in systems with random attributes, part 2," *Automat. Rem. Contr.*, vol. 22, pp. 1141–1146.

70. N, N. Krasovskii and E. A. Lidskii, "Analysis design of controller in systems with random attributes, Part 1," *Automat. Rem. Contr.*, vol. 22, pp. 1021–1025, 1961.

71. F. Kratz, S. Bousghiri, and G. Mourot, "A finite memory observer structure for robust residual generation," in *Proc. 32nd IEEE Conf. Decision and Control*, vol. 2, December 1993, pp. 1247–1249.

72. F. Kratz, S. Bousghiri and W. Nuninger, "A finite memory observer structure of continuous descriptor systems," in *Proc. American Control Conf.*, vol. 5, June 1995, pp. 3900–3904.

73. G. A. Kurina and R. Marz, "On linear quadratic optimal control problems for time-varying descriptor systems," *SIAM J. Control Optim.*, vol. 42, no. 6, pp. 2062–2077, 2004.

74. L. Kuzmina, "Stability theory methods and mechanics singular systems," in *Proc. IEEE Int. Conf. Systems, Man, and Cybernetics*, vol. 1, October 1999, pp. 62–67.

75. W. Lan and J. Huang, "Semiglobal stabilization and output regulation of singular linear systems with input saturation," *IEEE Trans. Automatic Control*, vol. 48, no. 7, pp. 1274–1280, July 2003.

76. F. L. Lewis, "A survey of linear singular systems," *Circuits Syst. Signal Process.*, vol. 5, pp. 3–36, 1986.

77. D. J. N. Limebeer, B. D. O. Anderson, and B. Hendel, "A Nash game approach to mixed $\mathscr{H}_2/\mathscr{H}_\infty$ control," *IEEE Trans. Automatic Control*, vol. 39, no. 1, pp. 69–82, January 1994.

78. J. Lin and S. Chen, "Exact bounds for stability robustness of uncertain singular systems via LFT-based method," in *Proc. 36th IEEE Conf. Decision and Control, 1997*, vol. 5, December 1997, pp. 4896–4901.

79. J. Lin and S. Chen, "LFT approach to robust stability bounds of uncertain linear singular systems," *IEE Proc. D Control Theor. Appl.*, vol. 145, no. 2, pp. 127–134, March 1998.

80. J. Lin and S. Chen, "Robustness analysis of uncertain linear singular systems with output feedback control," *IEEE Trans. Automatic Control*, vol. 44, no. 10, pp. 1924–1929, October 1999.

81. X. Liu, "Robust stabilization of nonlinear singular systems," in *Proc. 34th IEEE Conf. Decision and Control*, vol. 3, December 1995, pp. 2375–2376.

82. X. Liu, "Input–output decoupling of linear time-varying singular systems," *IEEE Trans. Automatic Control*, vol. 44, no. 5, pp. 1016–1021, May 1999.

83. X. Liu, X. Wang, and D. Ho, "Input-output block decoupling of linear time-varying singular systems," *IEEE Trans. Automatic Control*, vol. 45, no. 2, pp. 312–318, February 2000.

84. Y. Liu, "Robust output regulation for linear singular systems subject to input saturation," in *Int. Conf. Control and Automation*, June 2002, pp. 97–97.

85. Y. Liu and Y. Li, "Stabilization of nonlinear singular systems," in *Proc. 1998 American Control Conf.*, vol. 4, June 1998, pp. 2532–2533.

86. Y. Liu and X. Xie, "On problem of stabilization by output feedback for linear singular systems with time delay: A new approach," in *Proc. IEEE Int. Conf. Industrial Technology (ICIT '96)*, December 1996, pp. 561–564.

87. G. Lu and D. W. C. Ho, "Generalized quadratic stabilization for perturbated discrete-time singular systems with delayed state," in *The Fourth Int. Conf. Control and Automation*, June 2003, pp. 56–56.

88. G. Lu, D. W. C. Ho, and L. Yeung, "Generalized quadratic stability for perturbated singular systems," in *Proc. 42nd IEEE Conf. Decision and Control*, vol. 3, December 2003, pp. 2413–2418.

89. R. Lu, H. Su, and J. Chu, "Robust $\mathscr{H}_\infty$ control for a class of uncertain Lurie singular systems with time-delays," in *Proc. 42nd IEEE Conf. Decision and Control*, vol. 6, December 2003, pp. 5585–5590.

90. R. Lu, H. Su, and J. Chu, "Robust $\mathscr{H}_\infty$ filtering for a class of uncertain Lurie time-delay singular systems," in *IEEE Int. Conf. Systems, Man and Cybernetics*, vol. 4, October 2003, pp. 3176–3181.

91. S. Ma, "Robust stabilization for a class of uncertain discrete-time singular systems with time-delays," in *Fifth World Congr. Intelligent Control and Automation*, vol. 2, June 2004, pp. 970–974.

92. S. Ma and Z. Cheng, "An LMI approach to robust stabilization for uncertain discrete-time singular systems," in *Proc. 41st IEEE Conf. Decision and Control*, vol. 1, December 2002, pp. 1090–1095.

93. S. Ma and Z. Cheng, "Mixed $\mathcal{H}_2/\mathcal{H}_\infty$ control for linear singular systems," in *Proc. 4th World Congr. Intelligent Control and Automation*, vol. 1, June 2002, pp. 283–287.

94. X. Mao, "Stability of stochastic differential equations with Markovian switching," *Stoch. Process. Appl.*, vol. 79, pp. 45–67, 1999

95. M. Mariton, *Jump Linear Systems in Automatic Control.*   New York: Marcel Dekker, 1990.

96. M. Mariton, "Control of nonlinear systems with Markovian parameter changes," *IEEE Trans. Automatic Control*, vol. 36, pp. 233–238, 1991.

97. V. Mehrmann, *The Autonomous Linear Quadratic Control Problem, Volume 163 of Lecture Notes in Control and Information Sciences.*   Berlin: Springer, 1991.

98. R. Nikoukhah, S. Campbell, and F. Delebecque, "Kalman filtering for general discrete-time linear systems," *IEEE Trans. Automatic Control*, vol. 44, no. 10, pp. 1829–1839, October 1999.

99. Z. Palmor and Y. Halevi, "On the existence of an optimal observer in singular measurement systems," *IEEE Trans. Automatic Control*, vol. 31, no. 7, pp. 683–685, July 1986.

100. P. Paraskevopoulos and F. N. Koumboulis, "Unifying approach to observers for regular and singular systems," *IEE Proc. D Control Theor. Appl.*, vol. 138, no. 6, pp. 561–572, November 1991.

101. P. Paraskevopoulos and F. N. Koumboulis, "Observers for singular systems," *IEEE Trans. Automatic Control*, vol. 37, no. 8, pp. 1211–1215, August 1992.

102. S. Sastry, *Nonlinear Systems: Analysis, Stability and Control*, New York: Springer, 1999.

103. C. Scherer, P. Gahinet, and M. Chilai, "Multiobjective output feedback control via LMI optimization," *IEEE Trans. Automatic Control*, vol. 42, no. 7, pp. 896–911, July 1997.

104. U. Shaked, "Explicit solution to the singular discrete-time stationary linear filtering problem," *IEEE Trans. Automatic Control*, vol. 30, no. 1, pp. 34–47, January 1985.

105. P. Shi and E. K. Boukas, "$\mathcal{H}_\infty$-control for Markovian jumping linear systems with parametric uncertainty," *J. Opt. Theor. Appl.*, vol. 95, pp. 75–99, 1997.

106. W. Shu and Q. Zhang, "$\mathcal{H}_\infty$ control for singular systems with time-delay," in *Fifth World Congr. Intelligent Control and Automation*, vol. 1, June 2004, pp. 773–777.

107. W. Shu and Q. Zhang, "Robust $\mathcal{H}_\infty$ control for interval singular systems with time-delay in state," in *Fifth World Congr. Intelligent Control and Automation*, vol. 1, June 2004, pp. 765–768.

108. Q. Su and V. Syrmos, "Robust stabilization of singular systems with $\mathcal{H}_\infty$-bounded uncertainty," in *Proc. 1998 American Control Conf.*, vol. 5, June 1998, pp. 2725–2729.

109. K. Takaba, N. Morihira, and T. Katayama, "A generalized Lyapunov theorem for descriptor system," *Systems Control Lett.*, vol. 24, pp. 49–51, 1995.

110. W. Terrell, "The output-nulling space, projected dynamics, and system decomposition for linear time-varying singular systems," in *Proc. 32nd IEEE Conf. Decision and Control*, vol. 4, December 1993, pp. 3294–3299.

111. Y. Uetake, "Adaptive observer for continuous descriptor systems," *IEEE Trans. Automatic Control*, vol. 39, no. 10, pp. 2095–2100, October 1994.

112. G. Verghese, B. Levy, and T. Kailath, "A generalized state-space for singular systems," *IEEE Trans. Automatic Control*, vol. 26, no. 4, pp. 811–831, August 1981.

113. D. Wang and P. Bao, "Robust impulse control of uncertain singular systems by decentralized output feedback," *IEEE Trans. Automatic Control*, vol. 45, no. 3, pp. 500–505, March 2000.

114. D. Wang and P. Bao, "Robust impulse control of uncertain singular systems by decentralized output feedback," *IEEE Trans. Automatic Control*, vol. 45, no. 4, pp. 795–800, April 2000.

115. D. Wang and C. Soh, "On regularizing singular systems by decentralized output feed-back," *IEEE Trans. Automatic Control*, vol. 44, no. 1, pp. 148–152, January 1999.

116. R. Wang and Y. Liu, "Asymptotic stability and robustness for discrete-time singular systems with multiple time-delays," in *Proc. 3rd World Congr. Intelligent Control and Automation*, vol. 5, June/July 2000.

117. R. Wang and Y. Liu, "Conditions for d-stability of discrete singular systems with time-delays," in *Proc. 3rd World Congr. Intelligent Control and Automation*, vol. 4, June/July 2000.

118. W. Wang and Y. Zhou, "The detectability and observer design of 2-D singular systems," *IEEE Trans. Circuits and Systems I: Fundam. Theor. Appl.*, vol. 49, no. 5, pp. 698–703, May 2002.

119. X. Wang, X. Liu, and D. Ho, "Input–output block decoupling of nonlinear time-varying singular systems," in *Proc. 37th IEEE Conf. Decision and Control*, vol. 1, December 1998, pp. 349–354.

120. X. Wang, X. Liu, and Y. Jing, "Input–output block decoupling of linear time-varying singular systems," in *Proc. 1998 American Control Conf.*, vol. 6, June 1998, pp. 3737–3741.

121. Y.-Y. Wang, P. Frank, and D. Clements, "The robustness properties of the linear quadratic regulators for singular systems," *IEEE Trans. Automatic Control*, vol. 38, no. 1, pp. 96–100, January 1993.

122. S. Xu, P. V. Dooren, R. Stefan, and J. Lam, "Robust stability and stabilization for singular systems with state delay and parameter uncertainty," *IEEE Trans. Automatic Control*, vol. 47, no. 7, pp. 1122–1128, July 2002.

123. S. Xu and J. Lam, "Robust stability and stabilization of discrete singular systems: An equivalent characterization," *IEEE Trans. Automatic Control*, vol. 49, no. 4, pp. 568–574, April 2004.

124. S. Xu, J. Lam, W. Liu, and Q. Zhang, "$\mathcal{H}_\infty$ model reduction for singular systems: Continuous-time case," in *IEE Proc. Control Theor. Appl.*, vol. 150, no. 6, November 2003, pp. 637–641.

125. S. Xu, J. Lam, and L. Zhang, "Robust d-stability analysis for uncertain discrete singular systems with state delay," *IEEE Trans. Circuits and Systems I: Fundam. Theor. Appl.*, vol. 49, no. 4, pp. 551–555, April 2002.

126. S. Xu, J. Lam, and Y. Zou, "$\mathcal{H}_\infty$ filtering for singular systems," *IEEE Trans. Automatic Control*, vol. 48, no. 12, pp. 2217–2222, December 2003.

127. S. Xu and C. Yang, "Stabilization of discrete-time singular systems: A matrix inequalities approach," *Automatica*, vol. 35, pp. 1613–1617, 1999.

128. S. Xu and C. Yang, "$\mathcal{H}_\infty$ state feedback control for discrete singular systems," *IEEE Trans. Automatic Control*, vol. 45, no. 7, pp. 1405–1409, July 2000.

129. A. Youssouf and M. Kinnaert, "Observer based residual generator for singular systems," in *Proc. 35th IEEE Conf. on Decision and Control*, vol. 2, December 1996, pp. 1175–1180.

130. R. Yu and D. Wang, "Algebraic properties of singular systems subject to decentralized output feedback," *IEEE Trans. Automatic Control*, vol. 47, no. 11, pp. 1898–1903, November 2002.

131. R. Yu and D. Wang, "On impulsive modes of linear singular systems subject to decentralized output feedback," *IEEE Trans. Automatic Control*, vol. 48, no. 10, pp. 1804–1809, October 2003.

132. M. Zasadzinski, H. Ali, H. Rafaralahy, and E. Magarotto, "Disturbance decoupled diagnostic observer for singular bilinear systems," in *Proc. American Control Conf.*, vol. 2, June 2001, pp. 1455–1460.

133. M. Zasadzinski, M. Darouach, and S. Nowakowski, "A transfer function approach to the linear discrete stationary optimal filtering for singular systems," in *Proc. 31st IEEE Conf. Decision and Control*, vol. 1, December 1992, pp. 979–980.

134. G. Zhang and Y. Jia, "New results on discrete-time bounded real lemma for singular systems: Strict matrix inequality conditions," in *Proc. American Control Conf.*, vol. 1, May 2002, pp. 634–638.

135. H. Zhang, L. Xie, and Y. C. Soh, "Optimal recursive filtering, prediction, and smoothing for singular stochastic discrete-time systems," *IEEE Trans. Automatic Control*, vol. 44, no. 11, pp. 2154–2158, November 1999.

136. L. Zhang, B. Huang, and J. Lam, "LMI synthesis of $\mathcal{H}_2$ and mixed $\mathcal{H}_2/\mathcal{H}_\infty$ controllers for singular systems," *IEEE Trans. Circuits and Systems II: Anal. Digit. Signal Process.*, vol. 50, no. 9, pp. 615–626, September 2003.

137. Q. Zhang, V. Sreeram, G. Wang, and W. Liu, "$\mathcal{H}_\infty$ suboptimal model reduction for singular systems," in *Proc. American Control Conf.*, vol. 2, May 2002, pp. 1168–1173.

138. S. Zhang, "Generalized functional observer," *IEEE Trans. Automatic Control*, vol. 35, no. 6, pp. 733–737, June 1990.

139. K. Zhou, K. Glover, B. Bodenheimen, and J. Doyle, "Mixed $\mathcal{H}_2/\mathcal{H}_\infty$ performance objectives I: Robust performance analysis," *IEEE Trans. Automatic Control*, vol. 39, no. 8, pp. 1564–1574, August 1994.

140. S. Zhu and Z. Cheng, "A singular-system method for singularly perturbed linear systems with perturbed quadratic cost," in *7th Int. Conf. Control, Automation, Robotics and Vision, ICARCV*, vol. 2, December 2002, pp. 869–873.

141. S. Zhu, L. Sun, and Z. Cheng, "Input estimation for uncertain linear singular systems and robust stabilization," in *Proc. 40th IEEE Conf. Decision and Control*, vol. 3, December 2001, pp. 2856–2857.

142. S. Zhu, L. Sun, and Z. Cheng, "Input estimation for uncertain linear singular systems and robust stabilization," in *Proc. 4th World Congr. Intelligent Control and Automation*, vol. 4, June 2002, pp. 2917–2921.

Index

Printing: Krips bv, Meppel, The Netherlands
Binding: Stürtz, Würzburg, Germany